THE ROLE OF ASCORBIC ACID IN GROWTH, DIFFERENTIATION AND METABOLISM OF PLANTS

ADVANCES IN AGRICULTURAL BIOTECHNOLOGY

Related titles previously published

Akazawa T., et al., eds: The New Frontiers in Plant Biochemistry. 1983. ISBN 90-247-2829-0

Gottschalk W. and Müller H.P., eds: Seed Proteins: Biochemistry, Genetics, Nutritive Value. 1983. ISBN 90-247-2789-8

Marcelle R., Clijsters H. and Van Poucke M., eds: Effects of Stress on Photosynthesis. 1983. ISBN 90-247-2799-5

Veeger C. and Newton W.E., eds: Advances in Nitrogen Fixation Research 1984. ISBN 90-247-2906-8

The Role of Ascorbic Acid in Growth, Differentiation and Metabolism of Plants

by the late

J.J. CHINOY
Gujarat University, India

edited by

N.J. CHINOY
Gujarat University, India

co-edited by

I.C. DAVE
Bhabha Atomic Research Centre
Bombay, India

Y.D. SINGH
Saurashtra University
Gujarat, India

O.P. SAXENA
Gujarat University, India

A.V. VYAS
M.G. Science Institute
Ahmedabad (Gujarat), India

1984 **MARTINUS NIJHOFF/DR W. JUNK PUBLISHERS**
a member of the KLUWER ACADEMIC PUBLISHERS GROUP
THE HAGUE / BOSTON / LANCASTER

Distributors

for the United States and Canada: Kluwer Boston, Inc., 190 Old Derby Street, Hingham, MA 02043, USA
for all other countries: Kluwer Academic Publishers Group, Distribution Center, P.O.Box 322, 3300 AH Dordrecht, The Netherlands

Library of Congress Cataloging in Publication Data

Library of Congress Cataloging in Publication Data

Chinoy, J. J., d. 1978.
The role of ascorbic acid in growth, differentiation, and metabolism of plants.

(Advances in agricultural biotechnology)
Includes index.
1. Plants, Effect of Vitamin C on. I. Chinoy, N. J. II. Title. III. Series.
QK753.V58C48 1984 581.19'26 83-25104

ISBN-13: 978-94-010-3715-0 e-ISBN-13: 978-94-010-3713-6
DOI: 10.1007/978-94-010-3713-6

CONTENTS

PREFACE

There is a paucity of information on the dynamics of Ascorbic Acid (AA) turnover in relation to germination, metabolism, growth, differentiation and development of a plant and in those undergoing stress of various types, in presowing treatment of seeds etc. The turnover of AA plays an important role during the juvenile phase of growth of a plant and has a significant bearing on its subsequent growth, development and maturation. The beneficial effect of presowing treatment of seed with Ascorbic Acid (AA) + $H_2 O_2$ highlights the validity of the AA-nucleic acid - protein metabolism concept of growth and development of plants.

During the course of the last 30 years, work has been undertaken by the author and his collaborators on the metabolic drifts of regulatory substances during juvenile, vegetative, reproductive and senescent phases. The most important of these growth regulatory substances was found to be Ascorbic Acid. The dynamic role of AA turnover is revealed by its control of rates of metabolic processes as well as those of enzymic reactions which paves the way to "New Genetics".

A reference to the role of AA in metabolism and growth of the animal cell has been specially made with a view to emphasize the oneness of the fundamental reactions in both plants and animals. Fundamental processes viz. the onset of reproductive phase, microsporogenesis, megasporogenesis, embryogenesis, and others in the plant as well as spermatogenesis, embryogenesis, muscle metabolism and those of a number of other tissues in the animal reveal identical metabolic machinery powered by the electron energy of the free radicals formed by the falling apart of the complexing partners of the charge transfer complex between AA and macromolecules. Thus at the molecular and submolecular levels, the plant and the animal systems are very similar in their ascorbate turnover pattern.

The foregoing monograph on - 'THE ROLE OF ASCORBIC ACID IN GROWTH, DIFFERENTIATION AND METABOLISM OF PLANTS' is, therefore, a special treatise on a theme which has not been hitherto fully exploited in detail. Morphological, biochemical, biophysical and histochemical evidences have been presented in this compilation to elucidate the electron donating role of AA in many a plant process.

To the best of our knowledge, this book should rank as one amongst the few available monographs elucidating the role of Ascorbic Acid in various facets of plant life. Moreover, the work incorporated in the book has considerable practical importance for obtaining higher yield of crops, better growth of plants and for producing those which are able to withstand stress conditions.

Ahmedabad, 1983

Editor
Co-Editors

ACKNOWLEDGEMENTS

I place on record my sincere gratitude to the University Grants Commission, New Delhi, India, for allowing me to compile this monograph and for offering all necessary facilities for carrying out extensive research on multifarious roles of Ascorbic Acid in plant growth and productivity.

My grateful thanks are due to Professors O.P. Garg of Department of Botany, Punjab University, Chandigarh, India, and M.M. Laloraya of Department of Life Sciences, Indore University, Indore (MP), India, for providing timely material for compilation of Chapters 7 and 10 respectively.

I deem it a pleasure to express my appreciation to Professors F.A. Loewus, M. Michniewicz, D.I. Arnon, L.W. Mapson, A. Szent Györgyi, B. Aberg, K.K. Nanda, G.S. Sirohi, K. Mothes, M. Reid, M. Kutacek, O. Stocker, I. Yamazaki, P. Schöpfer, S. Tonzig and many others for making available their reprints.

My special thanks go to Mr. R.K. Modi for painstakingly typing out the monograph; Mr. N.A. Shah for the expert handling of photographic prints; the artists Mr. Somabhai Dholakia and Mr. N.V. Shah for their untiring efforts; and all the staff of the Department of Botany and The Gujarat University for helping me in various ways during the compilation of this monograph.

Finally, I would like to express profound thanks and appreciation to my dear wife, Roda for her dedication, silent forbearance, help and encouragement without which this work would have remained incomplete.

Professor J.J. Chinoy
Professor of Botany and Director
University School of Sciences
Gujarat University
Ahmedabad - 380 009

Ahmedabad, 1978

The Editor and Co-editors of this monograph acknowledge with sincere gratitude the help rendered by all research students of Professor J.J. Chinoy, especially Dr. J.N. Patel, Dr. P.V. Vaishnav, Dr. K.C. Bhatt, Dr. D.B. Jadeja, Dr. N.S. Sharma and Dr. S. Verghese after his sad and sudden demise on 12th May, 1978, in the ardous task of proof reading and checking of references. The willing assistance of Dr. Kersi Fanibunda of Newcastle-Upon-Tyne, UK; Dr. Lovji D. Cama of New Jersey, USA; Dr. M.V. Rao and Dr. R.J. Verma, Department of Zoology, Gujarat University, in proof reading is highly appreciated.

The timely help rendered in typing out the manuscript by Mr. V.T. Viswanathan is greatly appreciated. The Editor and Co-editors express grateful thanks to Madam Roda J. Chinoy for permitting the free and full access to all the relevant papers, data and literature of Late Professor J.J. Chinoy to enable us to finalize the monograph. We also hasten to mention that the reference to work on Ascorbic Acid Metabolism in Animals is the research work carried out by Dr. (Miss) N.J. Chinoy. We have made extensive use of Dr. O.P. Saxena's review paper on 'Presowing Treatment of Seeds' while compiling Chapter 8 for which we owe grateful thanks to him.

And finally, we would like to thank one and all in our respective families and from the Departments of Botany and Zoology, Gujarat University, Ahmedabad, India, for helping us in various ways.

Editor

Dr. (Miss) N. J. Chinoy
Gujarat University, India

Co-Editors

Dr. I. C. Dave
Bhabha Atomic Research Centre
Bombay, India.

Dr. O. P. Saxena
Gujarat University, India

Dr. Y. D. Singh
Saurashtra University
Gujarat, India.

Dr. (Miss) A. V. Vyas
M.G. Science Institute
Ahmedabad, Gujarat, India

ABBREVIATIONS

AA, AH_2	=	Ascorbic acid
A^0	=	Free Radical of ascorbic acid
AA - FR	=	Ascorbic acid free radical
AA - MM	=	Ascorbic acid - macromolecule complex
AAO	=	Ascorbic acid oxidase
AAU	=	Ascorbic acid utilization
$AgNO_3$	=	Silver nitrate
ASG	=	Ascorbic acid bound or ascorbigen
ATP	=	Adenosine triphosphate
Bz	=	Benzimydazole
C	=	Control (non-vernalized)
Ce	=	Cerium
Chlorophyll*	=	Light excited chlorophyll
Co	=	Cobalt
Col	=	Colchicine
Cs	=	Cesium
CTC	=	Charge transfer complex
Cu	=	Copper
cv	=	Cultivar
DCMU	=	-3-(3'-4' -dichlorophenyl)-1, 1-dime - thylurea
DCPIP	=	2, 6-Dichlorophenol indophenol
$DCPIPH_2$	=	2, 6 Dichlorophenol indophenol reduced
DDC	=	Diethyldithiocarbamate
DHA	=	Dehydroascorbic acid
DKG	=	2, 3 - diketogulonic acid
DNPH	=	2, 4 - Dinitrophenylhydrazine
DPPH'	=	a, a - diphenyl - B - picryl hydrazyl
DW	=	Distilled water
e. value	=	Extinction value
E, ε	=	Enzyme, Energon
E_o^1	=	Redox potential
EDTA	=	Ethylene diamine tetra acetic acid
EMH	=	Enzyme mobilizing hormone
EPR	=	Electron paramagnetic resonance

ES	=	Enzyme substrate
FAD	=	Flavine adenine dinucleotide
Fe	=	Iron
FMN	=	Flavin mononucleotide
FR	=	Free radical
GA, GA_3	=	Gibberellic acid
GLC	=	Gas Liquid Chromatography
g. fr. wt.	=	Gram Fresh Weight
GSH	=	Glutathione (reduced)
GSSG	=	Glutathione (oxidized)
Ha	=	Hectare
Hg	=	Mercury
H_2O_2	=	Hydrogen peroxide
HPLC	=	High performance liquid chromatography
HPO_3	=	Metaphosphoric acid
H_2S	=	Hydrogen sulphide
hv	=	Photon energy
IAA	=	Indole - 3- acetic acid
iP	=	Inorganic phosphate
Kr	=	Kilorad
^oK	=	Degrees Kelvin
LD	=	Long day
m	=	Metabolon
M	=	Molar
MDHA	=	Monodehydroascorbic acid
mM	=	Millimole
mmhos/cm	=	Millimhos per cm.
mg/1	=	mg per litre
Mn	=	Manganese
Mo	=	Molybdenum
MP	=	Maturation Phase
n	=	Nucleon
NADP	=	Nicotinamide adenine dinucleotide phosphate
$NADPH_2$	=	Nicotinamide adenine dinucleotide phosphate reduced
ND	=	Normal day

NDC	=	Normal day unvernalized (control)
nm	=	Nanometer
NDV	=	Normal day vernalized
O.D.	=	Optical density
ODNB	=	Ortho-dinitrobenzene
OPD	=	Ortho-phenylene diamine
~ P	=	High energy phosphate bond
PCMB	=	p- Chloromercuribenzoic acid
PGA	=	Phosphoglyceric acid
PMS	=	Phenazine methosulphate
ppm	=	Parts per million
r	=	rad
Ro	=	Reproductive organ
RP	=	Reproductive phase
rd	=	Reproductive differentiation
S	=	Substrate
SD	=	Short day
SH	=	Sulphydryl
SP	=	Senescence (Maturation) phase
Sr	=	Strontium
SS	=	Disulphide
TBA	=	t-butyl alcohol
TIBA	=	Tri-iodobenzoic acid
TLC	=	Thin layer chromatography
TMPD	=	(Peschek, 1980)
T.S.	=	Transverse section
µg	=	Microgram
µm	=	Millimole
µP	=	Microgram phosphate
µS	=	Microsecond
V	=	Vernalized
Vn	=	Vanadium
VP	=	Vegetative phase

DEDICATION

This monograph is dedicated to my revered Teachers:

Late Professors : F.G. GREGORY) Formerly at Biological Laboratories, V.H. BLACKMAN) Imperial College of Sciences and Technology, Kensington, London, U.K. (for my Ph.D. 1932-1935).

R.H. DASTUR - formerly at Royal Institute of Science, Bombay, India (for my M.Sc. 1929-1931).

1. STRUCTURE AND PROPERTIES OF FREE AND BOUND ASCORBIC ACID

The relation between the type of diet and incidence of scurvy was recognized early, however the notion of accessory food factors or vitamins had been clearly formulated in 1912 by Hopkins and Funk (see: Aberg, 1958). This disease was recognized due to deficiency of a specific dietary factor, 'Vitamin C' only in 1917 (Rangaswami and Seshadri, 1952; Aberg, 1958; Florkin and Stotz, 1963; Fieser and Fieser, 1965). Attempts to isolate this factor began in 1920 by Drummond. Later, Zilva and Tillmans made intensive studies but the decisive step was taken by Szent Györgyi (1928), who isolated a strongly reducing substance of the molecular formula $C_6H_8O_6$ from adrenal cortex, oranges, and cabbage and named it as 'Hexuronic acid'. The new name, 'Ascorbic acid (AA) was later coined (Szent Györgyi and Haworth, 1933) and the following terminology was adopted:-

Vitamin C = Ascorbic acid (AA) + Dehydroascorbic acid (DHA).

The study of the chemical constituents of AA began immediately after its discovery and in 1933 the correct structural formula was determined (see: Aberg, 1958). Chemically, AA is known as l-threo-2, 4, 5, -6 - pentoxyhexene -2-carboxylic acid lactone, with a molecular weight of 176.13 daltons. It is a white, crystalline solid which melts at 192° C. AA is soluble in water and has a specific rotation in water $(\alpha)_D + 23^\circ$. AA shows an intense single band absorption spectrum at 245 nm and an ionizing constant of pK = 4.25 in water (Lewin, 1976) and behaves like a monobasic acid.

1. STRUCTURE

L- AA when treated with strong acid yields furfural and carbon dioxide, suggesting that its molecule is unbranched. There are four hydroxyl groups in the AA molecule of which two are enolic in nature, viz., at C-2 and C-3 which has been confirmed by several reactions.

That the ring structure is a lactone has also been confirmed by several studies, viz., physiochemical consideration, magnetic measurements and molecular orbital considerations. There occurs a double bond between the C-2 (α) and (β) carbons and establishes the enolic nature of the two hydroxyl groups. The keto group of AA was capable of undergoing enolization (Fig. 1.1).

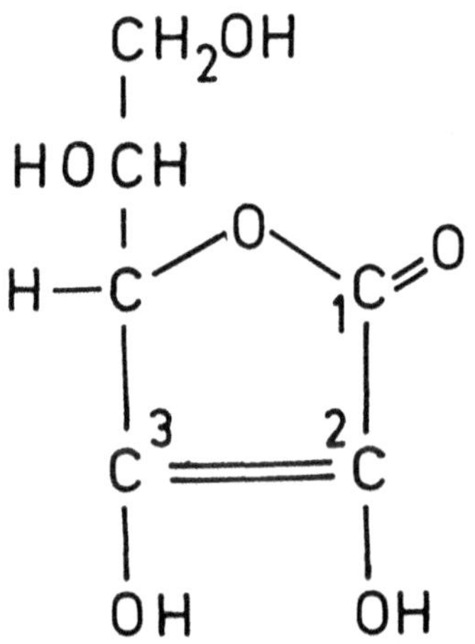

FIGURE 1.1. Structure of AA.

Methyl iodide and silver oxide were used to determine the precise nature of the ring in L-AA and tetra-O-methyl derivative of AA was obtained. As UV absorption spectrum of the product was the same as that of the original substance, it was inferred that the ring structure of L-AA remained intact during methylation. Experiments gave an unequivocal proof of the unsubstituted nature of C-2 hydroxyl of the L-threonic acid derivative and thus establishing the structure of L-AA as a five membered ring in which C-1 and C-4 were linked through the oxygen atom of the lactone. The formula given to AA was thus fully confirmed. X-ray crystallographic data also clearly indicated that the molecular model of L-AA built on the basis of its structural formula is quite flat (Planta, 1961; Hvoslef, 1964).

2. PROPERTIES

2.1. Optical absorbance

AA has optical absorbance and optical rotation properties due to the presence of the double bond between C-2 and C-3 and a ring structure. The optical absorbance spectrum of the ascorbate anion differs from that of AA. The optical absorbance of ascorbate has a peak value at between 265 nm and 266 nm. Lewin (1976) has determined the value to be 265.5 ± 0.3. The E_{max} value ranges between 7500 and 16,650 (Hewitt and Dickes, 1961). Lewin (1976) demonstrated that under strict anaerobic conditions, a continuous drop in absorbance occurred in aqueous solution at low ionic strengths of 10^{-4} to 10^{-3}. This drop was attributed to auto-oxidation of AA catalysed by traces of metal ions like Cu and Fe and the effect of light present during preparation of the solutions. Lewin (1976) recommends the use of freshly prepared solutions under N_2 free from O_2 and CO_2 in dim light. The highest E_{max} reported for AA in acidified solutions were at λ_{max} 244 nm. It was 10,500 (Hewitt and Dickes, 1961) whereas, Lawendel (1956, 1957) gave the value of 11,900 to 12,220. DHA is transparent in the region of 230 to 280 nm but possesses a weak absorption E_{max} = 720 at 300 nm (Mattock, 1965).

2.2. Optical rotation

Herbert et al. (1933) determined the optical rotation of ascorbate as $(\alpha)^{18°}_{5780} = +116°$ in neutral aqueous solution, and in acid solution (0.05N HCl) as $(\alpha)^{18°}_{D} = +22°$. This has not been rechecked and published for various other temperatures for the optical rotatory dispersion spectrum of ascorbate or ascorbic acid. However, Lewin et al., (unpublished observations, see: Lewin, 1976) determined the optical rotatory dispersion spectrum of ascorbate using spectropolarimetry and solutions prepared in DW and deionized water under anaerobic conditions (using N_2 free from O_2 and CO_2). This method was more sensitive than that of optical absorbance. The optical rotation values were used to compare relative stabilities of ascorbate solution under anaerobic conditions at different concentrations and ionic strengths and under aerobic conditions. The initial value for optical rotation of 10^{-2} M sodium ascorbate solution at 37° C and 365 nm, under anaerobic conditions was determined as + 0.0376° and this value dropped to + 0.0357° after 21 hours. Using a 10^{-3} M ascorbate solution under the same experimental conditions, the initial value of + 0.038° dropped to + 0.012° after 21 hours. This was a far greater drop in optical rotation than that shown by the 10^{-2} M sodium ascorbate solution. This data and numerous similar results obtained by Lewin and his group revealed that decreased ascorbate concentration (or decreased ionic strength) resulted in increased instability of the ascorbate solution under anaerobic conditions. However, aerobic conditions resulted in more marked changes in optical rotation values, in accordance with oxidative processes taking place, e.g., using a 10^{-3} M ascorbate solution, under the same experimental conditions as before, the initial reading was + 0.023° which fell to + 0.003° in 21 hours.

2.3. Ascorbic acid as reducing agent

Szent Györgyi (1928) had shown that ascorbic acid could reduce Fehling's solution at room temperature. Later investigators have revealed that reactions involved in the reduction of ionic copper and silver are of a complex nature yielding different types of substances under different conditions of the reaction. Two equivalents of argentous ion are required to form dehydroascorbic acid when the medium is acidic. On the other hand, in an alkaline medium, four equivalents of argentous ions are reduced (Constantinescu and Oteleanu, 1959). A number of cations catalyze the stepwise degradation of AA. Some of these are magnesium, manganese, selenium dioxide, selenious acid, methylene blue, ferron, platinum carbons, redox resins, acid permanganate, 1, 10-

phenanthroline and related ligands. Some of the antibiotics also catalyze the oxidation of AA such as streptomycin, terramycin and gordicin (Holker, 1955; Romanchuk, 1957; Chapon et al., 1959; Deshmukh and Bapat, 1955; Pihar 1955a, 1962; Manecke et al., 1959; Inczedy, 1961; Butt and Hallaway, 1961; Younathan and Frieden, 1961; Ichinose, 1962; Dudani and Krishnamurti, 1954; Markhel and Novel'nov, 1961). Quinone, phenolindophenol and its dichloro derivatives also stimulate oxidation (Nomura and Uchara, 1958; Gero, 1955).

Cupric ion is a more sensitive catalyst for the oxidation of AA in aqueous solution than ferrous and ferric ions (Vleeschauwer et al., 1959). In both the cases the first product of oxidation is DHA (Manecke et al., 1959; Arndt et al., 1952). Besides ferrous and cupric ions, hydrogen peroxide also oxidizes AA giving rise to CO_2 (Kamiya and Nakabayashi, 1957) and a large number of oxidation products including L-threonic, glyceric, glycolic, oxalic, and hydroxypyruvic acids (Herrman and Andrae, 1963). Two equivalents of halogens, chlorine, bromine and iodine are able to effectively oxidize AA in the cold, both in neutral or acid solution. AA can be recovered in the original form by the addition of hydrogen sulphide to the oxidized solution. In the case of treatment with iodine merely evaporating the reaction mixture *in vacuo* can reverse the reaction yielding AA and iodine.

Irradiation of different types oxidize AA both in aerobic and anaerobic conditions yielding different products (Douzou and Gallon, 1956). The effect of X- and γ- irradiation is better than that of ultraviolet radiation in producing DHA and hydrogen peroxide (Babin et al., 1955; Babin and Delmon, 1955; Pukhav, 1957; Douzou, 1956; 1958). Presence of metal ions, especially iron as well as oxygen accelerated the photochemical reaction (Baker, 1955; Baker et al., 1955; Tomana et al., 1963).

Barr and King (1956) have postulated that photo-oxidation of AA may be brought about by the generation of H and OH radicals. These radicals are known to oxidize AA (Habermann and Vernon, 1958). Perhydroxyl radical is also considered as the active species in such a photochemical reaction by a number of workers (Babin et al., 1955; Barr and King, 1956; Lampitt et al., 1956). This perhydroxyl radical is considered to be involved in the hydroxylation of aromatic compounds in aqueous medium containing AA, oxygen and iron (Norman and Radda; 1962; Cier et al., 1959 a,b; Udenfriend et al., 1954). A wide variety of oxidation which includes formation of carbon-carbon bond (Geyer et al., 1957), opening of N-heterocyclic ring (Moriyama, 1957), decarboxylation of aliphatic compounds (Mazdis, 1961), hydroxylation of steroids (Revol et al., 1958) and cleavage of carbon nitrogen bond are known to be affected by the above system. Nitrous acid also oxidizes AA in which monodehydro-ascorbic acid, its free radical, is considered to be an active intermediate (Dahn et al.,1960; 1960 a, b, c).

2.3. Effect of pH

An appropriate pH is essential for the continuation of aerobic oxidation of AA at an optimum rate. As the oxidation can proceed both in acidic and alkaline medium, two maxima were observed, one at pH 5.0 involving reaction with one equivalent of base and the other at pH 11.5, which corresponds to a reaction with two equivalents of base (Csüros and Petro, 1955; Burger and Becker, 1951).

The products of oxidation also depend to a large extent on the pH. At pH 4.0 oxalic acid is not produced, whereas, it is readily formed during auto-oxidation of AA in alkaline medium (Burger and Becker, 1951). The nature of the alkali used for the reaction has also a profound effect upon the rate and the extent of oxidation (Csüros and Petro, 1957; 1958; 1958 a). Oxidation in alkaline medium, however, leads to a greater acceleration in oxidation and the extent of degradation is also much more than in acid solution. Using aqueous solutions of the alkali or alkaline earth metals as well as ammonium hydroxide, Csüros and Petro (1957) were able to degrade AA to glycolate and oxalate. On the other hand, Kamiya and Nakabayashi (1957) obtained glycerate, formate and L-threonate also by using alkaline peroxide in the medium. Imanaga (1955) has shown that imidazole compounds evolve ammonia in the presence of AA and ferric ion in conditions of mild alkalinity to give acyclic amino acids after acid hydrolysis. This investigator has suggested the formation of a complex with MDHA as an intermediate in the above mentioned reaction.

A number of different compounds such as furfural and carbon dioxide, L-xylose and different organic acids are formed during anaerobic oxidation of AA in acidic aqueous medium (Norman and Radda, 1962; Heulin, 1953; Anet and Reynolds, 1955; Cier et al., 1959 a, b).

2.4. <u>Oxidation - reduction and its potential</u>

It was shown by a number of workers that AA could be readily oxidized with two atomic proportions of iodine to DHA with the removal of two hydrogens and the elimination of the conjugated system without disturbing the main structure in any way. Without oxidation, AA displayed strong selective absorption of light in the UV region of the spectrum having a band at nm = 2450 A moving to nm = 2650 A upon addition of alkali. These experimental findings lead one to conclude that group -CHOH- CO-COOR yields a system of conjugated double bonds. Such a conjugated double bond system is also found in dihydroxymalic acid, which has more or less a similar absorption spectrum in the UV region like that of AA. It is also easily oxidized to a compound which can also be brought back into the reduced form by treatment with H_2S.

2.4.1. <u>Oxidation</u>. Oxidation of AA by various oxidizing agents like oxygen, air, H_2O_2, $FeCl_3$, iodine and other halogens in acid or neutral solution, DCPIP in acid solution, quinones, copper acetate and even ultraviolet irradiation results in formation of DHA. At the time of its formation, DHA is a lactone, whose (lactone) ring can be easily hydrolyzed in aqueous solution with the formation of free carboxylic acid group. The above mentioned findings point to the fact that AA does not derive its acidic property from a carboxylic acid group, but from its enolic hydroxyl groups. AA cannot be obtained by treating the open chain DHA with H_2S alone. However, if H_2S treatment is combined with evaporation of the system in presence of hydriodic acid, the lactone formation again takes place and AA is reformed. However, opening of the lactone ring may lead to a complete rearrangement of the molecule as seen from changes in optical rotation and absorption spectrum. Prolonged oxidation of AA with H_2O_2 yielded oxalic acid suggesting the presence of a Keto group in α-position of the carboxyl group, whereas, its oxidation with sodium hypo-iodite in alkaline solution yielded oxalic acid and L-threonic acid in almost quantitative yield. The olefinic nature of the bond between the second and third carbon is revealed by the oxidative cleavage of the carbon chain of AA into L-threonic acid and oxalic acid. The unquestioned identification of L-threonic acid during oxidative cleavage of AA, left no room for doubt about the configuration at the asymmetric carbons (C-4 and C-5) of L-AA, and pointed to its structural relationship with L-glucose. This is quite surprising in view of the fact that practically all naturally occurring hexoses are members of the D-series.

On the basis of halogen oxidation of AA to DHA, the γ-lactone structure for DHA was suggested. The acidic property of L-AA was however ascribed to the enediol system of the ring structure. The two enolic hydroxyls situated on C-2 and C-3 differ considerably in their acidic property, the hydroxyl of C-3 being relatively more acidic than that on C-2 as shown by titration with 1 equivalent of ethereal diazomethane which gives rise to exclusively 3-O-methyl L-AA. Haworth et al. (1934) have shown that in the presence of alkali the lactone ring remained unimpaired. Only a proton was removed from the -OH at C-3. The high rotation of the sodium salt can be explained on the basis of the subsequent transformation of the ionized system into a state of resonance, as well as the shifting of the absorption band toward longer wavelength (Jonaitis and Krisciuniene, 1962; Lawendel, 1957; Nebbia, 1958).

The chemical properties as well as absorption of light by dihydromaleic acid are similar to those of L-AA and DHA. These facts suggest structural similarities between these compounds. The fact that DHA easily formed hydrazone like compounds with O-phenylenediamine, phenylhydrazine and substituted phenylhydrazines lent further support to the structure of these compounds.

2.4.2. Inhibitors of oxidation. There are a number of substances, both inorganic and organic, which directly inhibit the oxidation of AA. These are thioureas, thioacetals, N-heterocyclics, and others (Akagi and Aoki, 1957). In addition, there are certain chelating or complexing compounds which also inhibit or reduce the oxidation of AA by forming complexes with it, e.g., EDTA (Peel and Loughman, 1957; Rao et al., 1959; Flesch et al., 1960), flavonoids (Novotel'nov et al., 1955; 1959); O-diphenols (Gero, 1949), metaphosphoric acid (Morse, 1953), rubeanic acid (Smoczkiewiczowa and Grochmalicka, 1961), aryl thioureas (Inagaki et al., 1955) and acidic polysaccharides (Erickson and Gasparetto, 1953; Janecke, 1954).

In the case of bisulphite or with amino acids there was at first extensive degradation and decarboxylation of AA (Kamiya, 1960; Jackson et al., 1960), after which polymerization of oxidation products possessing lower molecular weight took place resulting in the production of coloured compounds (Kamiya, 1960; Ellis, 1959; Dulkin and Friedmann, 1956). Most of the proteins as well as amino acids are known to inhibit the oxidation of AA by the formation of stable AA - protein or cupric - protein complexes (Neiadas and Robert, 1958; 1958 a; Shamrai et al., 1959; Ichimonji, 1955; Rakshit et al., 1957).

The most important property of oxidation-reduction of AA and DHA has been made use of in the field of polarographic analysis (Kern, 1954; Heimann and Wisser, 1962; Wasa et al., 1961; Ono et al., 1953; 1958). Methods for the determination of AA alone, in the presence of DHA, other vitamins as well as in plant extracts have been devised (See Chapter 2). Use of AA has also been made as a supporting electrolyte in polarographic determination of a large number of elements especially uranium (Šušič, 1954; Lopez, 1961) and organic compounds (Shkodin and Tikhomirova, 1955). Similarly, many elements, viz., copper, phosphorus, lead, bismuth, halogens and tellurium have been estimated using the reaction with AA (Rao and Rao, 1955; Kolthoff and Elving, 1962; Maksimovic and Šušič, 1956; Campbell and Mellon, 1960; Hsien-Feng, 1958; Deshmukh and Bapat, 1955).

3. FREE RADICAL OF ASCORBIC ACID

It has been suggested that biological oxidation-reduction reactions occur by way of one electron transfer. As a consequence of univalent transfer of electrons in the oxidation of organic molecules free radicals are produced at the end of the first step. With the help of electron paramagnetic resonance spectrometer it has been established that several enzymic oxidation reactions involve one-electron transfer processes.

In the oxidation of AA to DHA both by enzymic and non-enzymic reactions, an intermediate state, the free radical of AA called monodehydro-ascorbic acid (MDHA) has been shown to occur (Yamazaki et al., 1960; Piette et al., 1961; Yamazaki and Piette, 1961; Lagercrantz, 1964; Foerster et al., 1965). This free radical has been identified as a characteristic doublet with a g-value of 2.004 and with a splitting of about 1.8 gauss having the formula, $C_6H_7O_6$ and with a molecular structure given below (Fig. 1.2):

An isozyme of peroxidase has been found in the actively growing and differentiating parts of plants and animals which catalyzes the formation of the free radical of AA, which is a much more powerful electron donor than L-AA and can reduce an inert substance like O-dinitrobenzene to dinitrophenyl hydroxylamine. AA alone is quite ineffective in this reaction (Gurevich, 1963).

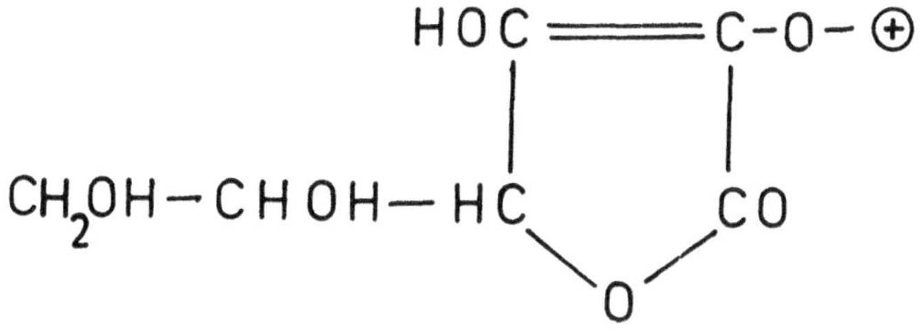

FIGURE 1.2. Structure of ascorbate free radical, monodehydroascorbic acid (MDHA)

Ascorbic acid complexes with macromolecules like DNA and RNA only when the free radical (FR) of AA is generated in the medium. Szent Györgyi (1960) has assigned important roles to free radicals and charge transfer complexes in the flow of electrons during active metabolism, growth and differentiation in plants and animals. Some of the biological aspects of this unique molecule are dealt with here. The position of the unpaired electron has been subject to argument. One group of workers view it as an ascorbate free radical (Lagercrantz, 1964; Duke, 1968; Russel et al., 1966) and have allocated it to the Oxygen on the C-2 position, while others (Foerster et al., 1965) have considered it as an ascorbic acid free radical and allocated the unpaired electron to the oxygen on the C-3 site. It is likely that the unpaired electron of MDHA or ascorbic free radical formed at acid pH values is located at the C-3 site, while that formed in the alkaline pH range is located at the C-2 site. However, the present data is inadequate to decide on the locations unequivocally (Lewin, 1976). The free radical has been characterized by Lavandoski et al. (1964) and complexing with AA was also indicated by Kluge et al. (1967).

3.1. Formation and elimination

The free radical can be formed in several types of reactions and eliminated in others. Its concentration at a given time thus depends upon its rate of formation and elimination. The reactions in which it is formed are: (1) By mixing of AA with DHA resulting in equilibrium (Foerster et al., 1965), (2) oxidation of AA with molecular oxygen (Lagercrantz, 1964), (3) reactions of AA with oxidants and (4) irradiation with light (Lewin, 1976). Reactions (1) to (4) are non-enzymatic reactions whereas, the enzymatic reactions are the redox reactions involving AA (Udenfriend et al., 1954; Yamazaki et al., 1959; Chinoy et al., 1969; Barnes and Kodicek, 1972). The non-enzymatic elimination reactions are (1) Dismutation into AA and DHA and (2) transformation into AA by capturing an electron or an H atom from another substance and thereby causing the latter to become a free radical (Blumberg et al., 1965); the enzymatic methods are the enzymatic redox reactions.

3.2. Properties and reactivity

The semiquinone like free radicals possess considerable energy which could be contributed to the lowering of the energy of activation of an associated reaction. The free radical is an unstable molecule and hence very reactive. The ascorbic/ascorbate free radical involvement with water free radicals and electron donor/acceptor trends are reported (Lewin, 1976).

As mentioned earlier, the electron paramagnetic resonance (EPR) signal of MDHA using pure AA + H_2O_2 (15 drops) is a doublet with a g-value of 2.004 and a splitting of about 1.8 gauss (Fig. 1.3). The signal heights were considerably increased when DNA solutions of different concentrations were added to pure AA + H_2O_2 (Figs. 1.4, 1.5). In Figure 1.5, the standardized signal heights (cm) vs time (in minutes) are plotted. A similar plot for different

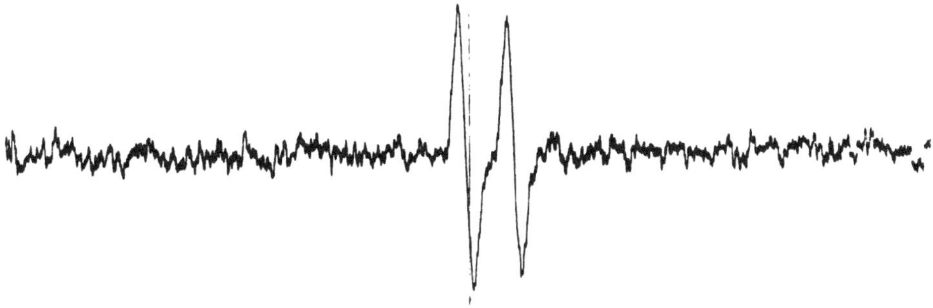

FIGURE 1.3. EPR Spectrum of pure AA + H_2O_2.

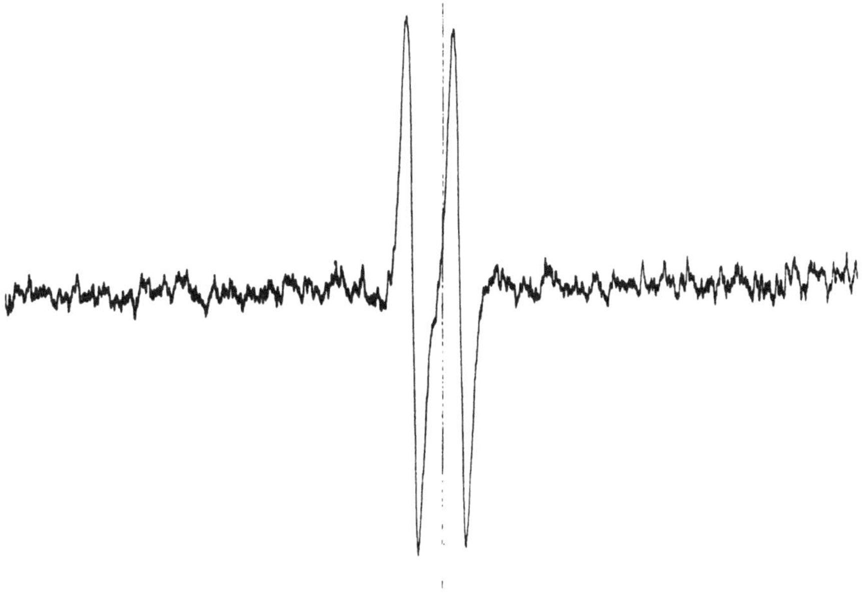

FIGURE 1.4. EPR Spectrum of AA + H_2O_2 + DNA.

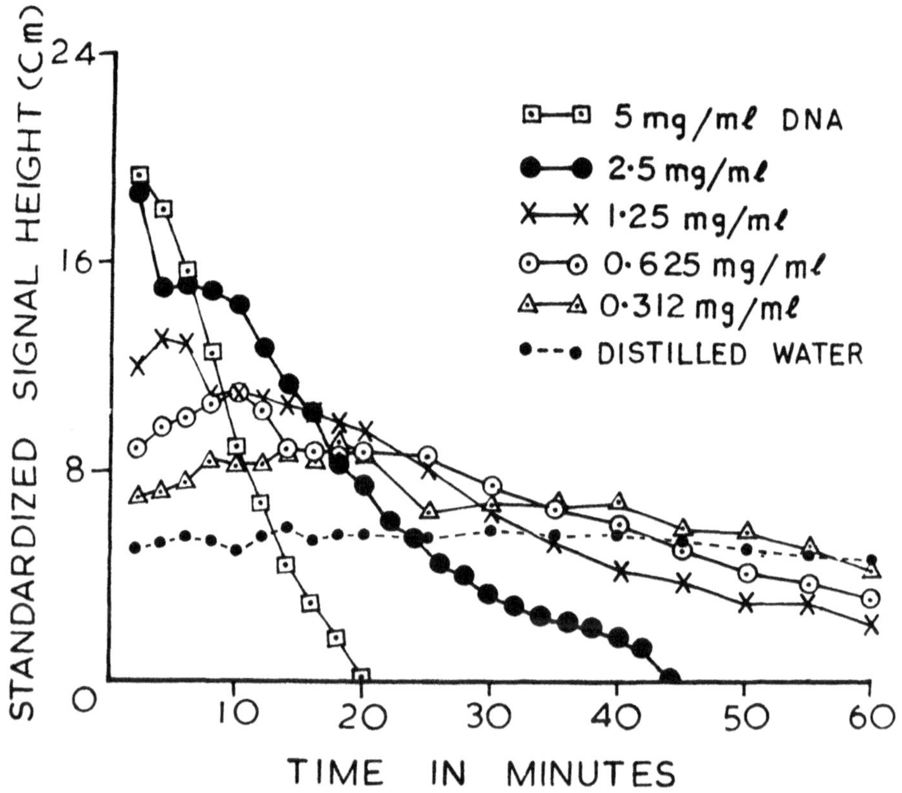

FIGURE 1.5. Standardized signal height (cm) vs. time (min) of the above at different concentrations of DNA.

concentrations of RNA is depicted in Figure 1.6, whereas, the amounts of AA-RNA complex (μg) is shown in Figure 1.7. AA is known to form charge transfer complexes (CTC) with macromolecules i.e., proteins, nucleic acids and steroids (Chinoy et al., 1972; 1974; Chinoy and Saxena, 1978; Chinoy, N.J. et al., 1978). In these complexes electrons may trespass between borderlines of two complexing molecules. Within a complex, an electron of one of the two complexing molecules may be transferred to the orbital of the other. This phenomenon is referred to as charge transfer complexing (CTC).

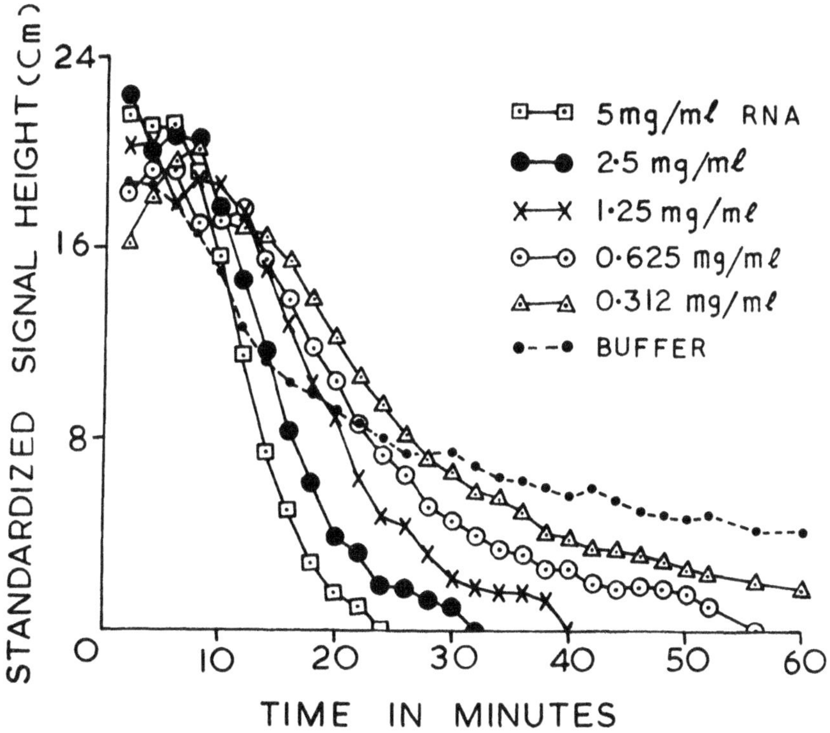

FIGURE 1.6. Standardized signal height (cm) vs. time (min) of AA + H_2O_2 + different concentrations of RNA.

The difference between CTC and oxidoreduction is that, in the latter, an electron pair is transferred from one molecule to another and two new closed shell molecules are formed. The molecules then part, one becoming richer and the other poorer by two electrons and the structure gets rearranged. In charge transfer however, only one electron is transferred. The donor and acceptor usually remain together and when they part, they do so not as closed shell molecules but as free radicals with an unpaired electron. Charge transfer means that the electrons of donor molecule are capable of using under certain conditions, orbitals of acceptor molecule.

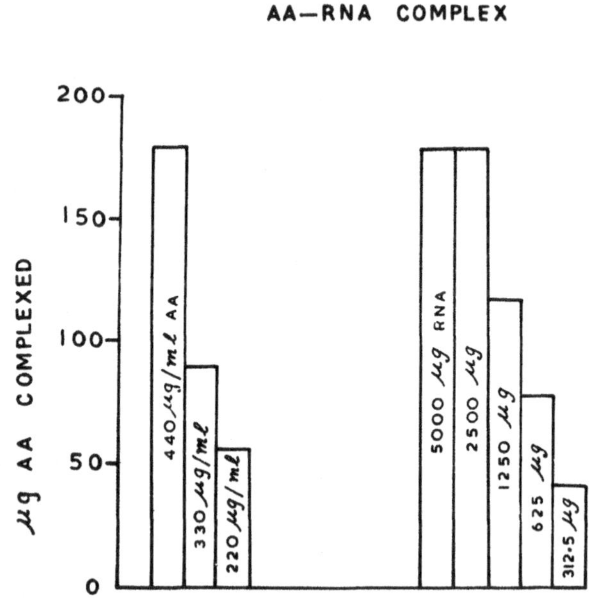

FIGURE 1.7. Amount of AA complexed with RNA.

In other words, CTC is a simple transfer of one electron from the highest filled orbital of the donor to the lowest empty orbital of the acceptor without any further rearrangement, subject to the condition, that the two molecules of the donor and acceptor should be so close to each other as to allow a passage of an electron from an orbital of the donor to an orbital of the acceptor and the electron clouds must overlap. The ionizing potential of the donor should be low, whereas, the electron affinity of the acceptor ought to be high. The electrons then pass from donor to acceptor and resonate between the two molecules contributing resonating energy to the binding forces. One or two parted electrons could also impart a dipole moment to the molecule and may compensate each other's magnetic moment, thus making the complex paramagnetic. In extreme cases the two molecules, acceptor and donor may even part as free radicals. CTC is an important and fundamental reaction which allows us to transfer an electron from donor to acceptor without any major loss in energy as no rearrangement is needed in the molecular structure and by formation of CTC, a relatively inactive molecule may acquire a high reactivity.

3.3. CTC of ascorbate with nucleic acids and protein

Hendry et al. (1977) have proposed the involvement of stereochemical recognition in DNA which are small molecules interaction in gene regulation processes. According to the authors, a close perusal of the chemistry of DNA and RNA indicates that while, the multiple nucleotide sequences may be able to code for the primary structural differences among proteins, a number of specific recognition sites cannot be explained on this basis alone. The authors maintain that specific proteins involved in molecular biosynthesis reflect the modifications in the template arising out of stereochemical recognition between the nucleic acids and tiny regulatory molecules, e.g., steroidal hormones, phytohormones and L-ascorbic acid, highlighting the importance of chirality in the stereochemical interaction of such molecules into DNA. The presence of at least one heteroatom appears in a common position. The exact location of such a heteroatom is of great significance in its biological activity in that the formation of hydrogen bondings between the DNA base pairs and an intercalated regulatory molecule would aid in positioning it in the helix (see: Hendry et al., 1977). Further, most if not all the chemicals could bind to at least one phosphate oxygen of DNA. This postulation is based on the feasibility study of three dimensional Corey-Pauling-Koltum (CPK) space-filling models. L-AA has been shown as a successful candidate for intercalation revealing remarkable specificity in hydrogen bonding with the bases. The intercalation of AA fits between the bases C-G and G-C in a right handed helix. It has five possible attachment sites. This model shows successfully that chirality of AA and its multiple attachments in base pair cavities indicate remarkable *in vivo* specificity for DNA. The extent of hydrogen bonding in terms of recognition of AA by helix is further governed by pH, ionic concentration and amount of chromosomal proteins. Such specific insertions described here may dictate the stereochemical specificity to be transcribed and translated, visualizing the protein formation "programmed" by tiny molecules, thereby involving the basic mechanisms regulating the processes leading to morphogenesis.

Jamaluddin et al. (1981) have presented the evidence for a direct interaction of L-ascorbate with rat liver DNA. They have proposed an additional direct mechanism by which L-AA could protect nucleophilic sites of DNA, namely, by a specific interaction with DNA. This proposition has been made based on the observed behaviour of L-AA in the presence or absence of rat liver DNA, when subjected to large-zone gel exclusion chromatography in a column of Sephadex G-50. Further, the initial rate of oxidation of AA in the absence of DNA which was a linear function of AA concentration became complex in the presence of a fixed concentration of DNA (3.6×10^{-6} M base pair of DNA). This complexity has been explained on the basis of the formation of a reversible complex between AA and DNA. The rate of AA is dependent on the relative proportions of bound and free AA. Amongst the bases, adenine could afford a better protective agent against AA oxidation than guanine. A lowering of the protective effect of adenine upon methylation at N^7 suggest that N^7 position of adenine may be involved in the interaction, through hydrogen bonding, specifically in native double helical DNA. Further, according to these authors, on a concentration basis, DNA is far superior to adenine in protection of L-AA hence invoking the importance of structural/conformational aspects of DNA in its interactions with AA.

Stich et al. (1976) have listed the properties of AA which include conversion of covalently - closed cDNA to open cDNA in the presence of molecular oxygen, causing single strand breaks in DNA, inactivation of transforming DNA of Pneumococcus, degradation of RNA of bacteriophage R 17, triggering of DNA repair synthesis, etc. According to these authors, all these properties are characteristic of most, if not, all chemical mutagens. They have further shown that the mutagenic action of AA is not manifested in absence of oxygen and that they are greatly enhanced in presence of Cu ions and H_2O_2. In reality, it is the free radical of AA which is involved in the mutagenic effects observed. What these authors have failed to notice is the nature of charge transfer generated in such *in vitro* experiments.

The milieu provided in such experiments is very conducive for the generation of a strong charge transfer. Szent Györgyi (1960) has described varieties of charge transfer. According to him, in extreme cases, when the difference between the electron affinity of the acceptor and ionizing potential of the donor (EA - IP) decreases, the molecular complex becomes more and more paramagnetic with 40% of the electrons being completely uncoupled. Under *in vivo* conditions, this could be considered as an extreme case and "such a strong charge transfer is ruled out as the existence of strong acceptors is incompatible with Life". The other way of visualizing the formation of the AA-macromolecule complex is the controlled generation of CTC subsequent to a controlled production of free radicals of AA enzymically *in vivo* inside the cell as and when required. *In vitro* synthesis of complexes between AA and DNA as well as RNA has been achieved in the author's laboratory as described earlier (Chinoy et al., 1972 a; 1974) and by the extensive data on concentration of AA, ASG, AAU, AA-macromolecule complexes, histones, AA - PR peroxidase, DNA, RNA, proteins, etc., during the period of juvenile phase, vegetative and reproductive differentiation and maturation of vernalized and non-vernalized plants raised under long day (LD) and normal day (ND) conditions (ref. subsequent chapters). Despite a high rate of AA turnover, AA-FR peroxidase activity and contents of free radicals, even during meiotic division, no mutation had occurred supporting thereby the compelling conviction that the endogenous AA or its free radical(s) are not mutagenic, but on the contrary, are inevitable for processes preparative to cell division and ontogenetic as well as morphogenetic differentiation.

3.4. CTC of ascorbate with metallic cations

The ascorbate anion possesses the ability to form a complex with metallic cations such as Fe^{+++} and Cu^{++} (Lewin, 1976).

3.5. CTC of ascorbate with steroids

The ascorbic acid - steroid CTC was demonstrated *in vitro* using various steroids for the first time by Chinoy, N.J. et al. (1978).

3.6. H--- bonding capacity of monovalent ascorbate anion

The monovalent ascorbate anion possesses both an H ---- donor group (OH) and is also capable of H--- acceptance and a $-C-O^-$ H---acceptor group and hence enables the molecule to participate in H--- bonding and simultaneous H... donation and H--- acceptance. However, DHA (anhydride form) has no H... donor groups in its ring, but, the hydrated form of DHA can, by virtue of its -OH groups on C-2 and C-3 sites, be presumed to possess activity in both H--- donation and H--- acceptance. These characteristics, the resonance potential due to the double bond between the C-2 and C-3 sites, as well as the negative end of the dipole of carbonyl group, impart enhanced potential for both unidirectional and complementary - two-dimensional double H--- bonding activities to ascorbate as described extensively by Lewin (1976) and recently also investigated by others (Stich et al., 1976; Hendry et al., 1977; Jamaluddin et al., 1981) in both plant and animal material (ref. earlier part of this chapter).

4. DEHYDROASCORBIC ACID (DHA)

As mentioned earlier in this chapter, DHA is obtained by oxidation of AA. It has an anhydrous form which lacks ionizable -OH groups on the lactone ring and thus does not have acidic properties, but the hydrated form exhibits acidity. DHA is highly reactive, unstable in aqueous solution, easily reduced and on hydrolysis yields 2;3:diketogluconic acid. DHA also reacts with thiol groups, amino and imino groups (Yano et al., 1976).

4.1. Redox potential

A mixture of AA and DHA exhibit redox potential which varies with pH and temperature. The chemical reactivity of the ascorbate/dehydroascorbate has been dealt with at length by Lewin (1976) and hence will not be discussed here.

5. CHEMICAL STRUCTURE OF BOUND ASCORBIC ACID OR ASCORBIGEN (ASG)

The chemical structure of ascorbigen was determined by Prochazka et al. (1957) using infra-red spectra and chemical reactions. Hydrogenation of ASG resulted in a compound $C_{17}H_{29}O_7N$ after taking up six molecules of hydrogen which remained unaffected by acid hydrolysis. Presence of one γ-lactone ring was detected in this compound and it consumed one equivalent of HIO_4 when oxidized. Saponification of the lactone ring in alkaline medium resulted in the consumption of three equivalents of HIO_4 and yielded two molecules of HCOOH. ASG yielded another compound with the chemical formula $C_{15}H_{19}O_6N$ in an alkaline medium. Through the action of warm alkali, this compound was partially converted into hetero-auxin and indole-3-carboxylic acid. A crystalline diazomethane derivative of ASG having the formula $C_{18}H_{19}O_7N$ was prepared which had an almost identical UV spectrum as that of ASG. Two alternative structures have been suggested for ASG by Prochazka et al. (Figs. 1.8, 1.9).

6. THE SIGNIFICANCE OF CHEMICAL PROPERTIES OF ASCORBIC ACID IN BIOLOGICAL PROCESSES

Although AA is relatively a small and a simple molecule, some of its chemical properties have a profound influence upon many biological processes.

It is universally present in almost all the organelles of cells of plants and animals. It is also present in the

I

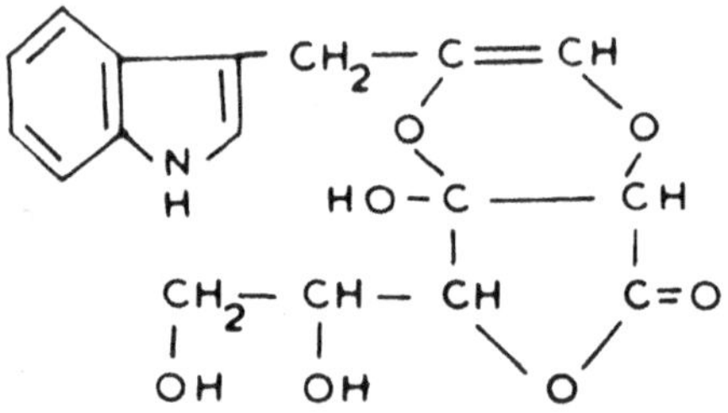

FIGURE 1.8. Two alternate structures of ascorbigen (ASG).

II

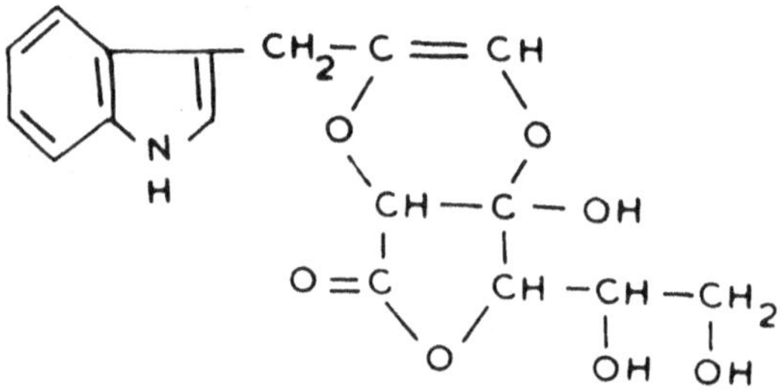

FIGURE 1.9

cytoplasm and nucleus both in the bound as well as free form. All actively growing and differentiating organs show high concentration of AA. It is constantly being utilized enzymatically. A number of enzymes like oxidases, peroxidases, catalases, dehydrogenases and laccases act upon it. The dienolic group is constituted by the presence of a double bond between the second and the third carbon, each having a hydroxyl group endowing AA molecule with a unique oxidation-reduction property. AA derives its acidic property not from a carboxylic group but its enolic hydroxyl groups. AA and its free radical are involved in reduction of disulphides, various biochemical and organic compounds viz., dichlorophenolindophenol and adrenochrome. The latter is a highly toxic substance and combines with -SH groups to form -SS. The involvement of AA in several enzymatic activities have been highlighted in Chapter 11. AA is also associated with hydroxylation of dopamine to noradrenaline, tryptophan to form serotonin and p-hydroxyphenyl-pyruvate to form homogentisic acid in animals. The dissolution of fat and cholesterol in the body are done by AA. AA is also known to increase membrane permeability and C-AMP by virtue of inhibiting the enzyme phosphodiesterase. An extensive review of the role of AA in animal and human tissues has been given elsewhere (Lewin, 1976; Chinoy, N.J. 1978).

AA is known to inactivate viruses (Murata and Uike, 1976) and antitumoric reductones like AA depolymerized nucleic acids and alter the priming action of DNA for DNA-polymerase (Omura et al., 1975; 1978).

7. METABOLISM OF D-AA

The D-isomer is distributed in animal tissues very much like L-AA but lacks vitamin activity. D-AA was able to replace some of the activities of L-AA when given in small doses and could substitute L-AA for some non-specific roles, e.g., tyrosine metabolism, wherein, L-AA could be replaced by various other structurally unrelated compounds (Burns, 1967).

Another metabolite, L-AA- 2- sulphate has been detected in the urine of animals. It is inactive (Tolbert et al., 1975; and cholesterol is sulphated *in vivo* by this compound (Verlangieri and Mumma, 1973). Numerous L-ascorbic acid phosphates have been prepared (Tolbert et al., 1975), but their biological significance is still not clear.

8. SUMMARY

The structure and properties of free and bound AA have been presented in the light of recent data. The importance of oxidation-reduction potential as well as of the free radical, MDHA in several biological processes has been elucidated. AA is utilized by cells and tissues via the formation of its free radical and CTC with macromolecules. Finally, some metabolic entities of AA have been indicated.

REFERENCES

1. Aberg B, 1958,Ruhland W (ed) Encycl Plant Physiol. Berlin: Springer-Verlag 6: 479.
2. Akagi M, Aoki I 1957,Yakugaku Zasshi 77: 1314.
3. Anet EFLJ, Reynolds TM 1955,Australian J Chem 8: 267.
4. Arndt F, Loewe L, Ayca E 1952,Chem Ber 85: 1150.
5. Babin R, Coustou F. Delmon G 1955,Bull Soc Pharm Bordeaux 94: 69.
6. Babin R, Delmon G 1955,Bull Soc Pharm Bordeaux 94: 72.
7. Baker LC 1955 Chem and Ind (London), p. 502.
8. Baker LC, Lampitt LH, Wittenburg E 1955,J Sci Food and Agri 6: 682.
9. Barnes MJ, Kodicek E 1972,In: Harris RS, Munson PL, Glover J, Dikzfalucy E (eds) Vitamins and Hormones. 30:
10. Barr NF, King CG 1956,J Amer Chem Soc 78: 303.
11. Blumberg WE, Goldstein M, Lauber E, Peisach J 1965, Biochim Biophys Acta 99: 187.
12. Burger M, Becker K 1951, Amer Soc Brewing Chemists Proc p. 78.
13. Burns JJ 1967 In: Greenberg DM (ed) Metabolic Pathways. New York. Academic Press. I: Chap 7.
14. Butt VS, Hallaway M 1961, Arch Biochem Biophys 92: 24.
15. Campbell RH, Mellon MG 1960, Anal Chem 32: 54.
16. Chapon L, Chapon S, Genty-Amie N, Urion E 1959, Bull Soc Chim France, p. 81.
17. Chinoy JJ, Saxena OP 1978, Botanical Progress 1: 6-75.
18. Chinoy JJ, Saxena OP, Abraham PG, Pandya RB, Dave IC, Gurumurti K, Sastri K 1969, Research Report, PL 480 Research Project, FG-in-182 (A7-CR-87) p. 1-350. Department of Botany, University School of Sciences, Gujarat University, Ahmedabad - 380009, India.
19. Chinoy JJ, Singh YD, Laloraya MM, Saxena OP 1972, Seminar: Current Trends in Plant Sciences, Delhi University, Delhi. Oct. 3-10, p. 47.
20. Chinoy JJ, Gurumurti K, Saxena OP, Singh YD 1972 a, UGC All India Symp: Biology of Land Plants, Meerut University, Meerut, June 18-20, p. 188.
21. Chinoy JJ, Singh YD, Laloraya MM 1974, Vth Int Symp Magnetic Resonance, Jan. 14-18, Bombay, India. Abstr 17 A 64.
22. Chinoy NJ, 1978, J Anim Morphol Physiol Silver Jubilee Volume: 68.
23. Chinoy NJ, Seethalakshmi L, Rao MV, 1978, J. Steroid Biochem 9: 824, Abstr 57.
24. Cier A, Nofre C, Ranc M, Lefier A, 1959 a, Bull Soc Chim France p. 1523.
25. Cier A, Nofre C, Drevon B, Lefier A, 1959 b, Bull Soc Chim France p. 74.
26. Constantinescu DG, Oteleanu R, 1959, Naturwissenschaften 46: 512.
27. Csüros Z, Petro J, 1955, Acta Chim Acad Sci Hung 7: 199.
28. Csüros Z, Petro J, 1957, Periodica Polytech 1: 3.
29. Csüros Z, Petro J, 1958, Acta Chim Acad Sci Hung 14: 95.
30. Csüros Z, Petro J, 1958 a, Acta Chim Acad Sci Hung 14: 381.
31. Dahn H, Loewe L, Burton CA, 1960, Helv Chim Acta 43: 303.

32. Dahn H, Loewe L, Burton CA 1960a, Helv Chim Acta 43: 317.
33. Dahn H, Loewe L, Burton CA 1960b, Helv Chim Acta 43: 320.
34. Dahn H, Loewe L, Luscher E, Menasse R 1960 c, Helv Chim Acta 43: 287.
35. Deshmukh GS, Bapat MG 1955, Z Anal Chem 147: 271.
36. Douzou P 1956, Compt Rend 243: 46.
37. Douzou P 1958, Actions Chim Biol Radiations 4: 86.
38. Douzou P, Gallon D 1956, Compt Rend 242: 3132.
39. Drummond JC, 1920, Biochem J 14: 660.
40. Dudani AT, Krishnamurti CR 1954, Biochem Biophys Acta 13: 505.
41. Duke PS 1968, Experimental and Molecular Pathology 8: 112.
42. Dulkin SI, Friedmann TE 1956, Food Res 21: 51.
43. Ellis GP 1959, Advan Carbohydrate Chem 14: 63.
44. Erickson LE, Gasparetto G 1953, Food Res 18: 178.
45. Fieser LF, Fieser M 1965, Advanced Organic Chemistry, 2nd Ed. Bombay: Asia Publishing House, p. 965.
46. Flesch C, Schuler W, Meier R 1960, Helv Chim Acta 43: 2014.
47. Florkin M, Stotz EH 1963, (eds) In: Comprehensive Biochemistry 5: 84.
48. Foerster GV, Weis W, Staudinger H 1965, Ann Chem 690: 166.
49. Geyer RP, Kydd S, Ryan H 1957, Arch Biochem Biophys 70: 129.
50. Gero E 1949, Bull Soc Chim Biol 31: 817 and 825.
51. Gero E 1955, Compt Rend 240: 1818.
52. Gurevich AA 1963, Doklad Biol Sciences 145: 674.
53. Habermann HM, Vernon LP 1958, Arch Biochem Biophys 76: 424.
54. Haworth WN, Hirst EL, Smith F 1934, J Chem Soc p. 1556.
55. Heimann W, Wisser K 1962, Ann Chem 653: 23.
56. Hendry LB, Witham FH, Chapman OL 1977, Perspectives Biol Med 21: 120.
57. Herbert RW, Hirst EL, Percival EGV, Reynolds RJW, Smith F 1933, J Chem Soc p. 1270, article no. 299.
58. Herrmann J, Andrae W 1963, Nahrung 7: 243.
59. Hewitt EJ, Dickes GJ 1961, Biochem J 78: 384.
60. Heyns K, Paulsen H 1962, Advances in Carbohydrate Chemistry 17: 183.
61. Holker JR 1955, J Chem Soc p. 579.
62. Hsien-Feng 1958, Ts'ai Hua Hsueh Shih Chieh 13: 462.
63. Huelin FE 1953, Food Res 18: 633.
64. Hvoslef J 1964, Acta Chem Scand 18: 841.
65. Ichimonji A 1955, Osaka Diagaku Igaku Zasshi 7: 389.
66. Ichinose Y 1962, Eiyo To Shokuryo 15: 186.
67. Imanaga Y 1955, J Biochem (Tokyo) 42: 657 and 669.
68. Inagaki C, Fukuba H, Kukai M, Toyosato S 1955, Nippon Nogeikagaku Kaishi 29: 416.
69. Inczedy J 1961, Acta Chim Acad Sci Hung 27: 185.
70. Jackson SF, Chicester CO, Joslyn MA 1960, Food Res 25: 484.
71. Jamaluddin M, Rajalakshmi S, Sarma DSR 1981, Biochim Biophys Acta 653: 423.

72. Janecke H 1954, Deut Lebensm Rundschau 50: 65.
73. Jonaitis H, Krisciuniene B 1962, Lietuvas Fiz Rinkinys TSR Mokslu Akad, Aukstosios Motyklos 2: 409.
74. Kamiya S 1960, Nippon Nogeikagaku Kaishi 34: 8 and 13.
75. Kamiya S, Nakabayashi T 1957, Bitamin 13: 250 and 384.
76. Kern DMH 1954, J Amer Chem Soc 76: 1011.
77. Kluge H, Rasch R, Brux B, Frunder H 1967, Biochim Biophys Acta 141: 260.
78. Kolthoff IM, Elving PJ 1962, Treatise on Analytical Chemistry, Vol. 1-9: Part II, New York: Wiley Interscience.
79. Lagercrantz C 1964, Acta Chem Scand 18: 562.
80. Lampitt LH, Baker LC, Wittenburg E 1956, J Sci Food Agri 7: 120.
81. Lavendoski NG, Baker EM, Canham JE 1964, Biochemistry 3: 1465.
82. Lawendel JS 1956, Nature 178: 873.
83. Lawendel JS 1957, Nature 180: 434.
84. Lewin S 1976, Vitamin C: Its Molecular Biology and Medical Potential. London: Academic Press.
85. Lopez JA 1961, Anales Real Soc Espan Fis Quim (Madrid) B 57: 733.
86. Maksimović ZB, Šušić MV 1956, Bull Inst Nucl Sci "Boria Kidrich" (Belgrade) 6: 137.
87. Manecke G, Bahr C, Reich C 1959, Angew Chem 71: 646.
88. Markhel EP, Novotel'nov NV 1961, Izv Vysshikh Uchebn Zavedenii, Pishchevaya Tekhnol 5: 31.
89. Mattock GL 1965, J Chem Soc 4728.
90. Mazdis M 1961, Nature (London) 189: 305.
91. Moriyama T 1957, Biochem (Tokyo) 44: 289.
92. Morse RE 1953, Food Res 18: 48.
93. Murata A, Uike M 1976, J Nutr Sci Vitaminol 22: 347.
94. Nebbia G 1958, Acta Vitaminol 12: 107.
95. Neiadas E, Robert L 1958, Bull Soc Chim Biol 40: 347.
96. Neiadas E, Robert L 1958a, Proc IV Internl Congr Biochem Vienna p. 97. Abstr 8. Pergamon (1960).
97. Nomura D, Uehara Y 1958, Hakko Kogagu Zasshi 36: 290.
98. Norman ROC, Radda GK 1962, Proc Chem Soc p. 138.
99. Novotel'nov NV, Golovkina MT, Bankovskaya VD 1955, Tr Leningr Tekhnol Inst Kholodil'n. Prom 7:12.
100. Novotel'nov NV, Golovkina MT, Stavrova ER 1959, Nauchn Dokl Vysshei Shkoly Biol Nauki No. 1, p. 137.
101. Omura H, Iiyama S, Tomita Y, Narazaki Y, Shinohara K, Murakami H 1975, J Nutr Sci Vitaminol 21: 237.
102. Omura H, Tomita Y, Fujiki H, Shinohara K, Murakami H 1978, J Nutr Sci Vitaminol 24: 263.
103. Ono S, Takagi M, Wasa T 1953, J Am Chem Soc 75: 4369.
104. Ono S, Takagi M, Wasa T 1958, Bull Chem Soc Japan 31: 356.
105. Peel JL, Loughman BC 1957, Biochem J 65: 709.
106. Piette LH, Yamazaki I, Mason SH 1961, In: Blois MS, Brown HW et al (eds) Free Radicals in biological systems. New York: Academic. p. 195.
107. Pihar O 1955, Coll Czech Chem Commun 20: 1151.
108. Pihar O 1955 a, Chem Listy 49: 340.

109. Pihar O 1962, Biochim Biophys Acta 56: 159.
110. Planta C Von 1961, Helv Chim Acta 44: 1444.
111. Prochazka Z, Sanda V, Sorm F 1957, Chem Listy 51: 1197.
112. Pukhav BI 1957, Tr Kirgizsk Gos Med Inst 9: 119.
113. Rakshit PC, Sen BC, Bhattacharya PK 1957, J Indian Chem Soc 34: 668.
114. Rangaswami S, Seshadri TS 1952, Chemistry of Vitamins and Hormones. 2nd edition, Andhra University Waltair: India. p. 183.
115. Rao GG, Rao VN 1955, Z Anal Chem 147: 338.
116. Rao MVL, Sastry LVL, Srinivasan M, Subrahmanyan V 1959, J Sci Food Agr 10: 436.
117. Revol A, Nofre C, Cier A (1958) Compt Rend 247: 2486.
118. Romanchuk PS 1957, Dokl Acad Nuak SSSR 117: 665.
119. Russell GA, Strom ET, Talaty ER, Chang KY, Stephens RD, Young MC 1966, Rec Chem Progress 27: 3.
120. Shamrai EF, Veremeenko KM, Khmelevskii YuV, Prikhozhan VL 1959, Ukr Biokhim Zh 31: 118.
121. Shkodin AM, Tikhomirova GP 1955, Urk Khim Zh 21: 265.
122. Smoczkiewiczowa Am Grochmalicka J 1961, Nature (Lond.) 192: 161.
123. Stich HF, Karim J, Koropatnick J, Lo L 1976, Nature (Lond.) 260: 722.
124. Šušič MV 1954, Bull Inst Nucl Sci "Boris Kidrich" (Belgrade) 4: 57 and 59.
125. Szent Györgyi A 1928, Biochem J 22: 1387.
126. Szent Györgyi A 1960, Introduction to a submolecular biology. New York: Academic.
127. Szent Györgyi A, Haworth WN 1933, Nature (Lond.) 131: 24.
128. Tolbert BM, Downing M, Carlson RW, Knight MK, Baker EM, 1975, Ann N Y Acad Sci (Vitamin C Conference, October 1974).
129. Tomana M, Svabensky O, Otta K 1963, Nahrung 7: 212.
130. Udenfriend S, Clark CT, Axelrod J, Brodie BB 1954, J Biol Chem 208: 731.
131. Verlangieri AJ, Mumma RO 1973, Atherosclerosis 17: 37.
132. Vleeschauwer A de, Tijtgat B, Hendrickx H 1959, Congr Mondial Rech Agron (Rome) p. 1.
133. Wasa T, Takagi M, Ono S 1961, Bull Chem Soc Japan 34: 518.
134. Yamazaki I, Mason HS, Piette L 1959, Biochem Biophys Res Commun 1: 336.
135. Yamazaki I, Mason HS, Piette L 1960, J Biol Chem 235: 2444.
136. Yamazaki I, Piette LH 1961, Biochim Biophys Acta 50: 62.
137. Yano M, Hayashi T, Namiki M 1976, Agricultural & Food Chem 24: 815.
138. Younathan ES, Frieden E 1961, Biochim Biophys Acta 46: 51.

2. BIOGENESIS, DETECTION AND DETERMINATION OF FREE AND BOUND ASCORBIC ACID

Ascorbic acid (AA) is an important biologically active reductant which is widely distributed in animal and plant cells and participates in their metabolism (Aberg, 1958, Mapson, 1958; 1958 a; Chinoy, 1962; 1977; Lewin, 1976; Chinoy, N.J. 1978; Chinoy, N.J. et al., 1982). Green plants are capable of biosynthesizing their own AA, but not all members of the animal kingdom appear to do so. Human beings and other primates, guinea pigs, marmot, Indian giant fruit bat and pipistrelle, bulbul and insects are incapable of synthesizing the vitamin in their bodies (Burns, 1957; Roy and Guha, 1958; Chatterjee et al., 1961; Dutta Gupta et al., 1972; 1973).

1. BIOSYNTHESIS OF ASCORBIC ACID (AA) AND ASCORBIGEN (ASG) IN PLANTS

1.1. Biosynthesis of AA

Several pathways have been discussed in studying the synthesis of AA in young seedlings, fully grown leaves or wounded tissues of various plants which are reviewed elsewhere (Smith, 1952; Aberg, 1953; 1958; Isherwood et al., 1954; Isherwood and Mapson, 1963; Mapson, 1955; 1958; Axelrod and Martin, 1961; Burns, 1958; 1961; 1967; Lewin, 1976; Counsell and Hornig, 1982).

Three major pathways have been proposed for the biosynthesis of AA in plants using radioactive tracer techniques by Isherwood et al. (1954) (two pathways) and Loewus and his coworkers (third pathway). Isherwood et al. (1954) administered a number of lactone and esters of sugar acids to cress seedlings and elucidated that only three lactones, viz., L-gulono- γ-lactone, L-galactono- γ-lactone and L-glucurono- γ-lactone as well as the methyl ester of D-galacturonic acid augmented their AA content. In one pathway, D-glucose was converted to D-glucuronic acid ----→ L-gulonic acid and finally L- AA; in the second pathway, D-galactose ----→ methyl D- galacturonate ----→ L-galactonolactone ----→ L-AA. These pathways involve respective inversion of C-1 of glucose and galactose to C-6 of L-AA. Loewus et al. (1956), Loewus and Jang (1957) and Loewus (1961) investigated the pathway of AA biosynthesis in ripening strawberry (Fragaria) and in germinating cress seedlings by administering D-glucose - 1-^{14}C and D-glucose - 6- ^{14}C and studied the distribution of ^{14}C label on the carbon atoms in crystallized AA, which clearly revealed that the major portion of ^{14}C label found in AA was on the same carbon atom as in the original glucose molecule (Table 2.1). With the use of D-galactose - 1- ^{14}C also, the resulting activity in the AA was again more in C-1 than C-6 (Table 2.2) from D-glucose (derived from sucrose) and galactose (from pectin). However, of the remaining label, an appreciable amount entered C-6. These data suggested that D-galactose was utilized for the biosynthesis of AA after its conversion to a derivative of D-glucose in the strawberry. Loewus and Kelly (1961 a, b) further concluded that higher plants were capable of converting D-glucose to L-AA by a pathway which preserved the carbon skeleton without inverting the carbon chain, but resulted in an epimerization at C-5 from the D- to the L- configuration. The labelling patterns of L-gulonic acid and L-AA in comparison with that of D-galacturonic acid in strawberry are shown in Table 2.3.

Shigeoka et al. (1979) reported that in Euglena gracilis, D-glucose was more effective for AA synthesis than D-galactose and a number of acid derivatives of both these sugars, excepting D-gluconolactone, could augment the AA formation considerably.

Considering the fact that an enzyme in rat kidney preparations readily converted myo-inositol to glucuronic acid (Charalampous and Lyras, 1957), attempts were made (Burns et al., 1956) to demonstrate the formation of L-AA in plants and animals from labelled myo-inositol. However, Loewus et al. (1962) completely ruled out the possibility of myo-inositol serving as an important precursor of L-AA.

Table 2.1. Distribution of ^{14}C in L-ascorbic acid recovered from ripened strawberries which were specifically labelled D-glucose treated (Loewus, 1963)

Labelled glucose given and berry colour	Period of metabolism (Hr)	^{14}C distribution (per cent) Carbon 1	2	3	4	5	6
Glucose-1-^{14}C							
Green	42	55	19	4	1	2	19
White	40	67	1	10	7	1	14
Pink	24	66	1	14	3	1	15
Red	24	56	9	18	2	-	15
Red	120	62	6	9	4	-	19
Glucose-2-^{14}C							
White	23		73		12	10	5
White	46		69	6	11	12	2
Glucose-6-^{14}C							
Green	47	24	1	1	2	-	73

Table 2.2. Distribution of ^{14}C into carbon 1 and 6 only of L-ascorbic acid, sucrose derived D-glucose and D-galacturonosyl residues of pectin from strawberries given D-galactose - 1- ^{14}C (Loewus, 1963)

Berry constituent	Experiment	^{14}C distribution Carbon 1	(per cent) Carbon 6
L-ascorbic acid	1	47	39
	2	65	23
	3	67	25
D-glucose (from sucrose)	1	78	14
	2	78	14
	3	80	15
D-galacturonate (from pectin)	1	67	21
	2	72	23
	3	68	25

Table 2.3. Distribution of radioactivity within the acidic constituents recovered from D-glucuronolactone -1- ^{14}C and -6 - ^{14}C labelled strawberries (Loewus, 1963)

	Site of radioactivity in glucuronolactone					
	Carbon - 1			Carbon - 6		
	% of ^{14}C administered to berry	% of ^{14}C incorporated into acid		% of ^{14}C administered into berry	% of ^{14}C incorporated into acid	
		Carbon 1	Carbon 6		Carbon 1	Carbon 6
D-galacturonic acid	2	99	1	2.5	-	99
L-gulonic acid	Approx. 5	-	100	Approx. 5	100	-
L-ascorbic acid	0.8	-	100	1.0	95	2

Table 2.4. Determination of ascorbic acid and its products by different methods.

Method	Determination of	Reference
1. Biological	Antiscorbutic activity of AA, Total AA	Oliver (1967); Bliss and Gyorgy (1967)
2. Chemical (titrimetric, colorimetric, spectrophotometric)	AA, DHA	See Roe (1954); Cooke and Moxon (1982)
a) methods based on oxidation of AA		Glick et al. (1953); Mapson and Moustafa (1956); Overbacher and Vines (1963); Kum Tatt and Leong, (1964); Oliver, (1967); Hubmann et al. (1969); Erdey and Svehla, (1973); Baraket et al. (1973); Cooke (1974); Omaya et al. (1979).
b) methods based on reactions of AA		Schmall et al. (1954); Shieh and Sweet (1979).
c) methods based on reactions of DHA		see Cooke and Moxon (1982).
d) methods based on reactions of 2,3-diketogulonic acid		Roe and Kuether (1943); Polk et al. (1960) and several others.
e) Using DCPIP;	Most widely used method; reduction of DCPIP to its Leuco form by AA	Aberg, 1958; Jensen, 1962; Lewin, 1976; Malik and Singh, 1980; Mahadevan and Sridhar, 1982.

Table 2.4. contd.

	Method	Determination of	Reference
	Using DCPIP;	AA, ASG, AAU, AA-MM complex	Chinoy et al. (1976 a) see chapter 2 for details. Saxena et al. (1974).
f)	2,4-dinitrophenylhydrazine; Diazotized 4-methoxy-2-nitroaniline; Methylene blue; Potassium ferricyanide; Ferri-a, a-dipyridyl, Phosphotungstic acid; Diazotized sulfanilamide; Silicomolybdic acid; Thionine; Phosphomolybdic acid; p-sulphophenyl hydrazine; Perinaphthindanetrione hydrate; Bromosuccinamide; titanium salts; uranium nitrate; Osmic acid; Chloramine T with variamine blue; Mercuric chloride and iodine.	AA	Same as under 2 e.
g)	P-chloromercuribenzoic acid (PCMB)	AA, specific, does not allow interference by other compounds	Owen and Iggo, 1956.
h)	Ferric chloride, pH 1.2.	Highly specific even in presence of reductone and other -SH compounds	Sullivan and Clark, 1953

Table 2.4. contd.

Method	Determination of	Reference
i) Formaldehyde	Highly specific	Aberg, 1958
j) 2,4-dinitrophenylhydrazine (DNPH)	Rapid assay, comparable in sensitivity and rapidity with other methods.	Roe and Kuether, 1943; Bessey et al., 1947; Polk et al., 1960; Dasgupta et al., 1962.
k) 1,2-Naphtho-quinone-4-sulphonic acid	Quantitatively reduced by AA	Hubmann et al., 1969.
l) Bromate	-	Sant and Mukherji, 1959.
m) Cupric sulphate or acetate	-	Baraket et al., 1973.
n) Using free radical a, a-diphenyl-B-picrylhydrazyl (DPPH)	-	Blois, 1958.
3. Enzymatic	AA, DHA	Mandels, 1953; 1953 a; Marchesini et al., 1970; Marchesini et al., 1972; Malik and Singh, 1980; List and Knechtel, 1980.
a) involving a fairly specific ascorbic acid oxidase (AAO)	AAO	
b) ascorbic acid peroxidase	AAO, Peroxidase	Yamaguchi and Ogata, 1978; Chinoy et al., 1969; Chinoy, N.J. 1973; Kelly and Latzko, 1980.
c) Involving C-AMP phosphodiesterase for hydrolyzing lactone structure of AA	AA	Lewin, 1976.

Table 2.4. contd.

Method	Determination of	Reference
4. UV absorption	AA	Chayen, 1953.
5. Paper, Column and Thin layer chromatography (TLC)	AA	Roe, 1954; Sweeley et al., 1963; Pfeilsticker, 1968; Allison and Stewart, 1971.
6. Gas liquid chromatography (GLC)	AA, DHA separation	For details see Cooke and Moxon, 1982 approximately 25 papers.
7. High performance liquid Chromatography (HPLC)	" "	Cooke and Moxon, 1982; approx. a dozen or more papers; Loetscher et al., 1978; Rouseff, 1979.
8. Automatic methods	AA	
a) Using DCPIP		Nesset et al., 1967; Beyer, 1967; Hoffman et al., 1970; Egberg et al., 1973.
b) Using O-phenylenediamine (OPD)		Roy et al., 1976; Egberg et al., 1977; Kirk and Ting, 1975; Dunmire et al., 1979.
c) Using 2,4-DNP		Pelletier and Brassard, 1975; 1977; Aeschbacker and Brown, 1972; Behrens and Madare, 1979.

Table 2.4. contd.

Method	Determination of	Reference
d) Diazotized 4-methoxy-2-nitro-aniline		Wilson and Guillan, 1969; Geller et al., 1969.
e) Polarographic technique using a tubular carbon electrode		Mason et al., 1972.
f) Amperometric procedure		Pasadka and Macholan, 1979.
g) Potentiometric procedure		AOAC, 1980; Davidek et al., 1971; Duetsch and Weeks, 1965.
h) Fluorometric	AA	Omaye et al., 1979.
i) Ion mass spectrometry		Benninghoven, 1979.
j) Isotachophoretic method	"	Rubach and Breyer, 1980.
k) Triple barrel C filament microelectrode and differential pulse voltameter scans	"	Plotsky, 1982
8. Radioactive tracer assay	"	Baker et al., 1963; Saari et al., 1967.

Table 2.4.A. Methods for the isolation and determination of ascorbigen in cruciferous and non-cruciferous plants

Method	Determination of	Reference
1. Biological	Anti-scorbutic property	Sumtsov, 1947; Kiesvaara and Virtanen, 1962, 1963; Virtanen and Kiesvaara, 1963; Terenteva, 1953.
2. Chemical Using titrimetric, colorimetric, spectrophotometric methods with DCPIP	ASG	See review by Guha 1961; Bentley, 1961; Kutacek et al., 1960; Malakar, 1963; Chinoy et al., 1969; 1969 a; 1976 a; 1970; 1970 a, b; 1972; 1972 a; 1973; Chinoy and Saxena, 1972; 1978; Saxena, 1969; Saxena et al., 1972; 1972 a; 1973; Jani, 1973; Mehta, 1973; Thakker, 1974; Chinoy N.J., 1978 (in animal tissues).
3. Paper and column Chromatography, Counter current extraction and fractional precipitation		Prochazka & Severa, 1960, Prochazka and Koristek, 1951; Prochazka, 1953 a, b; Prochazka et al., 1956; 1957; Prochazka and Sanda, 1958; 1960; Malakar and Guha, 1958; 1961; Kutacek et al., 1957; Kutacek, 1960; Valenta et al., 1958; Basu, 1961; Bose and Guha, 1963; 1963 a.
4. Paper ionophoresis		Sicho and Bradakova, 1955

The *in vitro* studies of Mapson and coworkers (Mapson, 1967; Mapson et al., 1954; Mapson and Breslow, 1958) and others (Bublitz and Lehninger, 1961), revealed that a mitochondrial enzyme, L-galactono- γ-lactone dehydrogenase of peas and pea seedlings and cauliflower florets could oxidize L-galactono- γ-lactone to L-AA in the presence of oxygen. This enzyme was highly specific for its substrate (Mapson, 1967; Mapson and Isherwood, 1958) and catalyzed AA synthesis in two steps as follows:

D-galacturonic acid derivatives $\xrightarrow{+ NADPH_2}$ L-galactonic acid derivatives + NADP

L-ascorbic acid $\xrightarrow[\text{L-galactono-}\gamma\text{-lactone dehydrogenase}]{\text{mitochondria}}$ L-galactono- γ-lactone

Although considerable quantity of the AA has been localized in the chloroplast, enzymic studies bring out the fact, that, the final stages of its synthesis are catalyzed by the enzymes of the mitochondria, whereas, microsomal enzymes in the soluble fraction of the cytoplasm control the initial stages of this biosynthesis (Mapson, 1967). The continued production of hexose sugars, NADPH, presence of oxygen and intact mitochondria are the necessary factors for sustaining the biosynthesis of AA at a high rate (Mapson et al., 1954; Mapson, 1967).

1.2. Biosynthesis of ASG

The biosynthesis of ascorbigen (ASG) using labelled/unlabelled precursors, viz., D, L- tryptophan-3-(^{14}C formyl) indolic acid, 3-hydroxy-methyl-indole and AA or from indole, formaldehyde and AA, has been studied in Brassica oleraceae L cv. Sabauda (Savoy cabbage) by a number of workers (Kutacek et al., 1960; 1962; Gmelin and Virtanen, 1961; Piironen and Virtanen, 1962; Veres et al., 1963).

2. BIOSYNTHESIS OF AA AND ASG IN ANIMALS

2.1. Biosynthesis of AA

The biosynthesis of L-AA in rat occurs from D-glucose and D-galactose, although the latter sugar was a more effective precursor of L-AA than D-glucose (Evans et al., 1960). The carbon chain of D-glucose was converted intact while undergoing inversion of configuration into L-AA (Burns, 1967; Lewin, 1976). However, the D-glucose had to be converted to D-glucuronic acid via the oxidation of uridine diphosphate glucose to uridine diphospho-glucuronic acid, the reaction being catalysed by a NAD-linked microsomal enzyme present in the liver (Storey and Dutton, 1955: reviews: Degkwitz et al., 1964; Burns, 1967; Chinoy N.J., 1978).

Most of the studies in animals support the fact that microsomal enzymes are involved in the conversion of L-glucuronolactone to L-AA. However, the use of D-glucuronic acid as substrate, requires its conversion to L-gulonolactone first and then to L-AA by microsomal enzymes (Chatterjee, 1970; Gupta et al., 1970; Lewin, 1976). Thus the biosynthesis of L-AA in rats occurs from D-glucose and uronic as well as aldonic acid derivatives. As in plants, lactone dehydrogenase enzymes from animal tissues have also been reported (see Mapson, 1967). Moreover, a number of lactonases have been found in animal tissues which use free L-gulonic or L-galactonic acid as substrates and convert them to L-AA via their γ-lactones (Hassan and Lehninger, 1956; Bublitz and Lehninger, 1961; Shimazono and Mano, 1961).

The loss of capacity to synthesize AA in primates and guinea pigs is an example of evolutionary loss of function and is related to the genetic absence of the enzyme, gulono-γ-lactone to AA. Ray Chaudhuri and Chatterjee (1969) and Chatterjee (1973) have shown a phylogenetic trend amongst animals on the basis of their ability to synthesize AA in various tissues. In the evolutionary ascent, the enzyme originally residing in the kidney, gradually passes into the liver and finally disappears from the liver also. However,

Rajalakshmi et al. (1967) confirmed the synthesis of AA in human placenta. Earlier, Bagchi (1952) had reported the absence of scurvy during pregnancy and lactation in poor Indian women. Unlike plants, the storage, distribution and synthesis of AA in rats and cockerals are regulated by hormones (Dieter, 1969; Majumder and Chatterjee, 1974; Chinoy, N.J. and Seethalakshmi, 1978; Chinoy, N.J. and Rao, 1979; Chinoy, N.J. et al., 1979). Apart from other factors, traces of Mn^{+2} and cobalt stimulate the synthesis of L-AA from L-gulonolactone (Sebrell and Harris, 1967; Sasmal et al., 1968; Kutsky, 1973).

2.2. ASG in animals

The existence of bound ascorbic acid was reported in the liver of the pig by Summerwell and Sealock (1952) but their method was criticized by Lewis et al. (1960). Later, Malakar (1963) found bound AA in goat liver and Chinoy, N.J. (1978) has reported its presence in several animal tissues by using the DCPIP method of Chinoy et al., (1976 a).

3. HISTOCHEMICAL LOCALIZATION

A number of workers have highlighted the role of AA in plant metabolism, growth and development (Chinoy, 1962), as well as in the general metabolism of animal tissues (Burns, 1960; Davidson and Passmore, 1963; Mitchell, 1964; Martin et al., 1981; Chinoy, N.J., 1969, 1969 a; 1970; Chinoy, 1978; Chinoy, N.J., et al., 1978; 1979 a; 1983). A need therefore arises for a specific method for the *in situ* localization of AA, so as to obtain information regarding its metabolic role at the cellular level.

Ascorbic acid localization presents difficulties which soluble compounds offer (Chayen, 1953; Jensen and Kavaljian, 1956). The localization procedure is based on the high reducing capacity of AA in acidic medium. The tissue is placed in an acidified silver nitrate solution in the dark and the black deposits of metallic silver are taken as sites of AA localization. No other substance normally found in the cell can cause the reduction of silver nitrate ($AgNO_3$) to Ag under these conditions (Chayen, 1953; Jensen, 1962).

Moreover, another difficulty in localization of AA is on account of its extreme sensitivity to oxygen (Chayen, 1953; Jensen and Kavaljian, 1956; Pearse, 1968). This was largely overcome by some of the previous workers who developed methods for AA localization where the material under study was microtomed or sectioned, treated with H_2S gas to reduce dehydroascorbic acid (DHA) and then brought in contact with silver nitrate reagent (see Pearse, 1968).

A histochemical procedure for the *in situ* localization of AA using freeze-dried material was developed by Jensen and Kavaljian (1956) in which sections of 10 μ were mounted by pressing them on slides coated with Haupts adhesive and allowed to remain overnight on a warming table at 40°C. The paraffin sections were then exposed to an atmpsohere of H_2S for 15 minutes to reduce all the DHA back to AA and subsequently high purity nitrogen was passed over them to remove the H_2S. The paraffin sections were then transferred to 10% aqueous silver nitrate solution containing 3% acetic acid for 4 - 24 hours, washed rapidly in water, dehydrated in ascending grades of ethanol, deparaffinated and mounted. From the time of transferring the tissue in silver nitrate, all following steps were carried out in the dark with only a weak red safety light. The sections were then stained by adding crystal violet in absolute alcohol to the xylene and at the same time removing the paraffin.

The three controls used were: (i) 3% acetic acid without the silver nitrate; (ii) sections were treated with copper sulphate solution for a short time in order to convert all the AA into DHA and then treated as before: (iii) the tissue was handled in the same manner as the other tissue, but was stained with azure B or Heidenhain's haematoxylin to show the effects of treatments on the morphology of the cells. The use of free-dried tissue and allowing the paraffin to remain in the tissue during the localization procedure, prevented AA from being displaced from its original *in situ* position.

On the contrary, the method has some drawbacks, viz., the use of aqueous silver nitrate solution either with or without the addition of acetic acid, incubation of the test material or sections at 37°C to 56°C, and exposing the sections of H_2S prior to treating them with aqueous solution of silver nitrate, not only increases the possibility of diffusion of AA from its original *in situ* position in the tissue, but also causes nonspecific staining of a number of other reductants besides AA.

These problems have been studied in detail in the writer's laboratory by a number of workers and suitable modifications made in the procedure to ensure the specificity of the silver reaction for the histochemical localization of AA in a variety of plant and animal tissues (Dave et al., 1968; Madhavan Unni and Shah, 1968; Dave et al., 1969; Chinoy, N.J., 1969; 1969 a; Chinoy, N.J. and Sanjeevan, 1978). The procedure outlined by Dave et al. (1969) is described here. The reagents A, B, used in this method were prepared as follows:

Reagent A: Five gram of silver nitrate was dissolved in 34 ml of glass distilled water. This solution was gradually added to 66 ml of absolute alcohol and five ml of glacial acetic acid and then stored in an amber coloured bottle at a low temperature (5 to 10°C) in the dark.

Reagent B: Five ml of liquor ammonia was added to 95 ml of 70% alcohol.

Fresh floral buds of Justicia repens L. and Tradescantia sp. were transferred to darkened specimen tubes containing reagent A and the tubes were immediately kept in a refrigerator at 0 to 3° C for about a week. The tissues were washed with three changes of reagent B for 15 minutes in order to remove the excess of silver nitrate. They were then dehydrated using t-butyl alcohol (TBA) series, embedded in paraffin and microtomed. The sections were deparaffinated in absolute xylol and mounted in Canada balsam.

Black, brownish-black or tan coloured granules or colloidal deposits of silver indicated the presence of AA. The above mentioned modified method was found to be very specific for AA as it was fixed before it had a chance to be displaced from its original position *in situ*. Alcohol in the reagent A helped in its quick penetration into the tissue. Another important modification in the method was the treatment of the tissue at low temperature (0 to 3° C) after the addition of the silver nitrate reagent. The low temperature not only effectively prevented enzymic oxidation of AA (which is often quite rapid at room temperature and above), but also prevented the reduction of silver nitrate by substances like glutathione, cysteine and others.

Using the above mentioned silver nitrate method, Dave et al. (1969) studied the AA content of different regions of the floral buds. The histochemical preparations (Figs. 2.1, 2.2, 2.3) clearly bring out the fact that cells of the procambial region, conducting zone, ovular region, archesporial cells of the anther primordia and chromosomes of the microspore mother cells at different divisional stages are all full of black silver granules indicating a high concentration of AA. Two distinct regions of localization, viz., cytoplasmic and nuclear, were observed. The silver granules were found to be thinly scattered in the cytoplasm of a mature cell, but considerably denser in the nuclear region of an actively dividing meristematic cell. During the different stages of meiotic division, the histochemical pattern varied considerably as metaphase and early anaphase showed lower density of AA, whereas, its intense localization occurred during the late telophase.

The above described histochemical method for the localization of AA was adopted by Chinoy, N.J. (1969; 1969 a) for a variety of animal tissues and she carried out extensive investigations for the confirmation and reappraisal of the

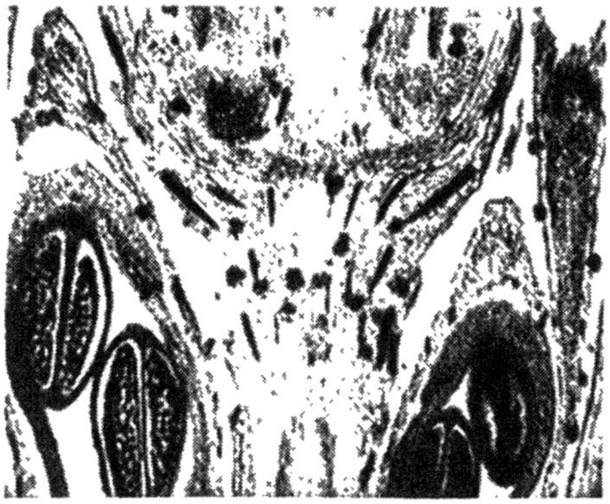

FIGURE 2.1. Longitudinal section of the inflorescence of *Justicia sp.* showing differential silver deposits in conducting tissues and anther lobes (Courtesy: Dave et al., 1969) x 100.

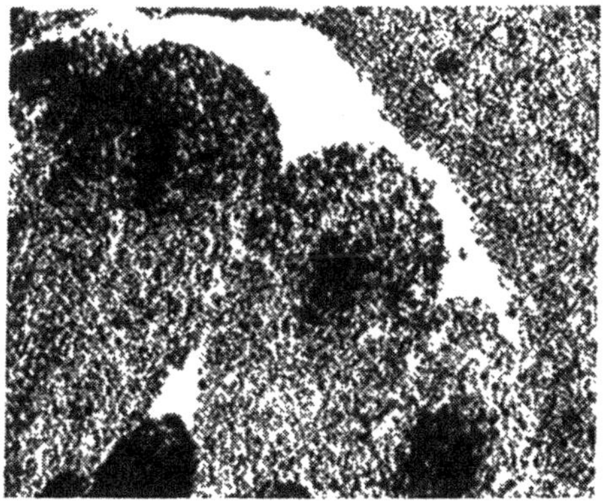

FIGURE 2.2. Vertical section of floral bud of *Tradescantia sp.* (Courtesy: Dave et al., 1969). x 100

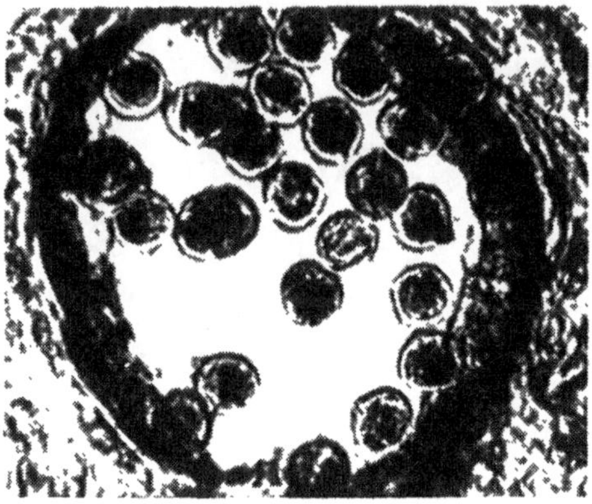

FIGURE 2.3. Microsporocytes of Tradescantia sp. at metaphase and telophase (Courtesy: Dave et al., 1969). x 300.

specificity of the silver reaction of AA (Chinoy, N.J. 1969 a: 1978; Chinoy, N.J. and Sanjeevan, 1978; Chinoy, N.J. et al., 1982; 1983) as well as developed an electron microscopic method (Chinoy, N.J., 1979).

The Blue Rock pigeon (Columba livia) and the common Pariah kite (Milvus migrans) Govinda Sykes), both maintained in the laboratory, as well as the Indian Roller bird (Coracias benghalensis Linn.) and pond Heron (Ardeola grayii Sykes) which were shot directly from the field were utilized. At autopsy of the animals, small pieces of M. pectoralis major were blotted free of blood and dropped into specimen tubes containing chilled, alcoholic silver nitrate reagent and treated as described earlier (Dave et al., 1969). After the requisite time in silver nitrate solution which varied with the experimental material, ranging from 24 hours or more, the reagent was decanted off and the tissue washed three to four times with alcoholic ammonia for 15 to 20 minutes, dehydrated, embedded in paraffin and sectioned (5 to 7 μ).

Fresh frozen sections of the muscle were also treated in the same manner, dehydrated and mounted as before in Canada balsam for comparison with the sectioned tissue. No stain was used other than treatment with alcoholic acidic silver nitrate. The control material was devitaminized by 10% formaline for 2 to 3 hours at room temperature and processed as already stated.

It was found (Chinoy, N.J., 1969) that the red muscle fibres had a comparatively greater deposition of silver granules than the white ones, indicating a higher AA content in the former (Fig. 2.4). Taking into consideration the presence of oxidative enzymes in the red muscle fibres besides fat (George and Berger, 1966), as well as a high content of AA in the same, Chinoy, N.J. (1969, 1970, 1978; Chinoy, N.J. et al., 1978; 1979 a; 1982; 1983) postulated that AA participates in the energy transfer mechanisms via its peroxidative transformation into its free radical, monodehydroascorbic acid (MDHA), which is a much more powerful electron donor on account of its unpaired electron. In a further study to establish the specificity of the silver reaction for AA, Chinoy, N.J. (1969 a) and Chinoy, N.J. and Sanjeevan (1978) have carried out investigations with a number of substances naturally occurring in plants and animals, which are likely to reduce silver nitrate under a given set of conditions. The following reactions were carried out at low temperature (0 to 3°C) for more than 24 hours using the alcoholic, acidic silver nitrate reagent described earlier.

(i) Two ml AA solution (3 mg/ml) + 3 ml reagent A, with or without 5% NH_3 or else with 3 ml ethanol gave a black precipitate (ppt). But, 2 ml glucose/sucrose (3 mg/ml) and 3 ml of reagent A yielded no black ppt.

(ii) Two ml AA solution + 2 ml 5% copper sulphate ($CuSO_4$), or a mixture of 2 ml 5% $CuSO_4$ + 2 ml water + 3 ml reagent gave a dirty white ppt or no ppt at all.

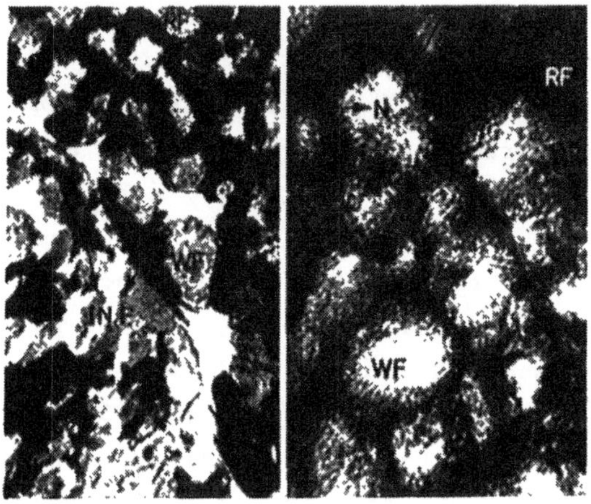

FIGURE 2.4 (1). Transverse section of M. pectoralis major of pigeon. The figure shows the interfiberal (IN. F.) localization of AA around the muscle fibres. WF = White fibres; RF = Red fibres. x 400. A similar localization was also obtained in the muscle of the other three birds studied (Courtesy: Chinoy, N.J., 1969).

FIGURE 2.4 (2). Transverse section of fresh frozen M. pectoralis major of pigeon. The red muscle fibres(RF) and their nuclei (N) show a greater deposition of reduced silver granules than the white ones (WF). x 428. (Courtesy: Chinoy, N.J., 1969).

(iii) When 2 ml of glutamic acid, cysteine, DL-valine, L-methionine, DL-alanine, L-leucine and L-tyrosine (3 mg/ml) each were reacted separately with 3 ml of reagent A, a milky white ppt was obtained. Similarly, 2 ml each of glutamic acid and 5% $CuSO_4$ and 3 ml of reagent A also yielded a white ppt. On the contrary, a mixture of 2 ml of glutathione (3 mg/ml) + 3 ml reagent A gave no ppt even after 24 hours at low temperature.

It was reported that at an acidic pH between 2 and 2.5 (Chinoy, N.J. 1969), did not cause the reduction of silver nitrate by reducing sugars and milder reductants such as cystein and glutathione. According to Roe (1954), the interference of polyphenols and sulphydryl compounds was diminished by carrying out the reduction at a pH below 4. Thus the interference of polyphenolic compounds is ruled out prima facie. Shah ahd Dalal (1978) have conducted *in vitro* experiments wherein 2 ml of various types of phenolic compounds were allowed to react with 3 ml of silver nitrate reagent. The phenolic compounds gave an extremely weak reaction only at room temperature ($27 \pm 2^{\circ}$ C), as opposed to the instantaneous reaction of AA with silver nitrate (Chinoy, N.J. and Sanjeevan, 1978). Besides, some of the phenolic compounds were found not to interfere even at room temperature after prolonged incubation.

The use of an appropriate washing solution is also important. The alcoholic ammonia (reagent B) used not only facilitates the removal of excess silver, but also the subsequent dehydration of the tissue.

In the histochemical methods outlined by various workers, no proper mention is made of the controls. Besides, the above mentioned reactions establishing the specificity of the silver reaction for AA, Chinoy, N.J. (1969 a) and Chinoy, N.J. and Sanjeevan (1978) have indicated a number of precautions to be observed in using the method which are as follows:

1. The preparation of the alcoholic, acidic silver nitrate reagent should be undertaken using only glass double distilled water saturated with CO_2.
2. Fresh blocks of tissues should be directly treated with alcoholic, acidic silver nitrate reagent in the dark at 0 to 3° C, for 24 hours or longer in a refrigerator. The reagent contains fixatives like acetic acid and alcohol which penetrate into the tissues rapidly and the silver nitrate in it fixes AA *in situ*.
3. At the end of the reaction period, the experimental material was thoroughly washed two to three times in alcoholic ammonia for 10 to 15 minutes (5 ml liquor ammonia + 95 ml 70% alcohol), dehydrated in ethanol series, embedded in

paraffin and microtomed. After the ribbons were dry, they were deparaffinated in xylol and mounted in Canada balsam or DPX.

4. The temperature of the reaction is maintained at a low level (0 to 3° C) in order to prevent oxidation of AA as well as to eliminate the possibility of reduction of acidic silver nitrate by substances like glutathione. Sharma and Sharma (1965) as well as Chayen (1953) have also recommended the maintenance of low temperature.

The following three controls were also used by the investigators:

1. The treatment of tissues or sections with 6 to 10% formalin prior to the application of alcoholic, acidic silver nitrate reagent. This treatment devitaminized the tissue within 3 to 4 hours.

2. Alternately, the tissues could be treated with 5 to 10% $CuSO_4$ solution for 24 hours.

3. The silver nitrate in reagent A was replaced by sodium chloride according to the method of Willis and Kratzing (1974).

3.1. Measurement of intensity of histochemical reaction

After the histochemical processing, the intensity of the staining could be measured by a simple, cytophoto-electrometer devised in the author's laboratory (Chinoy et al. 1971) (Fig. 2.5) and which is subsequently improved upon (Chinoy et al., 1976). A pointolite microscope lamp (A) produces a beam of light which after passing through the condenser (B) and a blue filter illuminates the preparation on the slide kept on the stage (C). This light passes through the objective (E) and the prism (F) and emerges through the eye piece (G), which protrudes out of the dark chamber and is reflected by the mirror (H) on the platform (I) with a very small aperture below which is a photo-electric cell (J-Photronic Cell, Weston, Model S. 123) is fitted. The photocell is connected with a mirror galvanometer which in turn is fitted with a lamp and a scale. The reading is taken by bringing the image of the cell or the nucleus concerned in alignment with the aperture on platform and noting the deflection on the galvanometer scale (M). The difference between the transmission value of the histochemical preparation of a cell, nucleus or a nucleolus and that of the corresponding control, gives the extinction value. At least ten such readings were taken for every tissue. (Fig. 2.5).

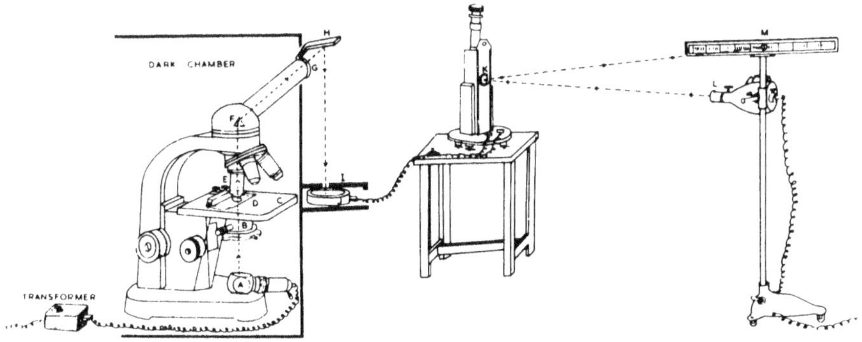

FIGURE 2.5. The cytophotoelectrometer (Courtesy: Chinoy et al., 1971, 1976).

The extinction value (e. value) was multiplied by the cross sectional area of the cell, nucleus or the nucleolus to obtain the AA content per cell. The extinction value was also divided by the cell area to obtain the concentration of AA per unit area of the cell.

Utmost precaution should be taken to maintain uniformity in cutting the sections for the histochemical preparations as well as its control. Photographing of both should also be regulated by maintaining equal distance between the lens and the object as well as during making prints.

4. BIOCHEMICAL DETERMINATIONS

Ascorbic acid is a very strong reducing agent and is highly soluble in aqueous solutions. It can be quantitatively extracted from the tissue by 0.5% oxalic acid, 5% metaphosphoric acid (HPO_3) and 10% acetic acid. The determination of AA is complicated by the readiness with which it is oxidized to dehydroascorbic acid (DHA) and diketogulonic acid (DKG). This oxidation must be prevented if true AA content of the tissue is to be measured (Jensen, 1962).

The accuracy of a biological assay of ascorbic acid (AA) depends upon a number of factors and great care as well as precautionary measures are needed to maintain the validity of such a test. It is more so in the case of AA than in any other vitamin or hormone, on account of its high reactivity with oxygen of the atmosphere as well as by interference of ferrous, cuprous salts, glucose amine type of compounds, sulphite salts, pigments, phenols, -SH compounds, reductone, reductic acid and tartronaldehyde (Elizabeth, 1963) with metal ions, especially copper, stannous and iron. Besides that, the AA content of different organs of the plant, such as fruits, flowers, leaves, growing spices and others vary considerably not only with age but also with changes in light, temperature, conditions of cultivation and genetic constitution. AA is also susceptible to autoxidation and enzymatic oxidations. There is also active complexing of AA with macromolecules and other smaller molecules during metabolism (Chinoy, 1962; Prochazka et al., 1956; 1957), and therefore it is necessary to use freshly cut material for feeding experiments.

The various experimental procedures that have been developed for its determination in different biological materials could be conveniently divided into bioassay or biological methods, chemical methods, radioactive tracer determinations, enzymatic techniques, paper, column, thin layer, gas chromatography, high performance liquid chromatography (HPLC) and automated methods (ref. Table 2.4 for references to these techniques and their grouping). The methods for determination of ascorbigen have also been listed separately in Table 2.4. A.

4.1. A method for the simultaneous determination of AA, ASG, AAU and AA - MM complex (Chinoy et al., 1976 a).

A 2, 6 - dichlorophenol indophenol colorimetric method has been developed by Chinoy et al (1969; 1969 a; 1976 a) for the simultaneous determination of ascorbic acid (AA), ascorbigen (ASG), ascorbic acid utilization (AAU) and ascorbic acid - macromolecule complexing (AA- MM complex) from a given plant sample. The solutions required are:

1. Metaphosphoric acid (HPO_3) ; 3% and 15% solutions. These can be stored in a refrigerator for four days.
2. Buffer solution A: 10.55 gram of citric acid dissolved in 100 ml of 1 N NaOH. The buffer solution B (31.65 gram of citric acid dissolved in 100 ml of 3 N NaOH) was used for estimations of ASG and AA - MM complex.
3. Buffered HPO_3: Two volumes of 3% HPO_3 were mixed with one volume of buffer solution A (pH 3.6). This solution was freshly prepared before use daily.
4. Ascorbic acid solution for AAU: 10 mg of ascorbic acid powder was dissolved in glass distilled water, which was previously boiled, cooled and saturated with CO_2. The volume of the solution was made upto 100 ml and stored in an amber coloured bottle at 3 to 5° C, till used. This solution was also freshly prepared daily.
5. Standard dye solution: 10 mg of 2, 6 - dichlorophenol - indophenol (DCPIP) (BDH) was dissolved in deionized water at 80° C, cooled, and its volume made upto 200 ml. Its strength was standardized by the addition of either deionized water or more concentrated dye solution until 5 ml of the dye solution when mixed with 2 ml of deionized water, gave a reading of 390 on the scale of the Klett-Summerson colorimeter or an optical density (O.D.) of 0.78. This standardization of the dye solution was done every time just prior to starting of estimations.

Standard Curve: Ascorbic acid solutions of concentrations ranging from 0.01 to 0.09 mg/ml were prepared from a stock solution by diluting with cold, CO_2 - saturated glass distilled water to the requisite concentration.

One ml aliquot of each AA solution was mixed with 1.0 ml of buffered HPO_3 solution and 5 ml of standardized dye solution and the readings were recorded on a Klett-Summerson colorimeter using a green filter. Ten to 20 replicates were taken for each concentration of AA solution and these were converted into O.D. by multiplying with 0.002 (calibration factor of Klett Summerson colorimeter).

The regression of AA concentration on O.D. was calculated and the position of the trend line determined from the following regression equation:

$$Y = 0.1103 - 0.14 X$$

where Y = concentration of AA in mg and
X = optical density

Extraction: The weighed sample of plant material was placed in a mortar and covered with 1 - 2 ml of ice cold CO_2 saturated glass distilled water (DW) and a pinch of acid-digested and thoroughly washed silica sand free from any chemical impurity. It was then quickly homogenized and the contents were transferred to a test tube. The pestle and mortar were rinsed two to three times with 1 to 2 ml of cold CO_2 - saturated glass DW, transferred to the test tube and the volume of the homogenate was made upto 16 ml with DW. The homogenate was divided into three parts:

i) 4 ml for the estimation of AA,
ii) 4 ml for the estimation of ASG, and
iii) 8 ml for the estimation of AAU and AA - MM complex respectively.

4.2. Determination of AA

Four ml cooled buffered HPO_3 was added to an equal volume (4 ml) of the original homogenate and after thorough shaking, a 2 ml aliquot of the mixture was diluted with 5 ml of DW, and the pointer on the colorimeter scale was adjusted to '0' for the turbidity factor. Five ml of standard dye solution was added to another 2 ml aliquot of the same mixture and the colorimetric reading was noted.

Calculation: The colorimetric reading was converted into optical density (O.D.) by multiplying with 0.002 (the calibration factor of Klett Summerson colorimeter). Using the regression formula, Y = 0.1103 - 0.14 X, the concentration of AA in one ml of the original extract and then AA/g fresh weight of the material was calculated.

Considerable evidence has now accumulated to demonstrate the presence of ascorbigen in plants (see reviews: Guha, 1961; Bentley, 1961). Its chemical structure has been determined and two types of ASG recognized (see Chapter I for details). It is also known that the ASG content of the plant is affected during germination, vegetative and reproductive growth and differentiation, before and after flowering, as well as during various stress conditions, including drought as detailed elsewhere in this book. The comparatively high levels of ASG during reproductive differentiation of shoot apex of wheat and a number of other cereals, suggests that ascorbic acid complexes with macromolecules (Chinoy, 1962; Chinoy et al., 1958) and there is sufficient experimental evidence to support the above mentioned contention. Important contributions on ASG of different plants by various workers from the writer's laboratory are listed in Table 2.4 A.

4.3. Determination of ascorbigen (ASG)

Two ml of 15% HPO_3 (reagent no. 1) was added to 4.0 ml of the original homogenate and the mixture was kept in a water bath at 75° C for 15 minutes in order to hydrolyze the ascorbigen. After cooling, the system was buffered to pH 3.6 by adding 2.0 ml of the buffer solution B, thus increasing the volume of the mixture to 8.0 ml. To a 2.0 ml aliquot of this buffered - hydrolyzed extract, 5.0 ml of distilled water was added and the pointer on the colorimetric scale was adjusted to '0' for the turbidity factor. Five ml of the standard dye (DCPIP) solution was added to another 2.0 ml aliquot of the buffered hydrolyzed extract and the colorimetric reading was noted.

The concentration of AA in 2.0 ml of the buffered hydrolyzed extract was calculated from the regression formula and the value of free AA in 1.0 ml of the homogenate, which was determined previously, was subtracted from it to obtain the AA equivalent of ascorbigen (ASG) in 1.0 ml of the homogenate. The ASG content per gram fresh tissue weight was then calculated as follows:

$$B = \frac{v\ (b - a)\ .\ 1000}{w}$$

where, B = AA equivalent of ASG in mg/g fresh weight of the sample, b = AA (mg) in 2.0 ml of buffered-hydrolysed extract a, v and w are the AA in mg/ml of the original homogenate, volume of the original homogenate and weight (in mg) of the plant sample used for analysis respectively.

4.4. Ascorbic acid utilization (AAU)

A number of enzyme systems capable of catalyzing the oxidation of AA in plant extracts have been discovered. Ascorbic acid oxidase (AAO), a copper containing enzyme (whose prosthetic group contains six copper atoms per molecule) is of widespread occurrence in plant tissues (Aberg, 1958; Mapson, 1958; Price, 1974; Malik and Singh, 1980; Mahadevan and Sridhar, 1982). Szent-Györgyi (1931) first described AAO from cabbage, cucurbits and crucifers, which are rich sources. The enzyme is considered as a soluble oxidase, but an insoluble counterpart exists in some tissues, e.g., barley seedling (Butt and Hallaway, 1958), tobacco callus (Newcomb, 1951). A large fraction of enzyme activity was found to be associated with the cell wall fraction which is normally discarded during preparation of enzyme (Beevers, 1961).

AAO regulates the levels of reduced and oxidized glutathione and NADPH, due to the existence of a system linking NADPH to oxygen (Aberg, 1958; Mapson, 1958; Price, 1974; Malik and Singh, 1980; Mahadevan and Sridhar, 1982). AAO catalyses the direct oxidation of AA by molecular oxygen while polyphenol oxidase, laccase and peroxidase act via the intermediate formation of quinone derivatives, and cytochrome oxidase via cytochrome C (Bonner, 1957; Aberg, 1958; Mapson, 1958; Chinoy, 1962; Chinoy, et al., 1969; 1969 a; 1970). In some cases these enzyme systems may readily oxidize all AA to DHA upon grinding, while in other cases, natural inhibitors, e.g., oxalic acid, strongly retard this process (Aberg, 1958; Mapson, 1958). Hence a method for the determination of ascorbic acid utilization (AAU) was developed.

Determination of ascorbic acid utilization (AAU): Eight ml of AA solution for AAU (10 mg/100 ml) were added to 8.0 ml of the original homogenate and the mixture was incubated at 30° C $\pm$ 2° C with thorough shaking after every 10 minutes. Two ml aliquots were taken from each of these 8.0 ml portions immediately after incubation as well as after 1,2 and 3 hours of incubation respectively and analyzed for AAU and AA - MM complex as described below.

Two ml of buffered HPO_3 was added to two ml of the incubated solution immediately after incubation and after thorough shaking, a 2.0 ml aliquot was mixed with 5.0 ml of DW and the pointer of the colorimeter adjusted to '0' as before for the turbidity factor. Another 2.0 ml aliquot was mixed with 5.0 ml of the standard dye (DCPIP) solution and the colorimetric reading was noted, which was converted to O.D. as before, and the AA content in mg of 2.0 ml aliquot of the buffered incubated solution (i.e. 0.5 ml of the original homogenate) was calculated from the regression equation. The regression value was multiplied by 2 to obtain the value of AA in mg/one ml of the original homogenate left unutilized immediately after incubation as well as after one, two or three hours of incubation respectively. The calculation of AAU per g fresh weight or as percent of the total AA originally present in the incubated solution is as follows:

$$C_1 = \frac{V\ (a + 0.1) - 2\ c}{w} \times 1000$$

where, C_1 = AA in mg utilized/g tissue fresh weight during a given period of incubation, C = Amount of AA (in mg) left over in 2.0 ml of the buffered - incubated solution (i.e., 0.5 ml of the original homogenate).

4.5. Ascorbic acid - macromolecule complexing (AA - MM complex

Further, work in the writer's laboratory has shown that complexing of AA presumably with macromolecules is continuously occurring in all active centres of growth. This complexing is most intense during reproductive differentiation when cellular activity is at its highest (Chinoy, 1962, 1967). Also during juvenile differentiation, active AA-macromolecule complexing (AA - MM complex) occurs. Formation of this complex is enhanced by the exogenous application of AA during germination in a number of plants, such as Cicer, Arachis, Phaseolus, Zea etc., (see later chapters). Again, during the reproductive phase of Sesamum, AA - MM complexing was considerably accelerated as compared with that during the period of vegetative differentiation.

Recently, *in vitro* formation of complexes between A on the one hand and DNA as well as RNA on the other, have been achieved in the writer's laboratory (Chinoy et al., 1972 a; 1974). Similar results have been reported in the case of rat liver DNA by Jamaluddin et al. (1981). Based on their experimental findings, on AA and DNA, Hendry et al. (1977) have proposed models and postulated the role of tiny molecules like AA inserted in cavities in DNA, thereby dictating the stereochemical specificities for gene products. Chinoy, N.J. et al (1978 a) have also reported for the first time that AA complexes with a number of steroids including testosterone *in vitro*. Therefore, the method for determination of AA - MM complexing was developed.

Determination of AA - MM Complex: One ml of 15% HPO_3 was added to 2.0 ml aliquot of the incubated solution and the mixture was transferred to a water bath at 75° C for 15 minutes for hydrolysis of the complex. After cooling, the mixture was buffered at pH 3.6 by adding 1.0 ml of buffer solution B and bringing the total volume to 4.0 ml. Then 5.0 ml of distilled water was added to a 2.0 ml aliquot of this buffered - hydrolyzed extract and the pointer on the colorimetric scale was adjusted at '0' for the turbidity factor as before. Another 2.0 ml aliquot of the above solution was pipetted out and after the addition of 5.0 ml of the standard dye solution, colorimetric reading was noted which was converted into O.D., and the AA content in the aliquot was calculated from the regression equation. The value of AA (in mg) obtained from the regression equation was multiplied by 2, because 2.0 ml aliquot taken for colorimetric reading contained only 0.5 ml of the original homogenate. Subtracting the value of AA of 1.0 ml of the original homogenate utilized before hydrolysis from the above value, gave the value of AA released by hydrolysis of the AA - MM complex (in mg/ml). The amount of AA - MM complex per g fresh tissue weight could be calculated for the aliquot taken immediately after incubation as well as those taken after 1, 2 and 3 hours of incubation as follows:

$$D_1 = \frac{2v\ (d - c)\ .\ 1000}{w}$$

where, D_1 = mg AA released from the complex / g fresh weight of the material.

d = mg AA in 2 ml of the incubated- hydrolyzed- buffered solution (i.e. on 0.5 ml of the original homogenate).

or

$$D_2 = \frac{2\ (\ d - c\ )\ .\ 100}{2\ d}$$

where, D_2 = % AA complexed with macromolecules.

5. Future prospects

It is hoped that in the future, still better assay methods will be developed which would enable the differential assay of AA and its other oxidation products as initiated in the writer's laboratory.

6. CONCLUSIONS

Ascorbic acid (AA) synthesis occurs in all green plants but not all members of the animal kingdom are capable of synthesizing the vitamin including human beings and therefore, have to depend on the dietary source of AA. Several sugars, D-glucose, D-fructose, D-galactose, D-mannose, sucrose etc., are the precursors for AA synthesis via three major pathways in plants; whereas, in animals, the main intermediates for its synthesis are D-glucurono- γ-lactone and L-gulono- γ-lactone. Three lactones and the methyl ester of galacturonic acid are utilized in plants. The site of synthesis is the green tissue in plants, and the liver as well as the kidney are the major sites of synthesis in animals. Moreover, unlike plants, the storage, synthesis and distribution of AA in animals is under hormonal control.

The biosynthesis of ascorbigen (ASG) using labelled and unlabelled precursors has been studied by numerous workers in plants and its existence has also been demonstrated in animal tissues.

A specific histochemical method for the *in situ* localization of ascorbic acid in several plant and animal tissues has been developed using the chilled, alcoholic, acidic silver nitrate reagent at a pH of 2 to 2.5 and subsequently, a cytophotoelectrometric technique was also devised to measure the intensity of the silver staining.

Numerous experimental procedures have been utilized for determination of ascorbic acid (such as biological, chemical, radioactive tracer determinations, enzymatic techniques, paper, column, thin layer, gas chromatography, high performance liquid chromatography and automated methods). The difficulty in formulating a method is due to the presence of numerous interfering substances. A method for the simultaneous determination of AA, ASG, AAU and AA - MM complexing was developed in the writer's laboratory as well as those for the histochemical and quantitative assay of an AA-free radical forming peroxidase. These have been successfully used in both plant and animal materials. Nevertheless, it is hoped that in the future, still better assay procedures enabling the differential assay of AA and its oxidation products will be made available.

REFERENCES

1. Aberg B 1953. Kungl Lantbrukshogsk Ann Roy Agr Coll (Uppasala) Sweden 20: 125.
2. Aberg B 1958. In: Ruhland W (ed) Encyclopedia of plant physiol Berlin: Springer-Verlag, p. 479.
3. Axelrod AE, Martin CJ 1961. Ann Rev Biochem 30: 383.
4. Bagchi K 1952. Indian Med Gaz 87: 198.
5. Bublitz C, Lehninger AL 1961. Ann N Y Acad Sci 92: 87.
6. Burns JJ 1957. Nature 180: 552.
7. Burns JJ 1958. In: Umbreit W. Molitor H (eds) Proc 4th International Congress biochemistry, Vienna, London: Pergamon.
8. Burns JJ 1960. Ascorbic acid. In: Greenberg DM (ed) Metabolic pathways. New York: Academic p 341.
9. Burns JJ (ed) 1961. Symp Vitamin C. Ann N Y Acad Sci 92: 1.
10. Burns JJ 1967. In: Greenberg DM (ed) Metabolic pathways. New York: Academic. Vol. I. Chap. 7.
11. Burns JJ, Peyser P, Moltz A 1956. Science 124: 1148.
12. Charalampous FC, Lyras C 1957. J Biol Chem 228: 1.
13. Chatterjee IB 1970. Methods Enzymol 18: Pt. A. 28.
14. Chatterjee IB 1973. Science 182: 1271.
15. Chatterjee IB, Kar NC, Ghosh NC, Guha BC 1961. Ann N Y Acad Sci 92: 36.
16. Chayen J 1953. Internl Rev Cytol 2: 78.
17. Chinoy JJ 1962. Indian J Plant Physiol 5: 172 (F.G. Gregory Memorial Volume).
18. Chinoy JJ 1967. Indian J Plant physiol 10 (2): 202.
19. Chinoy JJ 1977. Indian J Plant physiol 20: 1.
20. Chinoy JJ, Shah CK, Neelakandan N, Madhavan Unni NV 1971. Indian J Plant physiol 14: 212.
21. Chinoy JJ, Shah CK, Bhatt PN 1976. Vidya, J Gujarat University 19 (1): 12.
22. Chinoy JJ, Singh YD, Gurumurti K 1976a. Indian J Plant physiol 19: 122.
23. Chinoy NJ 1969. Histochemie 19: 125.
24. Chinoy NJ 1969a. Histochemie 20: 105.
25. Chinoy NJ 1970. Stain Technol 45: 99.
26. Chinoy NJ 1978. J Anim Morphol Physiol, Silver Jubilee Volume: 68.
27. Chinoy NJ 1979. Proc IIIrd All India Cell Biology Conf. Bombay, Jan. 10 - 12, p.27.
28. Chinoy NJ, Sanjeevan AG 1978. Histochemistry 56: 275.
29. Chinoy NJ, Seethalakshmi L 1978. J Anim Morphol Physiol 25 (1,2): 235.

30. Chinoy NJ, Seethalakshmi L, Singh YD 1978. Indian J Exp Biol 16 (2): 162.
31. Chinoy NJ, Rao MV 1979. J Anim Morphol Physiol 26 (1,2): 36.
32. Chinoy NJ Rao MV, Seethalakshmi L 1979. Indian J Exp Biol 17 (11): 1171.
33. Chinoy NJ, Seethalakshmi L, Buch RP, Sanjeevan AG 1979a. In: Talwar GP (ed). Recent advances in reproduction and regulation of fertility. Amsterdam: Elsevier North Holland p. 259.
34. Chinoy NJ, Sharma JD, Chinoy MR 1983. Zeit Mikrosk. Anat Forsch. In Press.
35. Chinoy NJ, Sheth KM, Seethalakshmi L, Parmar PY, Sanjeevan AG, Rao MV, Sharma JD, Chinoy MR, Maithili RS, Trivedi DG, Asok Kumar R, Geetha Ranga M 1982. Comp Physiol Ecol 7: 325.
36. Chinoy NJ, Chinoy MR, Sharma JD 1983. under publication.
37. Counsell JN, Hornig DH (eds) 1982. Vitamin C Ascorbic acid London: Applied Science Publishers.
38. Dave IC, Saxena OP, Abraham PG, Pandya RB 1968. In: Proc 37th Ann Session Nat Sci Acad India Abstr. p. 54.
39. Dave IC, Saxena OP, Abraham PG, Pandya RB 1969. Vidya, J Gujarat University 12 (1): 199.
40. Davidson SS, Passmore R 1963. Human Dietetics and Nutrition. Baltimore: Williams & Wilkins.
41. Degkwitz Von E, Schnieder W, Staudinger HJ 1964/5. In: Ascorbinsaure Symp Mainz 2 - 3 April 1964. Darmstadt: Verlag, p. 17 - 60.
42. Dieter MP 1969. Proc Soc Exp Biol Med 1390: 210.
43. Dutta Gupta S, Ray Chaudhuri C, Chatterjee IB 1972. Arch Biochem Biophys 152: 889.
44. Dutta Gupta S, Choudhury PK, Chatterjee IB 1973. Int J Biochem 4: 309.
45. Evans C, Conney AH, Trousof N, Burns JJ 1960. Biochem Biophys Acta 41: 9.
46. George JC, Berger AJ 1966. Avian Myology, New York: Academic.
47. Gmelin R, Virtanen AI 1961. Ann Acad Sci Fenicae, Ser A II, Chemica No.107.
48. Gupta SD, Gupta CS, Chaudhuri CR, Chatterjee IB 1970. Analyt Biochem 38: 46.
49. Hassan MU, Lehninger AL 1956. J. Biol Chem 223: 123.
50. Isherwood FA, Mapson LW 1962. Ann Rev Plant Physiol 13: 329.
51. Isherwood FA, Chen YT, Mapson LW 1954. Biochem J 56: 1.
52. Jensen WA 1962. Botanical Histochemistry. San Francisco: W.H. Freeman & Co.
53. Jensen WA, Kavaljian LG 1956. J Biophys Biochem Cytol 2: 87.
54. Kutacek M, Prochazka Z, Grunberger D 1960. Nature (Lond) 187: 61.
55. Kutacek M, Prochazka Z, Grunberger D, Stajkova R (1962) Coll Czech Chem Commun 27: 1278.
56. Kutsky RJ 1973. Handbook of vitamins and hormones. New York: Van Hostrand Reinhold.

57. Lewin S 1976. Vitamin C, its molecular biology and medical potential. New York: Academic.
58. Lewis WR, Chiang JL, Gross S 1960. Arch Biochem Biophys 69: 21.
59. Loewus FA 1961. Ann N Y Acad Sci 92: 57.
60. Loewus FA, Jang R 1957. Biochim et Biophys Acta 23: 205.
61. Loewus FA, Jang R, Seegmiller CG 1956. J Biol Chem 222: 649.
62. Loewus FA, Kelly S 1961a. Arch Biochem Biophys 95: 483.
63. Loewus FA, Kelly S 1961b. Nature (Lond)191: 1059.
64. Loewus FA, Kelly S, Neufeld EF 1962. Proc Natl Acad Sci U S 48: 421.
65. Madhavan Unni NV, Shah CK 1968. Indian J Plant Physiol 11 (2): 172.
66. Majumder PK, Chatterjee GC 1974. Indian J Exp Biol 12: 387.
67. Malakar MC 1963. Nature (Lond) 198: 185.
68. Mapson LW 1955. Vitamins and hormones 13: 71.
69. Mapson LW 1958. Ann Rev Plant Physiol 9: 119.
70. Mapson LW 1958a. In: Umbreit W, Molitor H (eds). Proc 4th international congress biochem Vienna London: Pergamon. p. 1 - 18.
71. Mapson LW 1967. In: The vitamins. Sebrell WH, Harris HS (eds). vol. 1. New York: Academic.
72. Mapson LW, Breslow E 1958. Biochem J 68: 395.
73. Mapson LW, Isherwood FA 1958. Biochem J 64: 13.
74. Mapson LW, Isherwood FA, Chen YT 1954. Biochem J 56: 21.
75. Martin DW, Mayes PA, Rodwell VW (eds) 1981. Harper's Review of Biochemistry. Lange Medical Publications. Maruzen Asia (pvt) Ltd.
76. Mitchell HH 1964. Nutrition of man and domestic animals. New York: Academic.
77. Pearse AGE 1968. Histochemistry, theoretical and applied. Vol. 1. London: J. & A. Churchill Ltd.
78. Piironen E, Virtanen AI 1962. Acta Chem Scandinavica 16: 1286.
79. Rajalakshmi R, Subbulakshmi G, Ramakrishnan CV, Joshi SK, Bhat RV 1967. Curr Sci 36 (2): 45.
80. Ray Chaudhuri C, Chatterjee IB 1969. Science 164: 435.
81. Roe JH 1954. In: Glick D (ed) Methods of biochemical analysis. Vol. 1. New York: Interscience publishers p. 115.
82. Roy RN, Guha BC 1958. Nature (Lond) 182: 319.
83. Sasmal N, Mukherjee D, Kar NC, Chatterjee GC 1968. Ind. J Biochem 5: 123.
84. Sebrell WH, Jr, Harris RS 1967. (eds). The vitamins. New York: Academic, Vol. I.
85. Shah CK, Dalal SL 1978. Geobios 5: 101.
86. Sharma AK, Sharma A 1965. Chromosome technique, theory and practice. London: Butterworths.
87. Shigeoka S, Yoshihisa N, Shozaburo K 1979. J Nutr Sci Vitaminol 25 (4): 299.
88. Shimazono N, Mano Y 1961. Ann N Y Acad Sci 92: 91.
89. Smith FG 1952. Plant Physiol 27: 736.

90. Storey IDE, Dutton GJ 1955. Biochem J 59: 279.
91. Summerwell WN, Sealock RR 1952. J. Biol Chem 196: 753.
92. Veres K, Kozel K, Prochazka Z 1963. Coll Czech Chem Commun 28: 750.

BIBLIOGRAPHY OF BIOCHEMICAL METHODS

1. Aeschbacker HU, Brown RG 1972. Clin Chem 18: 965.
2. Allison JH, Stewart MA 1971. Analyt Biochem 43: 401.
3. Anon 1980. In: Official methods of analysis of the Association of Official Analytical Chemists. Horwith W (ed) 13th edn, Washington DC, AOAC.
4. Baker EM, Levandoski VG, Sauberlich HE 1963. Proc Soc Exptl Biol Med 113: 379.
5. Baraket MZ, Shenab SK, Darwish N, El-Zohary A 1973. Anal Biochem 53: 245.
6. Basu S 1961. Studies on ascorbigen, Ph.D. Thesis. Calcutta University, Calcutta, India.
7. Beevers H 1961. Harper & Row, New York.
8. Behrens WA, Madare R 1979. Analyt Biochem 92: 510.
9. Benninghoven A 1979. NBS Spec Publ (U. S) 519: 627.
10. Bentley JA 1961. In: Ruhland W (ed) Encycl Pl Physiol. Vol. 14. Berlin: Springer-Verlag.
11. Bessey OA, Lowry OH, Brock MJ 1947. J Biol Chem 168: 187.
12. Beyer WF 1967. J Pharm Sci 56: 526.
13. Bliss CI, Gyorgy P 1967. In: The Vitamins, Gyorgy P, Pearson WW (eds) Vol. VII, p 49. New York: Academic.
14. Blois MH 1958. Nature (Lond) 181: 1199.
15. Bonner WD 1957. Ann Rev Plant Physiol 8: 427.
16. Bose S, Guha BC 1963. Ann Biochem Exptl Med 23 (5): 149.
17. Bose S, Guha BC 1963a. Ann Biochem Exptl Med 23 (12): 589.
18. Butt VS, Hallaway M 1958. Biochem J 69: 20.
19. Chinoy JJ, Singh N, Sirohi GS 1958. Proc Sem "Modern Developments in Plant Physiology", University of Delhi, 1957, p. 45 - 48.
20. Chinoy JJ, Patel HD, Jani BM, John D, Raj KPS, Vyas AV 1969. Res Progress Report P.L. 480 (FG - In - 208) A 7 - CR - 108, Botany Department, Gujarat University, Ahmedabad - 380009, India. p. 1 - 456.
21. Chinoy JJ, Saxena OP, Abraham PG, Pandya RB, Dave IC, Gurumurti K, Sastri K 1969a. Res Progress Report P.L. 480 Research Project (FG - In - 182) A 7 - CR - 87, Botany Department, Gujarat University, Ahmedabad - 380009, India, p. 1 - 350.
22. Chinoy JJ, Gurumurti K, Saxena OP 1970. ISPP Symp Plant growth regulators, Kharagpur, India.
23. Chinoy JJ, Saxena OP, Gurumurti K 1970a. ISPP Symp Plant growth regulators, Kharagpur, India.
24. Chinoy JJ, Shah PN, Ghesani PI 1970 b. Final Res Progress Report, ICAR, Botany Department, University School of Sciences, Gujarat University, Ahmedabad - 380009, India, p. 1 - 92.
25. Chinoy JJ, Saxena OP 1972. Proc Int Seed Test Assoc 37: 903.

26. Chinoy JJ, Saxena OP 1978. Bot Prog 1: 6.
27. Chinoy JJ, Gurumutri K, Saxena OP, Singh YD 1972. Proc All India Symp Biology of Land Plants. Meerut University, Meerut, India, p. 189 - 197.
28. Chinoy JJ, Singh YD, Laloraya M 1974. Vth Int Symp Magnetic Resonance, Jan. 14 - 18, Bombay, India, Absts 17 A 64.
29. Chinoy JJ, Singh YD, Laloraya MM, Saxena OP 1972a. All India Seminar "Current Trends in Plant Sciences" Delhi University, Delhi, p. 47 - 48.
30. Chinoy NJ 1973. Indian Natl Sci Acad 39 B: (1): 119.
31. Chinoy NJ, Seethalakshmi L, Rao MV 1978a. Fifth Int. Congr. Horm Steroids. J Steroid Biochem 9(9): Abstr 57, p. 824.
32. Cooke JR 1974. In: Vitamin C. Recent aspects of its physiological and technological importance. Birch CG, Parker K (eds) p. 31, London: Applied Science Publishers.
33. Cooke JR, Moxon RED 1982. In: Vitamin C, Ascorbic Acid Counsell JN, Hornig DH (eds) p. 167 - 198, London: Applied Science Publishers.
34. Dasgupta PR, Prahlad KV, Dhar C 1962. Curr Sci 31: 332.
35. Davidek J, Velisek J, Nezbedova M, Sb Vys Sk Chem Technol Praze Potravimy 1971. E30, 17.
36. Deutsch MJ, Weeks CE 1965. J Assoc Off Analyt Chem 48: 1248.
37. Dunmire DL et al., J Assoc Off Analyt Chem (1979) 62: 648.
38. Egberg DC et al., 1973. J Sci Food Agric 24: 789.
39. Egberg DC et al., 1977. J Assoc Off Analyt Chem 60: 126.
40. Elizabeth P 1963. Carbohydrates, In: Florkin M, Stotz EH (eds) Comprehensive Biochemistry, Vol. 5, Chap. 2, Amsterdam: Elsevier p. 84.
41. Erdey L, Svehla G 1973. Ascorbimetric titrations, Budapest: Akademiai Kiado.
42. Geller M et al., 1969. J Pharm Sci 58: 477.
43. Glick D, Alpert M, Stecklein 1953. J Histochem Cytochem 1: 326.
44. Guha BC 1961. J Indian Chem Soc 38: 492.
45. Hendry LB, Witham FH, Chapman OL 1977. In: Persp Biol Med 21: 120.
46. Hoffman M et al., 1970. Analyt Biochem 38: 35.
47. Hubmann B et al., 1969. Clin Chim Acta 25: 16.
48. Jamaluddin M, Rajalakshmi S, Sarma OSR 1981. Biochim Biophys Acta 653: 423.
49. Jani RB 1973. M.Sc. Dissertation, Botany Department, University School of Sciences, Gujarat University, Ahmedabad, India.
50. Kelly GJ, Latzko E 1980. J Agric Food Chem 28: 1320.
51. Kiesvaara M, Virtanen AI 1962. Acta Chem Scandinavica 16: 510.
52. Kiesvaara M, Virtanen AI 1963. Acta Chem Scandinavica 17: 840.
53. Kirk JR, Ting N 1975. J Food Sci 40: 463.
54. Kum-Tatt L, Leong PC 1964. Analyst 89: 674.
55. Kutacek M 1960. Chem Abstr 55: 16704, 1961.

56. Kutacek M, Valenta M, Icha F 1957. Experientia 13: 284.
57. List D, Knechtel W 1980. Fluess Obst 47: 57.
58. Loetscher KM, Brander B, Heiner K 1978. Labor Praxis 2: 36.
59. Mahadevan A, Sridhar R 1982. Methods in physiological plant pathology. Madras: Sivakami Publications.
60. Malakar MC, Guha BC 1958. Science and Culture 24: 147.
61. Malakar MC, Guha BC 1961. J Indian Chem Soc 38: 939.
62. Malik CP, Singh MB 1980. Plant enzymology and histo-enzymology. A text manual, New Delhi: Kalyani Publications.
63. Mandels GR 1953. Arch Biochem 44: 164.
64. Mandels GR 1953a. Arch Biochem 44: 362.
65. Mapson LW, Moustafa EM 1956. Biochem J 62: 248.
66. Marchesini A, Polesselo A, Zoja G 1970. Agrochimia 14: 453.
67. Marchesini A, Manitto P 1972. Agrochimia 16: 351.
68. Mason WD et al., 1972. J. Pharm Sci 61: 1301.
69. Mehta BH 1973. M.Sc. dissertation, Botany Department, University School of Sciences, Gujarat University, Ahmedabad - 380009, India.
70. Nesset BL. et al., 1967. Analyt Biochem 19: 89.
71. Newcomb EH 1951. Proc Soc Exptl Biol Med 76: 504.
72. Oliver M 1967. In: The Vitamins, Vol. I, 2nd Edn. p.338, New York: Academic.
73. Omaye ST, Thurnbull JD, Sauberlich HE 1979. Methods Enzymol Vol 62, Part D, p. 3.
74. Overbacher MF, Vines HM 1963. Nature (Lond) 197: 1203.
75. Owen JA, Iggo B 1956. Biochem J 62: 675.
76. Pasadka P, Macholan L 1979. Collect Czech Chem Commun 44: 3395.
77. Pelletier O, Brassard R 1975. J Assoc Off Analyt Chem 58: 104.
78. Pelletier O, Brassard R 1977. J Food Sci 42: 1471.
79. Pfeilsticker K 1968. Fresnius' Z anal Chem 237 (2) 97.
80. Plotsky PM 1982. Brain Res 235: 179.
81. Polk A, Flanagan TL, Vanloon EJ 1960. Clin Chem 6: 558.
82. Price CA 1974. Molecular approaches to plant physiology, New Delhi: Tata-McGraw Hill Publishing Co., Ltd.
83. Prochazka Z, 1953a. Chem Listy 47: 1637.
84. Prochazka Z 1953 b. Chem Listy 47: 1643.
85. Prochazka Z, Koristek S 1951. Chem Listy 45: 415.
86. Prochazka Z, Sanda V 1958. Chem Listy 52: 2378.
87. Prochazka Z, Sanda V 1960. Coll Czech Chem Commun 25: 270.
88. Prochazka Z, Severa Z 1960. Coll Czech Chem Commun 25: 1100.
89. Prochazka Z, Sanda V, Sorm F 1956. Chem Listy 50: 167.
90. Prochazka Z, Sanda V, Sorm F 1957. Chem Listy 51: 1197.
91. Roe JF, Kuether CA 1943. J Biol Chem 147: 399.
92. Rouseff R 1979. Liq Chromatogr Anal Food Beverages. Proc Symp Anal Food Beverages 1: 161. Charalampous G (ed) New York: Academic.
93. Roy RB, et al 1976. J Assoc Off Analyt Chem 59: 1244.

94. Rubach K, Breyer C 1980. Btsch Lebensm-Rundsch 76: 228.
95. Saari JC, Baker EM, Sauberlich HE 1967. Analyt Biochem 18: 173.
96. Sant BR, Mukherji AK 1959. Curr Sci 28: 16.
97. Saxena OP 1969. Ph.D. Thesis, Botany Department, University School of Sciences, Gujarat University, Ahmedabad - 380009, India.
98. Saxena OP, Chinoy JJ, Ghesani PI, Shah HK 1972. Sorghum Newsletter 15: 31.
99. Saxena OP, Chinoy JJ, Mehta BH 1972a. In: Seminar, "Recent Trends in Biological Sciences", University of Udaipur, Udaipur, Dec. 4 - 6.
100. Saxena OP, Chinoy JJ, Shah HK, Mehta D, Ghesani PI 1973. Sorghum Newsletter 16: 36.
101. Saxena OP, Patel HD, Chinoy JJ 1974. 17th ISTA Congress, Warsaw (June 17 - 22) Reprint 57 - S IX: 1-5.
102. Schmall M et al., 1954. Analyt Chem 26: 1521.
103. Sicho V, Bradakova V 1955. Czeskoslov framicie 4: 451.
104. Shieh Hui-Hua, Sweet TR 1979. Anal Biochem 96: 1.
105. Sullivan MX, Clark HCW 1953. Fed Proc 12: 277.
106. Sumtsov BM 1947. Biokhimiya 12: 539.
107. Sweeley CG, Bentley R, Makita M, Wells W 1963. J Amer Chem Soc 85: 2497.
108. Szent Györgyi A 1931. J Biol Chem 90: 385.
109. Terenteva EL 1953. Biokhimiya 18: 296.
110. Thakker PS 1974. M.Sc. Dissertation, Botany Department, University School of Sciences, Gujarat University, Ahmedabad - 380009, India.
111. Valenta M, Kutacek M, Icha F, Holkova L 1958. Sbornik CS Akad Zemedelskych Ved rostlinna Vyroba 31: 1579.
112. Virtanen AI, Kiesvaara M 1963. Acta Chem Scandinevica 17: 848.
113. Wilson SS, Guillan RA 1969. Clin Chem 15: 282.
114. Willis RJ, Kratzing CC 1974. Stain Technol 49: 381.
115. Yamaguchi N, Ogata K 1978 J Japan Soc Hort Sci 47: 121.

3. PHOTOSYNTHESIS AND RESPIRATION

The role of ascorbic acid (AA) in photosynthesis and respiration has been discussed in detail by various workers (Aberg, 1958; Mapson, 1958; Beevers, 1961; Chinoy, 1962; San Pietro et al., 1967; Rabinowitch and Govindjee, 1970; Krogmann, 1973). Conclusive evidence has been presented to show that chlorophyll bearing tissues (including leaves, green tomatoes etc.,) are responsible for the greater accumulation of AA in light grown plants compared with that in the dark grown ones (Reid, 1937; Prokoshev and Dancheva, 1946; Popovskaya, 1950; Haggen, 1953; Asselbergs, 1957; Hoffmann and Albrecht, 1966). Significant correlations between seasonal variations in photosynthetic phosphorylation, photochemical transphosphorylation and the Hill reaction of spinach leaves harvested in different seasons and the ability of these leaf extracts to synthesize AA from its precursor was reported (Mitsui and Oi, 1961). These authors (Mitsui and Oi, 1961a) also found that all the photoproduct or the major portion of isolated chloroplast was AA, whereas, the heat labile precursor of AA was found in chloroplast fraction. Thus the chloroplast appears to be an important seat of AA in green tissues of plants.

1. ASCORBIC ACID AND THE MODEL PHOTOCHEMICAL REACTION

Krasnovsky (1948) discovered the model reaction for primary photochemical reaction. The components for simplest reaction system are chlorophyll - a in oxygen free pyridine solution and AA. In the presence of light, the chlorophyll spectrum bleaches and a red pigment with a broad absorption band at 523 nm appears. The reaction product is pink coloured and is named as Eiosinophyll.

$$\text{Chlorophyll in pyridine + ascorbate} \xrightarrow{\text{light}} \text{reduced Chlorophyll or Eiosinophyll + dehydroascorbic acid (DHA)}$$

When the light was turned off, the red pigment disappeared and chlorophyll reappeared.

Reduced chlorophyll $\xrightarrow{\text{dark}}$ chlorophyll + ascorbic acid + DHA.

The chlorophyll - a destruction was inhibited by iron, hydroquinone and ascorbate (Shimazaki et al., 1980). The reduced chlorophyll could be reoxidized by a number of oxidizing compounds. The net result was the photocatalyzed reduction of these compounds by AA. In many cases this reaction resulted in a net storage of chemical energy, e.g. riboflavin was reduced by AA in an illuminated chlorophyll solution overcoming an adverse potential gradient of 0.2 volts. It was suggested that light excited chlorophyll molecules were reduced by AA *in vitro*.

Chlorophyll + ascorbate ⟶ Chlorophyll*
Chlorophyll* + ascorbate ⟶ Chlorophyll. H
H + (E_0^1 = redox potential) dehydroascorbate.

In vivo, the chlorophyll molecules are reduced by water and restored by reducing an appropriate intermediate which in turn reduced CO_2 (Lynch and French, 1954; Rabinowitch, 1956). Bannister (1959) studied the reversible photochemical chlorophyll reduction and suggested that the triplet form of chlorophyll participated but not the singlet.

Commoner (1958) noted a signal after AA oxidation, belonging to monodehydroascorbic acid (MDHA). This signal and the one obtained in chlorophyll photoreduction were stucturally similar but MDHA had a greater life time than the radical of chlorophyll. Bubnov et al. (1960) and others (Krasnovsky and Umrikhina, 1958; Krasnovsky, 1960; Krasnovsky et al., 1980) have also shown this signal. However, the signal disappeared when the light was turned off and did not reappear without the reductant.

2. ROLE OF FREE RADICAL OF ASCORBIC ACID IN PHOTOSYNTHESIS

The EPR studies of Bubnov et al. (1959; 1960) in *Hordeum vulgare*, *Triticum vulgare* and *Avena sativa* leaves showed that upon illumination, a characteristic doublet centered at g = 2.004 and with a splitting of 1.8 gauss was observed supporting the earlier work of Commoner (1958), who reported the signal to be that of MDHA and not that belonging to some radicalized form of chlorophyll, since all pigments viz., chlorophyll a, b, pheophytin a, b, haematoporphyrin and magnesium phthalocyanin in the oxidized as well as the reduced form yielded negative results. Oxidation with molecular oxygen in pyridine or alcohol solution did not yield an EPR signal either at room or at lower temperature (-60°C). On the contrary, reduction of all the pigments with AA upon illumination produced a paramagnetic species showing an EPR spectrum much the same as those shown by *Hordeum vulgare* and *Triticum vulgare*. Further, solutions of AA in piperidine and different photosynthetic pigments, (upon illumination) yielded the same doublet with the same characteristics. The addition of any one of the previously used pigment did not alter the shape of this signal. The fact that upon photoreduction with AA all the pigment solutions yielded the same spectra, clearly indicated that the structure of the pigment is immaterial to the EPR signal recorded. Therefore, it was concluded that the observed EPR spectrum referred to the oxidized radical form of AA viz., MDHA. Reduction with other hydrogen donors failed to produce any EPR signal, which further confirmed that only MDHA was responsible for the recorded EPR spectrum.

Leblanc et al. (1981) observed the location of AA inside the chloroplast membrane. Similarly, Sechenska et al. (1968) have reported the localization of ascorbate oxidase in the grana of spinach chloroplasts which may indicate the possible close relationship of this enzyme to the photosynthetic transport of electrons. It may also form one of the possible enzymatic pathways for transfer of electrons to molecular oxygen in its light induced consumption by isolated chloroplasts. Ivanov and Povalyaeva (1979) observed that oxygen uptake in isolated pea chloroplasts recorded a significant upsurge in the presence of ascorbate during the photoreduction of ferredoxin and ferredoxin + NADP.

3. PHOTO - OXIDATION OF ASCORBIC ACID

Chloroplast suspensions supplied with AA and AA oxidizing agents (when illuminated) slowed down markedly, the rate of AA oxidation, indicating the possible participation of AA as an electron carrier in the oxidative phosphorylation chain functioning in the chloroplast (Marre and Arrigoni, 1958). An anaerobic photo-oxidation of ascorbate by chloroplast fragments in the presence of NADPH trapping system was observed by Marre et al. (1959) who also reported an equilibrium between oxidized ascorbate and reduced glutathione. Oxygen was consumed by ascorbate in the light, in presence of menadione or FMN but in the absence of DCPIP. In this reaction, 2 moles of O_2 were consumed per mole of ascorbate and the O_2 uptake was inhibited by O-phenanthroline (Ikeda, 1959).

Mapson (1963; 1964) (based on his experiments on illumination of intact strawberry leaves) showed that the rate of AA photo-oxidation depended on the intensity of the illumination, but when the leaves were poisoned with cyanide, DCMU, hydroxylamine or O-phenanthroline, both photo-oxidation and photoreduction were abolished. Similarly, since photo-oxidative consumption of ascorbate was inhibited by O-phenanthroline, Chiba and Okayama (1962) also proposed the role of AA as an electron donor for photosystem II. Phosphorylation and peroxide formation during ascorbate photo-oxidation in the presence of low redox potential was observed by others (Forti and Jagendorf, 1961; Trebst et al., 1963; Jacobi, 1964) who suggested that AA reacts with the electron transfer system at a site related to the O_2 evolution system.

Bohme and Trebst (1969) also suggested that AA donates electrons to the primary oxidant (Z) of system II trapping centre, thus by-passing a phosphorylation site located between the O_2 yielding reaction of Z of the system II trapping centre. It has also been shown that ascorbate serves as an electron donor at a site which is prior to photosystem II (Gozal and Avron, 1970). The addition of AA doubled the apparent rate of O_2 uptake.

Cedeno-Maldonado et al. (1972) have shown that a relatively high concentration of Cu^{2+} (> 25 µM) decreased the photosynthetic evolution of O_2 in isolated chloroplasts. Further, photosystem II exhibited a greater sensitivity to Cu^{2+} ions in comparison to photosystem I. This implies that the participation of AA is more vigorous in photosystem II. Yerkes and Babcock (1980) studied the reactions occurring on the oxidizing side of photosystem II by monitoring the decay kinetics of EPR signal II F from photoinduced Z, which is an intermediate in electron transport chain between P-680 and the water splitting enzyme. A detailed study of reduction of photosystem II by AA revealed the pH dependence of the decay kinetics of signal II F in the presence of the donor (both the neutral acid and the ascorbate mono-anion) serve as reducing agents to Z^+. The occurrence of a negative charge is indicated on the inner membrane surface in the vicinity of Z. The presence of poison inhibits the electron transfer from ascorbate to cytochrome via O_2. The inhibition of the ascorbate supported NADP photoreduction with high concentration of poison was incomplete, which suggests that AA may reduce endogenous electron carrier other than O_2. Kok (1967) also reported that when photosystem II was poisoned, AA served as an electron donor.

Oku and Tomita (1980) could locate the latent O_2 evolution centres in spruce chloroplasts which could be photoactivated by weak white light. This photoactivation of the latent O_2 evolving centres occurred at a pH optima of 7 - 8 and a strong stimulation could be achieved by L-AA.

The mechanism of photo-oxidation of AA was studied by Heelis et al. (1981) by employing a continuous and flash photolysis technique. An electron was abstracted from AA by flavin (FMN) singlet forming a triplet state of flavin - AA complex. Efficient photo-oxidation was observed in the presence of O_2.

4. ASCORBIC ACID AND PHOTOPHOSPHORYLATION

Arnon et al. (1954) studied the effect of AA on photosynthesis by supplying it to the illuminated chloroplasts. The addition of an enzyme system was unnecessary, which suggested that chloroplasts contained the enzyme system needed for the reaction. AA was the only exogenously added substance which consistently gave a large increase in photosynthetic phosphorylation and CO_2 fixation. Arnon (1955) put forward a scheme for photosynthesis by isolated chloroplasts and reviewed the work done on extracellular photosynthesis. He concluded that like Mg^{++}, FMN and vitamin K, ascorbate was a cofactor of photosynthetic phosphorylation and as it was one of the components of green cells, it served as an electron carrier, but as it was required in catalytic quantities, it could not be considered as a substrate (Arnon et al., 1955).

The work of several authors (Arnon, 1958; Arnon et al., 1957; 1958; Whatley et al., 1956; 1957; 1959; Wessels, 1964) suggest that AA functions as a protector or stabilizer of some essential components of the chloroplasts against inactivation rather than serving as a redox catalyst or cofactor of phosphorylation (Table 3.1). Confirmatory evidence was presented by Fuchtbauer and Simonis (1958) as well as Yu and Brand (1980) who demonstrated that under *in vitro* conditions, presence of L-AA was found to be inevitable for the stabilization of photophosphorylation activity which was carried out using photosynthetic membrane fragments isolated from Anacystis nidulans.

Jagendorf (1962) studied oxidation - reduction reactions and phosphorylation in isolated chloroplasts, and observed the electron flow pattern *in vitro* by observing oxidation of AA and redox dye. AA and particularly the redox dye, DCPIP were oxidized by chloroplast in light in the presence of air by at least three ways.

Table 3.1. represents the stabilization of phosphorylating capacity in chloroplast by addition of ascorbate (Arnon, 1958).

Duration of storage prior to test hr.	μgP esterified	
	Control	Ascorbate treated
0	5.5	7.7
0.5	2.4	7.1
1.0	1.7	8.1
2.0	-	8.0
4.0	-	8.0
24.0	-	3.0

1. If H_2O_2 was produced by consumption of O_2 under aerobic condition, the ascorbate (or reduced dye) might be peroxidized nonenzymatically.
2. The AA or dye might be oxidized by photo-oxidant.
3. AA or the dye might be oxidized in a direct reaction with O_2, catalyzed by an essentially nonenzymatically photodynamic action of chlorophyll.

There are some controversial reports about the site of ATP formation during light dependent transfer of electrons. It was found that ATP formation did not take place during the light dependent transfer of electrons from ascorbate or ascorbate + DCPIP (Trebst and Eck, 1961; Trebst, 1964; Avron, 1964; Wessels, 1964; Gromet-Elhanan, 1967). On the other hand, ascorbate + DCPIP catalyzed cyclic photophosphorylation in the absence of NADP (Keister, 1963; Gromet-Elhanan, 1968; Gromet-Elhanan and Avron, 1964). Thus there is no site of ATP formation in the chain where AA + DCPIP feed electrons (presumably to plastocyanin and NADP (Wessels, 1963; Avron, 1964; 1967). Keister (1965) and Trebst and Pistorius (1967) studied the effect of the uncoupler (NH_4Cl) in accelerating electron flow and the energy transfer inhibitor phloridzin in inhibiting the electron transfer process and suggested that ATP formation did exist on the path and AA + DCPIP donate electrons at two points, one before and one after the ATP forming site. Gorman and Levine (1965) studied the cyclic PMS photophosphorylation activity and the photo-induced electron transfer from ascorbate + DCPIP to NADP in cytochrome - f- less as well as plastocyanin- less mutant of Chlamydomonas. The cytochrome- f- less mutant possessed the above mentioned activity while the plastocyanin- less mutant did not.

A study on the kinetics of reduction of cytochrome -f by AA in Brassica rapa revealed the value of k (reaction rate constant of 2nd order) to be 250 M^{-1} S^{-1} at pH 7.0 with activation parameters ΔH = 8.5 K cal/mol and ΔS = -19 Cal / mol / degree C. These observations highlight the importance of photosynthetic electron transport in relation to L-AA as an electron donor (Ishikawa et al., 1981).

Itoh (1979) carried out a series of experiments on surface potential and reaction of membrane bound electron transfer components in sonicated chloroplasts with redox reagents. It was found that P-700 reduction rate depended on the concentration of ferricyanide, AA and PMS. The apparent first order rate of reduction depended on 0.6 - 0.7 th power of concentration of these reagents. The rate limiting step seemed to be electron transfer reaction at the surface. The reduction rate was considerably greater in presence of PMS + AA than in PMS alone. This indicated that the reduction rates of P-700 molecule were determined by membrane-surface and the reagents in turn depended upon the surface potential and valence of the reagent. P-700 in class II chloroplasts did not react directly with the reductants, but an electron transfer was mediated by some electron transfer components situated on the outer surface of the thylakoid membrane (Itoh, 1979a).

The photosensitization of methyl red reduction by AA was much less efficient when pigment molecules were bound to particles than when in solution (Widart et al., 1980).

5. ASCORBATE MEDIATED REVERSAL OF HIGH TEMPERATURE AND POISON

Katoh and San Pietro (1967) studied the NADP photoreduction in heated *Euglena* chloroplast in the presence of ascorbate, and which were suspended in the preparation medium (0.4 M sucrose, 0.01 M NaCl and 0.05 M phosphate buffer, pH 5). The suspension was heated in a small test tube in a waterbath at a desired temperature in the dark. The addition of ascorbate to heated chloroplast resulted in the resumption of NADP photoreduction.

The Hill reaction with ferricyanide was completely inhibited at 40° C, while with NADP, it was one half of that of the original. The photoreduction of NADP when carried out in presence of AA, did not appreciably affect its activity. However, when the chloroplast was heated at 40° C, the Hill reaction was stopped or it was very low, but addition of ascorbate restored it to about 70% of the original activity. The photoreduction of NADP with ascorbate and DCPIP was most resistant to heat treatment; upto 40° C the activity was unaffected or even slightly increased, while at 50° C it was reduced to one half. Katoh and San Pietro (1967) suggested that ascorbate may serve as an electron donor for NADP photoreduction or it may reactivate the oxygen evolution system of the heated chloroplast. This suggests that the actual electron donor for NADP photoreduction is water and oxygen is evolved during the reaction. They proposed the scheme showing heat inactivation of *Euglena* chloroplasts by ascorbate (Fig. 3.1). These authors also studied the effect of poisons on inhibition of the Hill reaction in presence of ascorbate. The low concentrations of Atrazine significantly inhibited the Hill activity in untreated chloroplast and slowed down the NADP photoreduction. When DCPIP and AA were used along with Atrazine, the rate of NADP photoreduction could be increased. They also found that cysteine replaced the ascorbate for the NADP photoreduction with heated chloroplast, but the concentration of cysteine required was double than that of AA (Katoh and San Pietro, 1967).

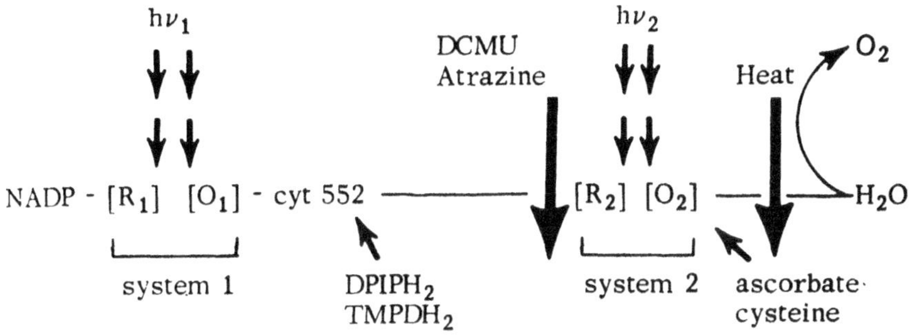

FIGURE 3.1. Mechanism of heat inacitvation of Euglena chloroplasts proposed by Katoh and San Pietro (1967).

Hendrich (1958) studied the effect of inorganic ions on the change of photopotential upon illumination of the chlorophyll - AA system. Copper and Nickel reacted with AA and prevented the appearance of the photopotential.

5.1. Digitonin and Tris buffer effect reversal by ascorbate

Digitonin treated chloroplasts were found to be fragmented and photophosphorylation did not take place in such particles, however, when AA was used as an electron donor, the illumination of such particles with added ferredoxin, ferredoxin-NADP reductase and plastocyanin resulted in the reduction of NADP (Wessels, 1964; Boardman and Anderson, 1964).

Yamashita et al. (1971) found that the Hill reaction of spinach chloroplast was inhibited by washing them with 0.8 M Tris buffer. This inhibition was further promoted by adding ferricyanide in the washing medium. If a reducing agent such as DCPIP + ascorbate system or the hydroquinine + ascorbate system was added to the Tris buffer, the Hill reaction activity was not affected but the activity recovered upto 70% of the initial level by rewashing the chloroplast with a preparation medium containing the above reducing agents. However, digitonin treated spinach chloroplast particles were capable of photoreduction of NADP if plastocyanin, ferredoxin, ferredoxin-NADP reductase and artificial electron donor system were added (Trebst and Erich, 1957). These workers suggested that ascorbate may be used as an electron donor but DCPIP/ascorbate or NAD/ascorbate gave higher rates. The DCPIP system donated electrons without plastocyanin as well but at a reduced rate.

5.2. Reversal of effects of various inhibitors by ascorbate

Wessels (1963) reported the suppression of inhibition of cyclic phosphorylation with PMS by DCMU when ascorbate and NADPH were added. The inhibition of phosphorylation with vitamin K_3 and anthraquinone disulphonate by DCMU could not be overcome by ascorbate but the addition of small amounts of DCPIP and ascorbate reinitiated ATP synthesis. (Fig. 3.2). Friend et al. (1966) found that AA and indophenol dye reduced some plastoquinones in darkness and even under illumination but in the presence of DCMU.

Vernon (1967) suggested that ascorbate may serve as an alternate source of electrons when pigment system II was inactivated by DCMU. AA was more effective when coupled with the dye, DCPIP and reduces the dye, which interacts directly with photosystem I.

Schwartz (1967) found that addition of ascorbate reversed the inhibition of PMS phosphorylation caused by DCMU. PMS when added, was active only in the red region where photosystem II photoreduced it to an advantageous redox level. In the presence of excess AA, the photophosphorylation catalyzed by PMS had a lower quantum required in the far-red region. Schwartz (1967 a) could also show that noncyclic photophosphorylation from water to NADP proceeded with a quantum of requirement of 7 quanta per ATP in the red region, which increased to about tenfold in the far-red region. Photophosphorylation with AA + DCPIP in the presence of DCMU and in the presence or absence of ferredoxin and NADP had a quantum requirement of about 30 quanta per ATP in the red region, and 15 quanta per ATP in the far red region. Avron

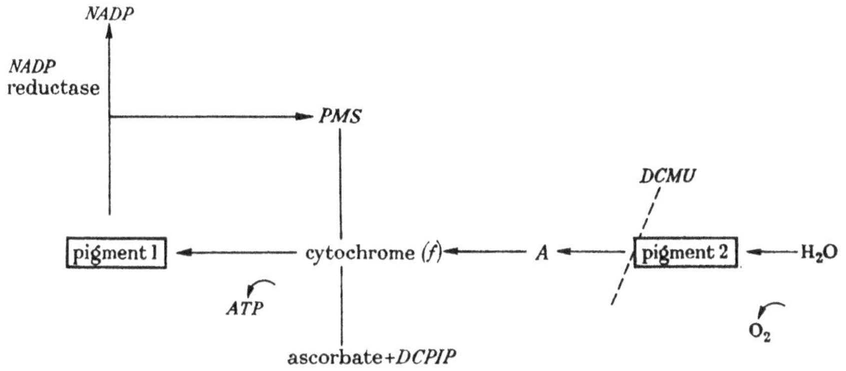

FIGURE 3.2. Scheme for photophosphorylation and photoreduction of NADP in chloroplasts. The straight arrows indicate the direction of the electron flow.

and Newmann (1968) interpreted these data as follows:

All the observed phosphorylation in the presence of DCMU + AA + DCPIP, with or without NADP, originate from a cyclic process cataylsed by ascorbate + DCPIP. The relatively low quantum yield observed in this system may have resulted from insufficient attention to the redox system considerations which in turn play a paramount role in this system.

6. PHOTO-EVOLUTION OF HYDROGEN

Recent reports have highlighted the importance of ascorbic acid in the process of photo-evolution of hydrogen. Porphyrin radical intermediates are active in the formation of molecular hydrogen during the photosynthetic processes (Marcer-Smith and Mauzerall, 1981). In fact, a new hydrogen generation system has been proposed consisting of a compound I - Zn^{2+}- tetraphenyl porphyrin p- sulphonic acid, methyl viologen, colloidal platinum and AA as an electron donor. This system was claimed to show satisfactorily a high recycling number for each constituent involved along with a reasonable quantum yield of molecular hydrogen (Tabushi and Yazaki, 1981). These models assume significance in the light of the elegant studies by Krasnovsky et al. (1980) pertaining to conditions for photo-evolution of hydrogen by chloroplasts in the presence of exogenous enzyme hydrogenase obtained from Clostridium butyricum and methyl viologen which could link the electron transfer from photosystem to the enzyme. According to these authors, the evolution of molecular hydrogen was limited in the absence of external electron donors, due to the occurrence of the cyclic electron flow short-circuiting the photosystem I. Ascorbic acid alone had limited efficiency in H-evolution because of the reverse reaction of the photoreduced dye. The presence of the di-thiothreitol could block the short circuiting and at this stage, the yield of H-photo-evolution was to the tune of $\leqslant$ 400 μ Mol H/ mg Chl / h.

7. PREVENTION OF CHLOROPLAST AGEING BY ASCORBATE

Siegenthaler (1969) found a beneficial effect of AA on the NADP photo-reduction of aged chloroplast within which it was markedly reduced, however, on addition of the AA + DCPIP electron donor system, the NADP photoreduction was reinitiated, thus suggesting that ageing may interfere with the photosystem II attacking the energy transducing mechanism in chloroplast. Vernon and Zaugg (1960) also reported recovery during suppression of NADP reduction by ageing chloroplast on addition of $DCPIPH_2$ - ascorbate system.

Rudolph and Bukatsch (1967) found that AA was also effective in the reduction of bleaching of chlorophyll solutions. In Rumex alpinus leaves fed with AA, the photoreduction of the latter was also reduced. The increased concentration of AA reduced the pressure of O_2 in the cell and moreover, the electron transfer was increased to pigments which also reduced the chlorophyll photoreduction.

8. TISSUE RESPIRATION AND ASCORBIC ACID

James and Cragg (1943) investigated the mechanism of oxidation in actively respiring barley. Rapid oxidation of AA occurred in spite of addition of H_2O_2 to barley extracts and its formation could not be detected in the tissue either from the action of AAO or any other source. They therefore, suggested the total reaction as:

$$AH_2 + \tfrac{1}{2} O_2 \xrightarrow{AAO} A + H_2O$$

without commitment as to its intermediary mechanism. Based on their studies on barley seedlings, these authors have attributed a specific respiratory function to AA as a part of the reduced NAD oxidizing system. James et al. (1944) also determined the possible role of the AA system as a producer of phosphoglycerate and outlined the linkage between the oxidation and glycolytic stages of respiration. Mathews (1951) had observed the enzymatic oxidation of NADH using a protein fraction from pea in the presence of AA and O_2.

It has been reported that respiratory particles from H_2 grown alga, Anacystis nidulans oxidized ascorbate + TMPD at a rate of 70 nmole O_2 taken up/mg protein/min, while the oxidation rate for H_2, NADPH, NADH and succinate was 28, 15, 6 and 2.5 respectively (Peschek, 1980). Thus, there occurs a preferential oxidation of ascorbate by the respiratory particles.

9. ENZYMES INVOLVED IN OXIDATION OF ASCORBIC ACID

Several enzyme systems are known to occur in plants which can oxidize AA. Transfer of electrons to O_2 can take place by the help of different oxidases varying in their prosthetic groups. Electron donating substances may also be different.

(1) Oxidases: These are divided into two classes in accordance with their oxidative action on AA which may be direct or indirect (Mapson, 1958):

(a) Ascorbic acid oxidase (AAO) which directly oxidizes AA in the presence of O_2 and has copper as its prosthetic group.

(b) Other oxidases which bring about indirect oxidation of AA during the oxidation of their own substrates, viz., polyphenol oxidases and cytochrome oxidases.

(2) Besides oxidases, laccases and peroxidases also oxidize AA (Aberg, 1958; Mapson, 1958; Chinoy, 1962).

9.1. Oxidases

9.1.1. Ascorbic acid oxidase (AAO). AAO was found by Dunn and Dawson (1951) to be a specific blue green copper protein with six copper atoms per molecule and which directly catalysed the oxidation of AA by molecular O_2 to dehydroascorbic acid (DHA). Recently, Newbury and Smith (1981) have presented an immunochemical evidence for specific activity of AAO enzyme in relation to the pigment phytochrome in mustard seedlings. An AAO which specifically oxidises L-AA only and not its analogues, was discovered earlier in *Myrothecium verrucaria* (Mandels, 1953) and *Physarum polycephalum* (Ward, 1955). Copper chelating agents like diethyldithiocarbamate (DDC), cyanide or 3-hydroxyquinoline did not inhibit the activity of this enzyme. Another special feature of this AAO is the production of H_2O_2 during oxidation, whereas in case of the common AAO, no H_2O_2 is formed (Steinmann and Dawson, 1942). Attempts have been made to assign the role of a terminal oxidase to AA in respiration, on the basis of an increase in respiratory rate under the influence of exogenous application of AA which brought about an increase in the rate of respiration in the cut tissues of apples and pea seedlings. This was interpreted as confirming the role of AAO as a terminal oxidase (Boswell, 1950; Waygood, 1950). These workers, however, did not take into consideration the fact that AA might be oxidized by polyphenol oxidase or cytochrome system.

James (1953 a, b) was of the view that more than one oxidase was participating in respiration and a shift in the main terminal oxidase was very often brought about by internal and external factors. According to Mayer (1958) AAO could be a terminal oxidase present in lettuce seeds and seedlings since there was very high initial content of the enzyme in the dry seeds which mediated the entire O_2 uptake of the seed. As germination proceeded, other terminal oxidases which could replace AAO, became more and more active (Mayer et al., 1957). Coumarin had no effect on AAO while thiourea completely inhibited it. Mayer (1958) believed that thiourea may thus cause an immediate and sudden shift in terminal oxidase.

Kiraly and Farkas (1957) observed a marked increase in O_2 consumption by wheat leaves after rust infection which is paralled by an increase in AAO activity of homogenates. Demonstration of participation of AAO in the O_2 uptake of sections of intact leaves was attempted with the use of inhibitors. Azide inhibited the O_2 uptake upto 85%, both in healthy and infected leaves, indicating the operation of a metal containing oxidase. The activity of polyphenol oxidase was not demonstrated by these tissues, but diethyldithiocarbamate inhibited the O_2 uptake by about 14% in non-infected leaves, while the inhibition in infected leaves ranged from 17 to 58%. On the basis of these results, Kiraly and Farkas (1957) suggested that AAO oxidase might be present in healthy wheat plants in an inactive state and this oxidase became active in infected plants thus acting as a major terminal oxidase in these plants.

From the evidence presented, it is still not very clear whether AAO acts as a terminal oxidase in respiration or not. Although it is generally proved beyond doubt that cytochrome oxidase is the terminal oxidase present in plant tissues, nevertheless, under certain stages of growth or environmental conditions, it appears likely that AAO contributes to a certain extent in the respiration by acting as terminal oxidase.

9.1.2. <u>Polyphenol oxidase</u>: Since the discovery of polyphenol oxidase by Bourquelot and Bertrand (1895), a number of studies have been carried out to determine its role in respiration with little success. A number of workers have reviewed the extensive literature on the chemistry, mode of action and role of polyphenol oxidase. The enzyme is a copper protein in which the copper moiety is associated with a specific protein and this accounts for its oxidase activity. The enzyme catalyzes the oxidation of O- and p-polyhydroxyphenols to the corresponding quinone which in turn oxidizes AA to DHA.

Polyphenol oxidase has been assigned the status of terminal oxidase in respiration by virtue of the fact that it catalyzes the aerobic oxidation of a variety of phenols, especially having orthodehydroxy grouping. The substrate used for this enzymatic reaction was mainly catechol which on oxidation gave orthoquinone. It may, however, be pointed out that the oxidation of catechol by polyphenol oxidase was found to be extremely rapid. On the other hand, no such vigorous O_2 uptake could be found in respiration of plant tissues in which this enzyme occurred. Moreover, the oxidation product, O-quinone was a very reactive oxidant which was likely to damage living tissue at higher concentrations. This fact led Szent-Györgyi (1939) to assign a dormant role to polyphenol oxidase in respiration. The enzyme was mobilized to activity purely to repair the damage done to the tissues by cutting or by pathogenic invasion by producing quinone which was toxic to the invader. However, the respiration rate of tissues was found to be many times higher than the mitochondrial preparation representing the equivalent tissue.

Two isozymes of polyphenol oxidase isolated from root tubers of <u>Dioscorea alata</u> were specific for only those polyphenols possessing O-dihydroxy configuration. Inhibition of the enzyme activity by a low concentration of AA was reported and reduction of O-benzoquinone to catechol was given as the possible explanation of AA action (Hsieh et al., 1977).

9.1.3. Cytochrome oxidase: During the oxidation of cytochrome oxidase, secondary reaction takes place viz., the oxidation of AA. Although Cytochrome C is a specific substrate for cytochrome oxidase, it can be reduced by a large number of reducing substances. A number of electron donors as well as enzyme systems can also bring about reduction of cytochrome C. As AA is a universal electron donor endogenously present in all tissues, it is not very surprising that cytochrome C also oxidizes it and gets itself reduced in the process.

Ljones and Skotland (1979) observed an acceleration in Cytochrome C reduction when AA was oxidized by dopamine - β-mono-oxygenase through one electron transfer, e.g. from AA to form MDHA. The reducing ability of AA alone was lesser in comparison to that of MDHA. Myer et al. (1981) have studied the cytochrome C: ascorbate reduction site and have traced the possible electron transfer path. They conclude that this reduction is a three step process for ferricytochrome C and is independent of the state of crevice-opening of the heme and the binding site for the ascorbate is one of the two arginyl side chains, i.e. arginine 8. Further, the electron transfer path has been indicated involving heme-Fe through electronic interaction between the indole group of tryptophan -59 of the porphyrin moiety.

9.2.

9.2.1. Laccase: The role of laccase in the respiratory mechanisms of the plant tissues still remains obscure. Taking into account the behaviour of laccase towards carbon monoxide and the characteristic blue colour, one can discern its close relationship with AAO. On the other hand, the nature of the substrate utilized by laccase entitled it to be classed as a close relative of polyphenol oxidase. Thus these three enzymes are closely related to one another.

The oxidative activity of laccase on AA diminishes on purification of the enzyme preparation, however, the addition of phenylenediamine (but not catechol) can restore the activity of the purified enzyme thus suggesting that the oxidation of AA is indirect.

9.2.2. Peroxidase: Peroxidases are also of general occurrence in plant suggesting their importance in respiration. Szent-Györgyi (1930, 1938) was the first to demonstrate that peroxidases catalyze the oxidation of AA. Purified peroxidase preparations with H_2O_2 could not catalyze this reaction and did so only after the addition of small amount of plant juice. This secondary reaction of peroxidase was found to be capable of quinone formation. The peroxidase activity can be represented as follows:

$$\text{Phenolic compound} + H_2O_2 \xrightarrow{\text{Peroxidase}} \text{quinone}$$

$$\text{Quinone} + \text{AA} \longrightarrow \text{Phenolic compound} + \text{DHA}$$

Yamazaki and his co-workers (Yamazaki et al., 1960; Yamazaki, 1962; Piette et al., 1961; Iyanagi and Yamazaki, 1969) have confirmed through EPR studies that a free radical of AA viz., MDHA was formed during its peroxidative oxidation and a free radical mechanism was the main pathway involved in the oxidation of AA by AA oxidase and peroxidase. Gurevich (1963) discovered a special isozyme of peroxidase in wheat and corn seedlings which catalyzed and formation of the free radical of AA and which in turn brought about the transfer of hydrogen from AA to an acceptor of a very low oxidation-reduction potential like O-dinitrobenzene (ODNB) to convert it into a yellow soluble product, O-nitrophenylhydroxylamine. Some work on AA-free radical producing peroxidase has been done in this laboratory showing a very high activity of this enzyme in metabolically active sites in both plants and animals (Chinoy, 1968; 1969; Chinoy et al., 1969; 1969 a; 1970; 1973; Chinoy N.J., 1970; 1973; 1978).

Similarly, thermolabile free radicals were measured in pea leaf discs at disc temperatures upto 47° C by EPR method. From the data and the nature of signals, the authors concluded that the free radicals participate directly in the biological processes of energy transfer which take place in the respiratory pathways (Sagatov et al., 1981). These *in vivo* studies support the concept of participation of AA in electron transfer through its 1 electron intermediates.

Ohnishi et al. (1969) showed that AAO catalysed one-electron oxidation and produced MDHA. According to them, the MDHA so produced from AA-AAO system could reduce cytochrome C. Yamazaki and his associates (Iyanagi and Yamazaki, 1969) have presented clear evidence to show that the MDHA once formed in the system could always oxidize NADH. The MDHA could be produced both enzymatically and non-enzymatically (see Ch. 1 for details).

In their studies on peroxide reaction, Ohnishi et al. (1969) further showed that certain phenols like p-cresol were very fast substrates for horse radish peroxidase.

Kow et al. (1982) studied the oxidation of NADPH in reconstituted spinach chloroplast preparations using ascorbate and H_2O_2 system. The generation of NAD(P) was monitored in the form of rate of PGA formation. They observed an enhanced rate of PGA formation, when AA was used with NAD as cofactor. Interestingly, addition of glucose oxidase, an enzyme generating H_2O_2 caused an additional increase in PGA formation. They concluded that oxidation of NADPH by a system utilizing AA and H_2O_2 could serve as a means of recycling NADPH, keeping thereby an option of another reaction involving AA and NADPH in chloroplast open.

H_2O_2 molecules generated during a number of reactions require to be scavenged as quickly as possible, as it poses a threat as a potential physiological toxin. Intact spinach chloroplasts were employed to study the scavenging of H_2O_2. The chloroplasts possessed specific peroxidase enzyme which used only AA as photoreductant, i.e., electron donor. AA was found to be localized in chloroplastic stroma which could be regenerated upon oxidation by DHA reductase localized also *in situ* (Nakano and Asada, 1981).

10. ASCORBIC ACID TURNOVER AND RESPIRATORY METABOLISM

Patel (1967) studied the effect of exogenous application of AA and sucrose in the respiratory activity of excised embryo axis of <u>Arachis hypogea</u>. During the I respiratory period (after 3 hours of incubation), enhanced respiratory activity was observed in substrates containing AA and sucrose alone, as compared with that in distilled water (DW). During the II respiratory period which was preceded by 20 hours of incubation period, the rate of respiration was appreciably higher compared with that in I respiratory period. Here also AA, sucrose and AA + sucrose significantly enhanced respiratory activity compared with that in distilled water.

During both I and II respiratory periods, there was acceleration in linear growth of the embryo axis cultured in AA, sucrose and AA + sucrose compared with that in DW (Table 3.2). Embryo axis cultured in AA imbibed more water compared with that in DW in both the respiratory periods at the end of which the AA concentration was higher in all the substrates compared with that in DW. In sucrose medium, AA content was the highest. At the end of the respiratory periods, ascorbigen (ASG) content was lower in all the three substrates compared with that in DW but on the other hand, the enzymic utilization of AA was higher.

Actively developing prefertilized carpels gave a higher rate of respiration compared with that of the developing Kernels of wheat cv. N.P. 18 (Fig. 3.3) (Abraham, 1969; Abraham et al., 1968; Chinoy et al., 1969). Studies of AA turnover in prefertilized carpels and developing kernels showed a similar trend. Concentration of AA and ASG remained at a higher level accompanied by a lower rate of AAU in the carpels (Figs. 3.4, 3.5). However, as soon as fertilization took place and the zygote began to develop, the concentration of AA and ASG fell to a lower level accompanied by a very high rate of AAU which persisted throughout the period of experimental observation.

Confirmatory evidence was obtained from an experiment in which the unfertilized carpels and developing kernels were incubated separately in AA and sucrose solutions respectively. Their respiratory activity, concentration of

Table 3.2. Showing the effect of exogenous application of ascorbic acid and sucrose on the respiratory activity of excised embryo axis of <u>Arachis hypogea</u> (Patel, 1967).

	I Respiratory period						II Respiratory period					
	Resp. μ ml.	Elongation	Imbibition	AA	ASG	AAU	Resp. μ ml.	Elongation	Imbibition	AA	ASG	AAU
DW	680	124	193	1.3	2.7	22.3	950	134	222	1.4	2.5	28.7
AA	795	138	210	1.9	1.6	29.1	1150	146	247	2.6	0.9	38.0
Sucrose	850	132	220	2.7	1.8	28.3	1450	151	252	2.0	1.1	39.8
Sucrose + AA	815	134	215	2.2	1.3	29.6	1365	155	252	2.9	2.4	39.4

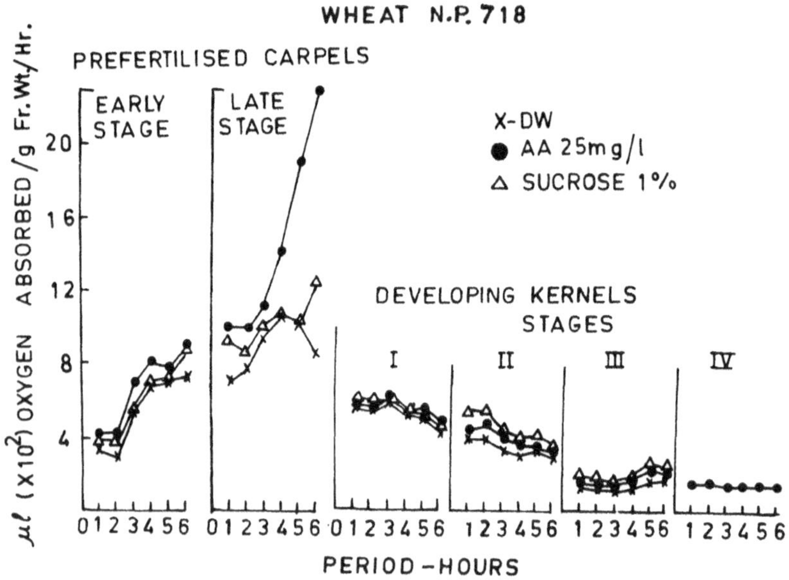

FIGURE 3.3. Oxygen consumption in prefertilized carpels and developing kernels of wheat incubated in 25 mg/l AA and 1% sucrose (Abraham et al., 1968; Abraham, 1969; Chinoy et al., 1969).

AA, ASG and AAU were found to be at a high level. The high respiratory activity in sucrose was firstly due to it being a suitable substrate for respiration and secondly, because sucrose acted as a precursor for the biosynthesis of AA as shown earlier by Chinoy (1962; 1968), Chinoy and Nanda (1959), Chinoy et al. (1965; 1967). A parallelism between the rate of respiration of AA metabolism has also been observed by a number of workers (Chinoy, 1962; Patel, 1967; Nanda and Krishnamurty, 1966). In another experiment on seeds of *Triticum aestivum* L. cv. Arnej 624, the rate of respiration was nearly the same in the case of AA as well as sucrose medium, but slightly higher than that in the control (Fig. 3.6) (Chinoy et al., 1969; Dave, 1969). From the second hour onwards, the sucrose effect became more pronounced

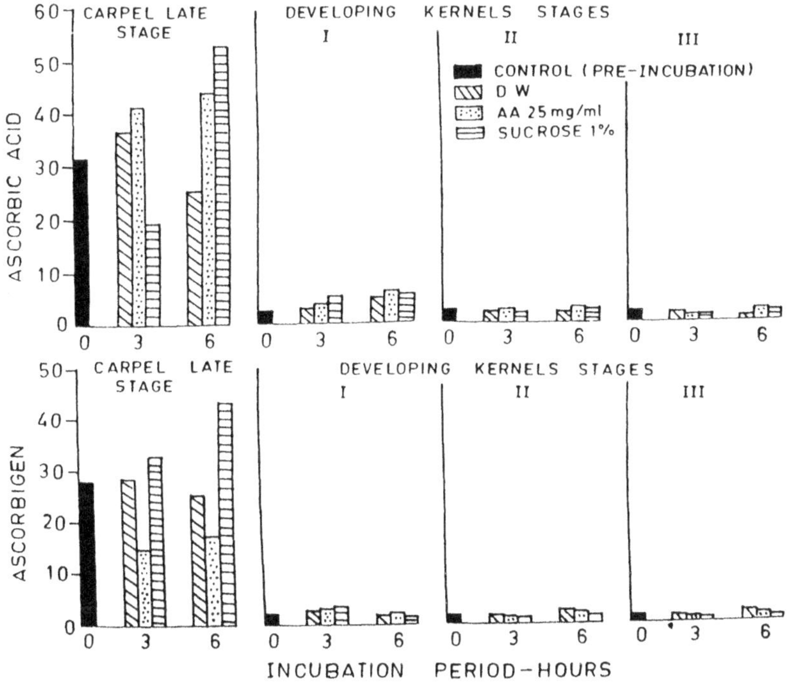

FIGURE 3.4. Ascorbic acid and ascorbigen (AA and ASG) contents of prefertilized carpels and developing kernels of wheat incubated in distilled water (DW), 25 mg/1 AA and 1% sucrose (Abraham et al., 1968; Abraham, 1969; Chinoy et al., 1969).

compared with that of the control (Fig.3.6). The parallelism between the effect of AA and sucrose was maintained upto the fourth hour. From this time onward, there was a significant shooting up in the respiratory activity of sucrose

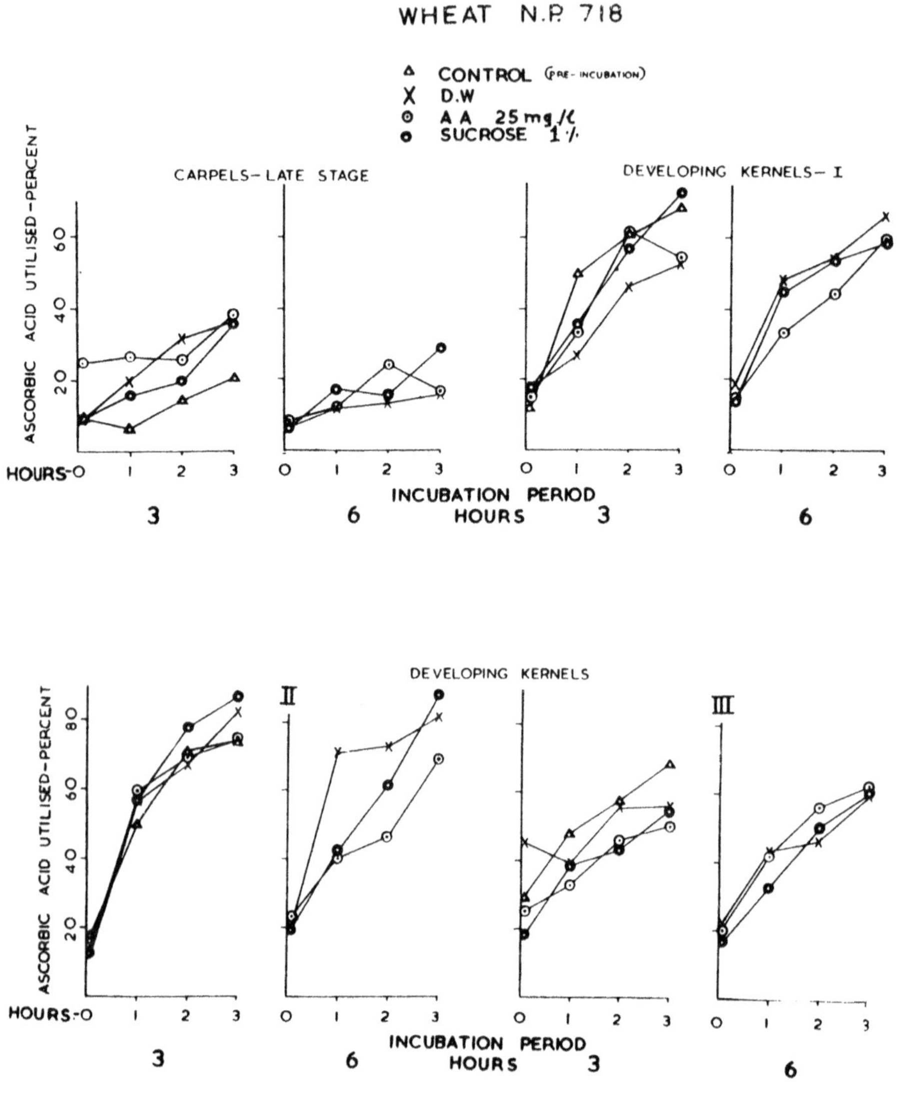

FIGURE 3.5.1,2. Ascorbic acid utilization in prefertilized carpels (late stage) and developing kernels (Stage I, II and III) of wheat after 3 and 6 hours of incubation respectively in DW, 25 mg/l AA and 1% sucrose (Abraham et al., 1968; Abraham, 1969; Chinoy et al., 1969).

III) of wheat after 3 and 6 hours of incubation respectively in DW, 25 mg/l AA and 1% sucrose (Abraham et al., 1968; Abraham, 1969; Chinoy et al., 1969).

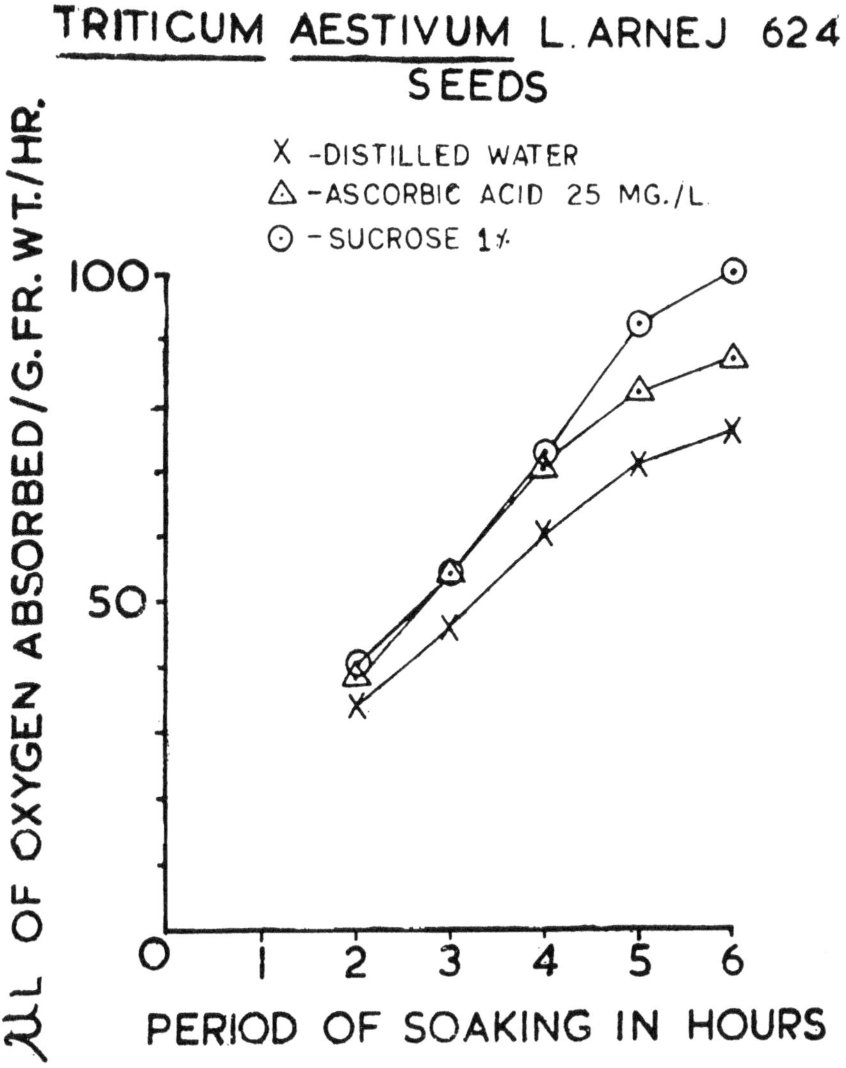

FIGURE 3.6. Effect of exogenous application of AA (25 mg/l) and sucrose (1%) to seeds of Triticum aestivum L. cv. Arnej 624 (Chinoy et al., 1969, Dave, 1969).

fed seeds followed by AA. The rate of respiration remained significantly lower in case of DW throughout the experiment and the effects of sucrose and AA were significantly higher.

From these results and the earlier biochemical studies, a new scheme of electron flow could be established, which may explain the high AAU and peroxidase activity and higher contents of AA phenols in the differentiating tissue with a very low oxygen uptake (Laloraya et al., 1972).

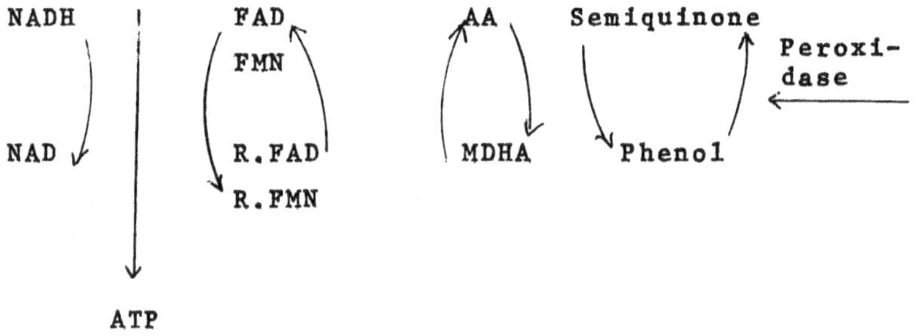

11. SUMMARY

The chloroplast of a plant cell is the chief centre of reduction of silver nitrate indicating the presence of AA in abundance. Chlorophyllous tissues like leaves have a high AA content. There is a close correlation between photosynthesis and AA formation as well as between AA content and surface area of leaves.

A model photochemical reaction involving chlorophyll - a in oxygen free pyridine solution and AA has been proposed and Eiosinophyll was the pink coloured reaction product. The photo-oxidation of AA was markedly enhanced by disrupting the chloroplasts. AA might act as an electron donor and DHA or more probably, MDHA as an electron acceptor in photosynthesis. Oxidation of AA results from the reaction with oxidized product of photolysis via a cyclic oxidation and reduction of either endogenous quinones or benzoquinones, added to the reaction mixture. It has also been suggested that AA replaces water as an electron donor for light reaction II. Further, that both the light reactions and the electron transport system between them including one ATP site participate in ascorbate photo-oxidation.

The higher AA content of the photosynthesizing organs of a plant suggests a possible role of AA as an intermediate or a catalyst in photosynthesis. AA was the only exogenously added substance which consistently enhanced photosynthetic phosphorylation along with Mg^{++}, FMN and vitamin K. The role of AA may be considered as beneficial for FMN photosynthetic phosphorylation and CO_2 fixation. AA is also assigned the function of a protector of some essential components of the chloroplasts against inactivation, rather than serving as a cofactor of phosphorylation. A specific respiratory function has been attributed to AA as a part of the reduced NAD oxidizing system and a linkage between the oxidation and glycolytic stages of respiration. A parallelism between the rate of respiration and AA turnover is now well established.

REFERENCES

1. Aberg B 1958. Ruhland W (ed) Encycl Pl Physiol 6: 479. Berlin: Springer Verlag.
2. Abraham PG 1969. Ph.D. Thesis, Botany Department, University School of Sciences, Gujarat University, Ahmedabad-380009, India.
3. Abraham PG, Dave IC, Pandya RB, Saxena OP 1968. Proc Nat Acad Sci Sec Biol. Abst. 98: 48.
4. Arnon DI 1955. Science 122: 9.
5. Arnon DI 1958. Chloroplast and Photosynthesis, Brookhaven Symp No. 11, p. 181.
6. Arnon DI, Allen MB, Whatley FR 1954. Nature (Lond) 174: 394.
7. Arnon DI, Whatley FR, Allen MB 1955. Biochem Biophys Acta 16: 607.
8. Arnon DI, Whatley FR, Allen MB 1957. Nature (Lond) 180: 182, 1325.
9. Arnon DI, Whatley FR, Allen MB 1958. Science 127: 1026.
10. Asselbergs EAM 1957. Plant Physiol 32: 326.
11. Avron M 1964. Biochem Biophys Res Commun 17: 430.
12. Avron M 1967. Current topics in bioenergetics Sanadi DR (ed) New York: Academic. II: p. 373.
13. Avron M, Newmann J 1968. Ann Rev Plant Physiol 19: 137.
14. Bannister TT 1959. Plant Physiol 34: 246.
15. Beevers H 1961. Respiratory metabolism in plants New York: Harper & Row.
16. Boardman NK, Anderson JM 1964. Nature (Lond) 203: 166.

17. Bohme H, Trebst A 1969. Biochim Biophys Acta 180: 137.
18. Boswell JG 1950. Ann Bot 14: 521.
19. Bourquelot E, Bertrand G 1895. C r Soc Biol Paris 47: 582.
20. Bubnov NN, Krasnovsky AA, Umrikhina AV, Tsepalov VG, Shliapintokh VI1960. Biofizika 5: 121.
21. Bubnov NN, Tsepalov VF, Shliapintokh VIa 1959. Izv Akad Nauk SSSR 10.
22. Cedeno-Maldonado A, Swader JA, Heath RL 1972. Plant Physiol 51: 698.
23. Chiba Y, Okayama S 1962. Plant & Cell Physiol 3: 379.
24. Chinoy JJ 1962. Indian J Plant Physiol 5: 172.
25. Chinoy JJ 1968. Vidya 11: 138.
26. Chinoy JJ 1969. Indian J Plant Physiol 12: 67.
27. Chinoy JJ, Gurumurti K, Singh YD 1973. Ind J Plant Physiol 16: 42.
28. Chinoy JJ, Nanda KK 1959. Proc. IX Internatl. Bot. Congr. 2: 69 (Abstracts).
29. Chinoy JJ, Patel Hemlata D, Jani BM, John D, Vyas AV, Raj KPS 1969 a. Report PL - 480 Research Project, FG-In-208 (A7-CR-108) p. 1-456, Botany Department, University School of Sciences, Gujarat University, Ahmedabad - 380009, India.
30. Chinoy JJ, Saxena OP, Abraham PG, Pandya RB, Dave IC, Gurumurti K, Sastri K 1969. Report PL - 480 Research Project, FG-In-182 (A7-CR-87) p. 1-350, Botany Department, University School of Sciences, Gujarat University, Ahmedabad - 380009, India.
31. Chinoy JJ, Saxena OP, Gurumurti K 1970. ISPP Symp "Plant Growth Regulators, Kharagpur, Jan 3-8, Abst. 1.
32. Chinoy JJ, Shah CK, Patel HT, Suthar HK 1965. Proc. Seminar "Tissue Culture", M.S. University, Baroda, p. 377.
33. Chinoy JJ, Shah CK, Patel HT, Suthar HK (1967) Biol Plant 9: 182.
34. Chinoy NJ 1970. Stain Tech 45: 99.
35. Chinoy NJ 1973. Ind Natl Sci Acad 39B: 119.
36. Chinoy NJ 1978. J Anim Morph Physiol Silver Jubilee Volume: 68.
37. Commoner B 1958. Personal Comm IV Internal Congress Biochem Vienna, Austria.
38. Dave IC 1969. Ph.D. Thesis, Botany Department University School of Sciences, Gujarat University, Ahmedabad - 380009, India.
39. Dunn FJ, Dawson CR 1951. J Biol Chem 189: 485.
40. Forti G, Jagendorf AT 1961. Biochim Biophys Acta 54: 322.
41. Friend J, Olsson R, Redfearn ER, 1966. In: Biochemistry of Chloroplasts, Goodwin TW (ed), London: Academic, II: p.537.
42. Fuchtbauer W, Simonis W 1958. Naturwissenschaften 45: 519.
43. Gorman DS, Levine RP 1965. Proc Natl Acad Sci U S 54: 1665.
44. Gozal B, Avron M 1970. Eur J Biochem 15: 155.

45. Gromet-Elhanan Z 1968. Arch Biochem Biophys 123: 447.
46. Gromet-Elhanan Z 1967. Biochim Biophys Acta 131: 526.
47. Gromet-Elhanan Z, Avron M 1964. Biochemistry 3: 365.
48. Gurevich AA 1963. Dokl Biol Sci 145: 674.
49. Haggen U 1953. Biol Abstr 29: 945.
50. Heelis PF, Parsons BJ, Phillips GO, McKellar JF 1981. Photochem Photobiol 33: 7.
51. Hendrich W 1958. Roczniki Chemii 32: 107.
52. Hoffmann P, Albrecht E 1966. Z Pflanzenphysiol 55: 292.
53. Hsieh Po-Tung, Su Yuan-Chi, Yin, Fwu-Shiow, Yang, Chien-Tse 1977. Hua Hsueh 2: 35 (In Chinese).
54. Ikeda S 1959. Mem Res Inst Food Sci Kyoto Univ 18: 57.
55. Itoh S 1979. Biochim Biophys Acta 548: 589.
56. Itoh S 1979 a. Ibid 548: 596.
57. Ishikawa H, Niew Satsuki, Takabe Teruhiro, Takenaka Kiyoshi, Nikai Sueki 1981. Meizo Daigaku Rikogakubu Kenkyu Hokoku 21: 220.
58. Ivanov BN, Povalyaeva TV 1979. Fiziol Rast (Moscow) 26: 276.
59. Iyanagi T, Yamazaki I 1969. Biochim Biophys Acta 172: 370.
60. Jacobi G 1964. Z Naturforsch 19 b: 470.
61. Jagendorf AT 1962. Survey of Biological Progress, Glass B (ed) Vol 4: 181.
62. James WO 1953 a. Biol Rev Cambridge Phil Soc 28: 245.
63. James WO 1953 b. Ann Rev Plant Physiol 4: 59.
64. James WO, Cragg NM 1943. New Phytol 42: 28.
65. James WO, Heard CRC, James GM 1944. Ibid 43: 62.
66. Katoh S, San Pietro A 1967. Arch Biochem Biophys 122: 144.
67. Keister DL 1963. J Biol Chem 238: PC 2590.
68. Keister DL 1965. Ibid 240: 2673.
69. Kiraly Z, Farkas GL 1957. Arch Biochem Biophys 66: 474.
70. Kok B 1967. In: Harvesting the Sun, Photosynthesis in Plant Life, San Pietro A, Greer FA, Army TJ (eds) New York: Academic, p. 29.
71. Kow Yoke Wah, Smith DA, Gibbs M 1982. Plant Physiol 69: 740.
72. Krasnovsky AA 1948. Compt Rend (Doklady) Akad Nauk USSR 60: 421.
73. Krasnovsky AA 1960. Ann Rev Plant Physiol 11: 363.
74. Krasnovsky AA, Umrikhina AV 1958. Doklady Akad Nauk SSSR 122: 1061.
75. Krasnovsky AA, Chan Van Nhi, Nikandrov VV, Brin GP 1980. Mol Biol (Moscow) 14: 287. Plant Physiol 66: 925.
76. Krogmann DW 1973. The Biochemistry of Green Plants. Prentice Hall of India, Pvt Ltd., New Delhi.
77. Laloraya MM, Srivastava HN, Bakshi N, Shah N, Pandya KJ, Singh YD 1972. All India Symp Biol Land Plants Meerut, India.
78. Leblanc RM, Du Pre Werson B, Dijkmans H, Aghion J 1981. Photosynthetica 15: 109.
79. Ljones T, Skotland T 1979. FEBS Lett 108: 25.
80. Lynch VH, French CS 1954. Carnegie Inst Washington Dept Plant Biol Ann Report 54: 166.

81. Mandels GR 1953 Arch Biochem Biophys 44: 362.
82. Mapson LW 1958. Ann Rev Pl Physiol 9: 119.
83. Mapson LW 1963. Proc. Roy Soc London B 157: 365.
84. Mapson LW 1964. Phytochem 3: 429.
85. Marre E, Arrigoni O 1958. Biochim Biophys Acta 30: 453.
86. Marre E, Arrigoni O, Rossi G 1959. Biochim Biophys Acta 36: 56.
87. Mathews MB 1951. J Biol Chem 189: 695.
88. Mayer AM 1958. Physiol Plantarum 11: 75.
89. Mayer AM, Poljakoff-Mayber A, Appleman W 1957. Physiol Plantarum 10: 1.
90. Mercer-Smith JA, Mauzerall D 1981. Photochem Photobiol 34: 407.
91. Mitsui A, Oi Y 1961. Plant Cell Physiol Tokyo 2: 45.
92. Mitsui A, Oi Y 1961 a. Plant Cell Physiol Tokyo 2: 99.
93. Myer YP, Pande A, Pande I, Thallam KK, Saturno AF, Verma BC 1981. Int J Quantum Chem 20: 513.
94. Nakano Y, Asada K 1981. Plant Cell Physiol 22: 867.
95. Nanda KK, Krishnamurthy HN 1966. Proc 53rd Indian Sci Congr III Abstr.
96. Newbury H, Smith JH 1981. Eur J Biochem 117: 575.
97. Ohnishi T, Yamazaki H, Iyanagi T, Nakamura T, Yamazaki I 1969. Biochim Biophys Acta 172: 357.
98. Oku T, Tomita G 1980. Physiol Plant 48: 99.
99. Patel HD 1967. Ph.D. Thesis, Gujarat University, Ahmedabad - 380009, India.
100. Peschek GA 1980. Arch Microbiol 125: 123.
101. Piette LH, Yamazaki I, Mason H 1961. In: Free Radicals in Biological Systems. Blois MS,J, Brown HW, Lemmon RM, Lindblom RO, Weissblutth M (eds) New York: Academic.
102. Popovskaya EM 1950. Biokhimiya 44: 10063.
103. Prokoshev SM, Dantsheva EI 1946. Biokhimiya 11: 481.
104. Rabinowitch EI 1956. Photosynthesis and related processes. New York: Intersciences Vol. II: Part I p. 603.
105. Rabinowitch E, Govindjee 1970. Photosynthesis New Delhi: Wiley Eastern Pvt Ltd.
106. Reid ME 1937. Am J Bot 24: 445.
107. Rudolph E, Bukatsch F 1967. Flora ABT A Physiol Biochem (Jena) 153: 443.
108. Sagotov ZS, Lakeev SG, Zhokevich VN 1981. Fiziol Rast 28: 150.
109. San Pietro A, Greer FA, Army TJ, (eds) 1967. Harvesting the Sun. New York: Academic.
110. Schwartz M 1967. Biochim Biophys Acta 131: 548.
111. Schwartz M 1967 a. Biochim Biophys Acta 131: 559.
112. Sechenska M, Tomova N, Dechev G 1968. Compt Rend Acad Bulg Sci 21: 277.
113. Shimazaki K, Sasaki T, Kondo N, Sugahara K 1980. Plant Cell Physiol 21: 1193.
114. Siegenthaler Paul-Andre 1969. Plant and Cell Physiol 10: 811.
115. Steinmann HG, Dawson CR 1942. J Amer Chem Soc 64: 1212.
116. Szent Györgyi A 1930. Science 72: 125.
117. Szent Györgyi A 1938. Zschr Physiol Chem 255: 126.

118. Szent Györgyi A 1939. On Oxidation, fermentation, Vitamins Health and Disease. Baltimore: Williams & Wilkins, p. 109.
119. Tabushi I, Yazaki A 1981. J Org Chem 46: 1899.
120. Trebst A 1964. Z Naturforsch 19 B: 418.
121. Trebst A, Eck H 1961. Z Naturforsch 16B: 455.
122. Trebst A, Erich E 1967. In: Proc NATO Adv Study Inst Biochem Chloroplasts Aug. 1965, Aberytswyth, Wales, U.K. London: Academic, Vol. 2, p. 531.
123. Trebst A, Pistorius E 1967. Biochim Biophys Acta 131: 580.
124. Trebst A, Eck H, Wagner S 1963. In: Photosynthetic Mechanism of Green Plants. Kok CB, Jagendorf AT (eds) Natl Acad Sci Natl Res Council Publ 1145, Washington, D.C. p. 174.
125. Vernon LP 1967. In: Harvesting the Sun. Photosynthesis in Plant Life. San Pietro A. Greer FA, Army TJ (eds) New York: Academic, p. 15.
126. Vernon LP, Zaugg WS 1960. Fed Proc 19: 329.
127. Ward JM 1955. Plant Physiol 30: 58.
128. Waygood ER 1950. Canad J Res Sect C 28: 7.
129. Wessels JSC 1963. Proc Roy Soc Lond 147 B: 345.
130. Wessels JSC 1964. Biochim Biophys Acta 79: 640.
131. Whatley FR, Allen MB, Arnon DI 1957. Plant Physiol 32: (Suppl) iii.
132. Whatley FR, Allen MB, Rosenberg LL, Capindale JB, Arnon DI 1956. Biochim Biophys Acta 20: 462.
133. Whatley FR, Allen MB, Arnon DI 1959. Ibid 32: 32.
134. Widart M, Dinant M, Techy F, Aghion J 1980. Photobiochem Photobiophys 1: 103.
135. Yamazaki I 1962. J Biol Chem 237: 224.
136. Yamazaki I, Mason HS, Piette L 1960. J Biol Chem 235: 2444.
137. Yamashita T, Junko T, Gitti T 1971. Plant Cell Physiol 12: 117.
138. Yerkes CT, Babcock GT 1980. Biochim Biophys Acta 590: 360.
139. Yu CM, Brand JJ 1980. Biochim Biophys Acta 591: 483.

4. GERMINATION, JUVENILE AND VEGETATIVE GROWTH

We are concerned here mainly, with the role of AA in germination, juvenile and vegetative growth. Reference to the work on the effect of auxins, gibberellins, and other regulators will be made only in relation to their interaction with AA.

Role of AA in germination of seeds and later stages of plant growth has been the subject of numerous reviews (Tonzig and Marre, 1955; Bonner, 1957; Mapson, 1958; 1960; Knox, 1960; Aberg, 1961; Chinoy et al., 1969 a, b; Chinoy, 1962, 1977).

Noting the parallelism between growth rate and AA content of higher plants, AA may be considered as a growth factor or even as an indispensable phytohormone. It also has growth promoting property. However, as pointed out by Aberg and Ekdahl (1948) such a parallelism is by no means a general one because an inverse correlation was found between growth rates and AA content of mature leaves in many species of plants. In the case of young growing tissues, however, there are better indications of a rise in AA content with increasing growth rate. Sreenivasan and Wandrekar (1950 a) observed an increase in the biosynthesis of AA in seeds germinating in the dark, which were previously soaked in cold water. Addition of hexose to nutrient media was found to double the AA formation in embryos. They considered leaves as the main seat of biosynthesis of AA because its highest values were obtained for leaves after cold treatment. In further studies, Sreenivasan and Wandrekar (1950 b) found an increase of 45% in AA content of 5 day old seedlings of <u>Phaseolus radiatus</u> sprouted in the dark or steeped in the cold at 5° C with a corresponding decrease of 37% in DHA. They also demonstrated a 25% fall in AA oxidase activity and a 158% increase in glutathione content. Other legumes also gave similar results. Goh (1950) found a parallel increase in glutathione and AA contents during germination which reached its maximum on the

15th day after germination.

A study of the role of ascorbic acid turnover in metabolism during germination and juvenile growth has been studied in three types of seeds having carbohydrate, protein and lipid as major reserve nutrients (Saxena, 1969; Gurumurti, 1971; Pandya, 1969; Chinoy; 1962; Chinoy et al., 1969 a, b; Abraham, 1969). As there is general agreement in the data, results of one of the workers (Gurumurti, 1971) are presented here. The AA content of the embryo axis and the endosperm of wheat var N.P. 718 remained more or less constant upto 72 hours of germination, after which there was a slight fall (Fig. 4.1). There was a gradual decline in AA utilization in the embryo axis, whereas, in the endosperm AAU showed a rhythmic rise and fall. ASG content of the embryo axis was also at a constant high level except in the determination made after 48 hours of germination. In the endosperm ASG content was at a much lower level and showed a fluctuating trend. AA - MM complexing in the embryo axis rose sharply after 24 hours of germination and reached its maximum after 72 hours, after which it was at a lower level (Fig. 4.1). AA - MM complexing was absent in the endosperm in this case. The peroxidase as well as AA - free radical peroxidase activities in both embryo axis and endosperm were enhanced with advance in germination (Fig. 4.2). This finding suggested the participation of AA in the form of its free radical, MDHA. DNA shows rhythmic trend in embryo but shows a net synthesis in endosperm (Fig. 4.2).

As a certain minimum level of AA is maintained in the endosperm, it is likely that it is synthesized locally in this organ from carbohydrates as shown earlier by Chinoy et al. (1958; 1961), to enable the redox balance to be maintained at an optimum level. The utilization of AA in the endosperm showed an inverse relation, i.e., with an increase in AAU activity there was a decline in the AA level (Fig. 4.1). Also, with increased AAU there occurred an increase in the bound form of AA, thus counteracting the

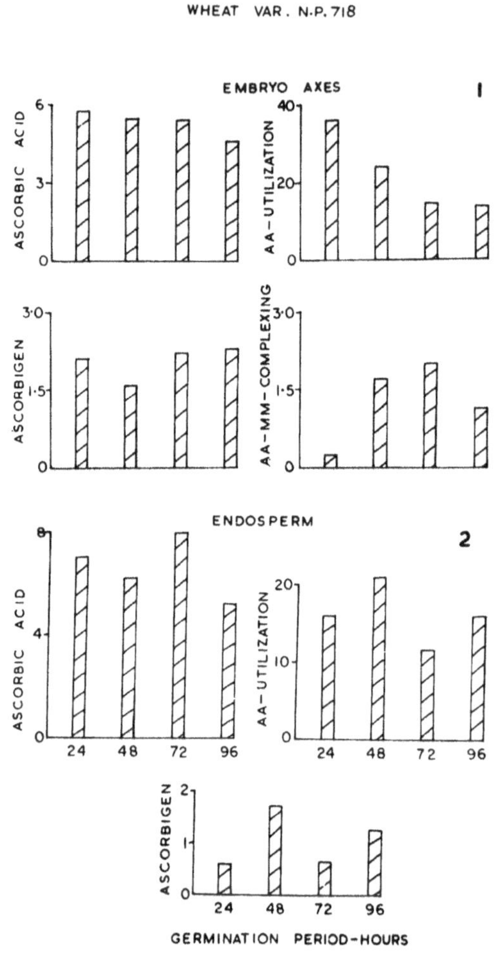

FIGURE 4.1. AA and ASG contents as well as AAU and AA - MM complexing in embryo axis and endosperm of wheat N. P. 718 during 96 hours of juvenile growth.

oxidative shift caused by AAU in the endosperm. This behaviour is consistent in the endosperm. In the embryo axis, on the other hand, the oxidative processes of AAU declined with a concomitant increase in ASG as well as a general increase in complexing of AA with macromolecules (Fig. 4.1).

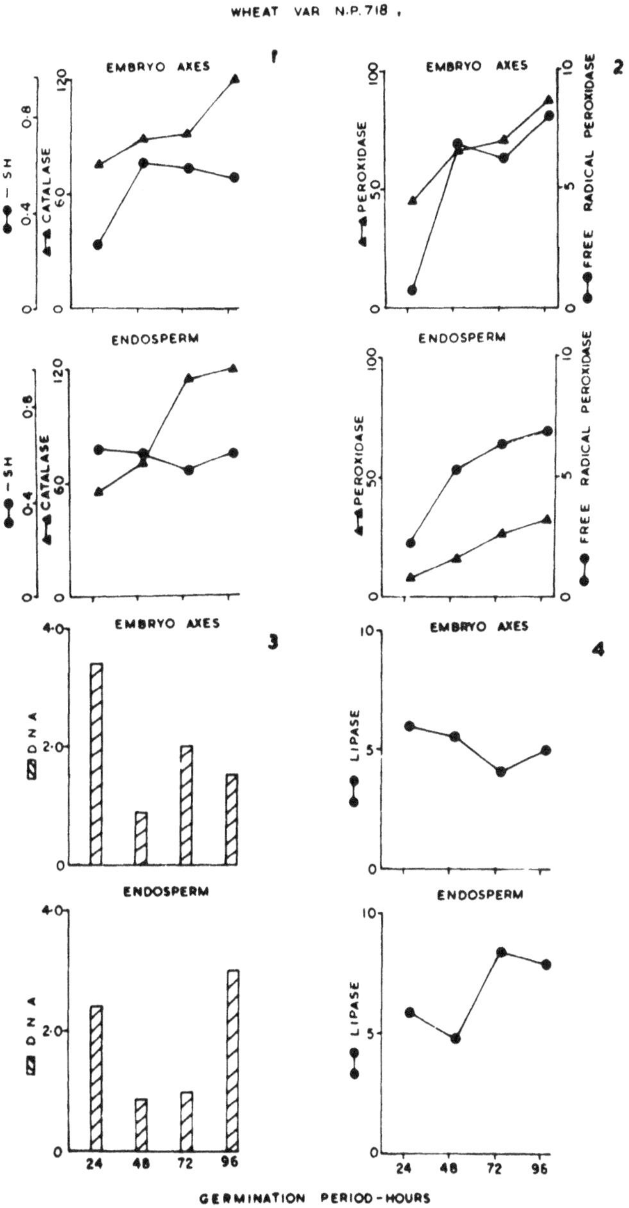

FIGURE 4.2. Catalase, peroxidase, AA-FR peroxidase and lipase activities as well as -SH and DNA contents of embryo axis and endosperm of wheat N.P. 718 during 96 hours of juvenile growth.

Thus the redox balance in the embryo axis was clearly shifted towards the reductive side aiding the synthetic processes. The more or less same level of -SH after 48 hours period inspite of increasing catalase and peroxidase activities was also indicative of the reductive atmosphere prevailing in the embryo axis. Thus it appeared that AA turnover goes on in the embryo axis at a faster rate as a result of which there was active growth (Pandya, 1969; Saxena, 1969; Gurumurti, 1971; Chinoy et al., 1972 a, b).

As in the case of wheat, the cotyledon of pea and gram also maintained a certain minimum level of AA (Fig. 4.3). AA utilization reached a maximum after 48 hours and the level was maintained throughout (Fig. 4.3). Mapson and Moustafa (1956) have also reported that AA oxidase does not attain its full activity during initial periods of germination. ASG content both of the embryo axis and cotyledon of pea and gram showed a fluctuating trend. AA - MM complex, on the other hand, increased in the embryo axis of pea upto 72 hours of germination after which there was a decline. The continuous increase of AA - free radical peroxidase in the embryo axis (Saxena, 1969; Gurumurti, 1971; Chinoy et al., 1972 a, b) is suggestive of the participation of AA in the form of its free radical in the synthetic processes of the embryo axis. (Fig. 4.4). SH and catalase showed an inverse correlation in endosperm while in the embryo, -SH showed a steep rise with increase in catalase activity (Fig. 4.4). Amylase activity rose with advancement of germination and showed a correlation with a decline in Starch (Fig. 4.4).

AA content decreased during the early period of germination and increased after 96 hours. Decline in AA content was presumably due to a considerable increase in its utilization, thus indicating active participation of AA in the germination process. Later on, AA content increased inspite of increase in AAU. This increased utilization of AA during juvenile differentiation is in confirmity with previous

PISUM VAR. PERFECTION

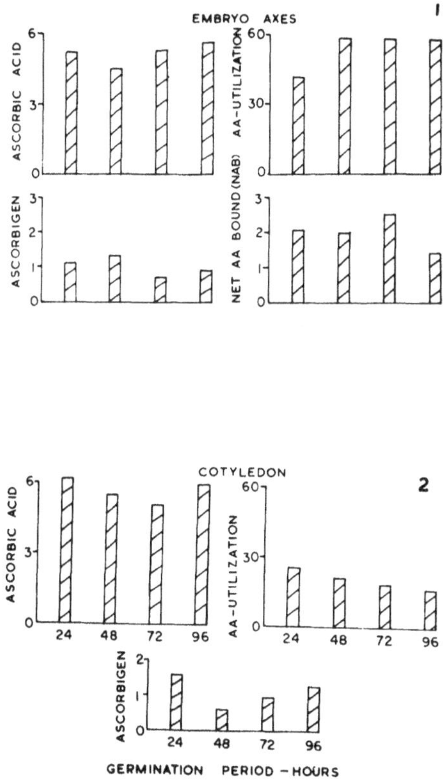

FIGURE 4.3. AA, ASG, AAU and AA - MM complexing in embryo axis and cotyledon of Pisum sativum L. cv. Perfection during 96 hours of juvenile growth.

findings in case of wheat (Abraham, 1969; Pandya, 1969; Saxena, 1969; Chinoy et al., 1969 a, b). ASG and AA - MM complexing showed fluctuating trends during germination which may be helpful in the release of AA when needed during active cell division and its complexing with macromolecules for future use. AA - FR peroxidase activity increased

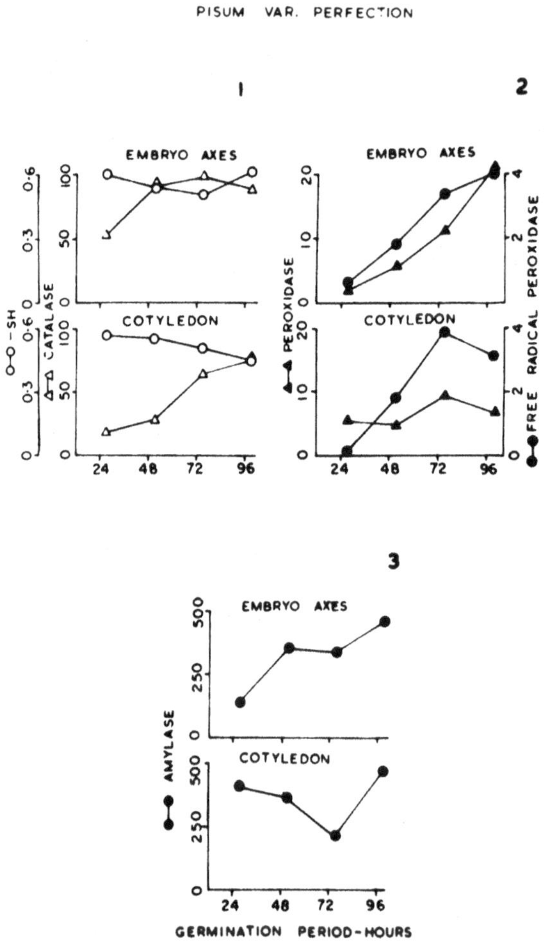

FIGURE 4.4. Catalase, peroxidase, AA - FR peroxidase and amylase activities and -SH content of embryo axis and cotyledon of Pisum sativum L. cv. Perfection during 96 hours of juvenile growth.

continuously with advance in germination in all the three types of plants. As the free radical of AA is formed due to the activity of this enzyme, and as this free radical is a more powerful reducing agent than AA itself, its increased production with progress in germination will enhance electron energy needed for the growth of embryo axis (Chinoy, 1962; Chinoy et al., 1969 a, b; Gurevich, 1963; Saxena, 1969).

1. ASCORBIC ACID AS ENZYME MOBILIZING HORMONE

Isolated endosperm of barley when incubated in different concentrations of ascorbic acid and gibberellic acid, it was found that AA proved to be superior to GA at all concentrations tried in accelerating the release of sugars (Saxena et al., 1969). The release of sugar in incubation medium containing 5 ppm of AA was 5 times greater than that in distilled water and 2.6 times more in the medium containing 5 ppm of GA. At the above mentioned concentration (5 ppm), both AA and GA showed optimum sugar release. In this connection it is well to note that Rowsell and Goad (1962) have demonstrated the accelerating effect of reducing substances on amylase activity. Sprent (1968) is of the opinion that GA cannot be considered as an enzyme - mobilizing hormone (EMH). Thus the information presented here conclusively points to the fact that AA is an enzyme mobilizing hormone because of its universal presence in plants as well as due to the free radical mechanism of its oxidative process.

On the basis of the foregoing evidence it appears that with the imbibition there was a rapid breakdown of various reserves into simpler and soluble substances by the activities of oxidative and hydrolytic enzymes. Besides operating as an enzyme mobilizing hormone, AA actively participates in cell division and cell differentiation processes by its active turnover (Chinoy, 1962). Moreover, by its oxidative transformation into its free radical form by the catalytic action of AA-FR peroxidase, it creates a strong reductive atmosphere which helps in the activation of synthetic processes at a faster rate by providing electron energy. Similarly, hydrolytic activities of lipase, protease and RNAse are also accelerated by AA turnover thus enhancing the breakdown of various reserves into simpler soluble and utilisable products which are translocated towards the growing points of embryo axis leading to its further growth and differentiation. According to Mapson and Cruickshank (1947) as well as Somers and Kelly (1951), AA contributes to the cation-anion balance in seedlings as well as in the tissues of mature leaves. In cress seedlings germinated in salt solution, AA content was markedly reduced by addition of ammonium sulphate and chloride. Potassium nitrate and malate increased the AA content. Havas and Gal (1936) were able to demonstrate favourable effect of low concentrations of AA on germination of wheat seed. Higher concentration on the other hand, retarded germination or proved lethal. Contradictory results were obtained in the case of oat germination. Davies et al. (1937) and Ruge (1947) were able to stimulate germination in oat to a slight-extent, whereas, Emmerich (1941) failed to observe any effect.

A number of workers have studied the effect of AA on *Avena* coleoptile sections singly as well as in combination with IAA and sucrose (Raadts, 1948; Scheuermann, 1952; Bitancourt, 1955; Chinoy et al., 1957 a, b). AA when used alone gave a slight stimulatory effect or none at all, but in combination with sucrose and IAA the response was very marked. The promotive effects of AA treatment have been reported in germination of cress and mustard (Davies et al., 1937), jute (Mitra and Datta, 1951), lupine and tomato (Lee, 1955) and tobacco (Dennison, 1940). The stimulatory effect of exogenous AA on seedling growth depends upon the concentration of AA (Havas and Gal, 1936).

Garg et al. (1958) reported that the retarding influence of colchicine on growth was considerably removed when it was applied in combination with AA. This observation strongly suggests that colchicine treatment brought about a decrease in the endogenous AA. A similar decrease in the endogenous AA of germinating seeds after treatment with colchicine has also been reported by Sreenivasan and Wandrekar (1950 a, b).

Gurumurti (1971) showed that the AA concentration was maintained at more or less the same level in the embryo-axis of <u>Arachis</u> but the utilization of AA increased continuously as germination advanced (Fig. 4.5). Further, the level of endogenous AA in AA - fed embryo-axis was maintained at a slightly higher level as compared to distilled water (DW) fed embryo axis. The more or less same level of AA with concomitant increase in utilization of AA showed that its synthesis occurred in the embryo axis from carbohydrates, especially sugars. This is supported by the data on total sugar content which shows a decreasing trend with the advance of germination upto 72 hours (Fig. 4.6). Here it must be noted that the level of total sugar in the embryo axis of AA fed media is considerably higher at almost all stages of germination compared with that in DW (Gurumurti, 1971). Further, it was observed that the sugar content of the AA fed cotyledon was gignificantly lower compared to the one fed DW. This showed that exogenous application of AA enhanced the transport of sugar to the embryo axis (Saxena, 1969; Gurumurti, 1971).

The activities of AA - FR peroxidase as well as general peroxidase of the embroy axis and the cotyledon increased generally with the advance of germination (Fig. 4.7). Exogenous AA also enhanced enzymic activity of lipase and RNAse (Fig. 4.7). The amylase activity showed an increasing trend in the embryo axis followed by a decline (Fig. 4.6). AA treatment slightly raised the levels of -SH, amylase and invertase in embryo as well as cotyledons (Fig. 4.6).

Exogenous application of AA (50 ppm) enhanced AAU as well as AA - FR - peroxidase activity concomitant with a decline in AA content in <u>Sesamum indicum</u> cv. Limbdi. In spite of the faster utilization of AA, its level was more in AA treated segments. This suggested that AA enhanced its own biosynthesis in the seedling by a feed-back mechanism.

AA and H_2O_2 not only increased the biosynthesis of AA, but also its utilization. Concomitant with increase in the AA content of AA + H_2O_2 treated seedlings, there was a

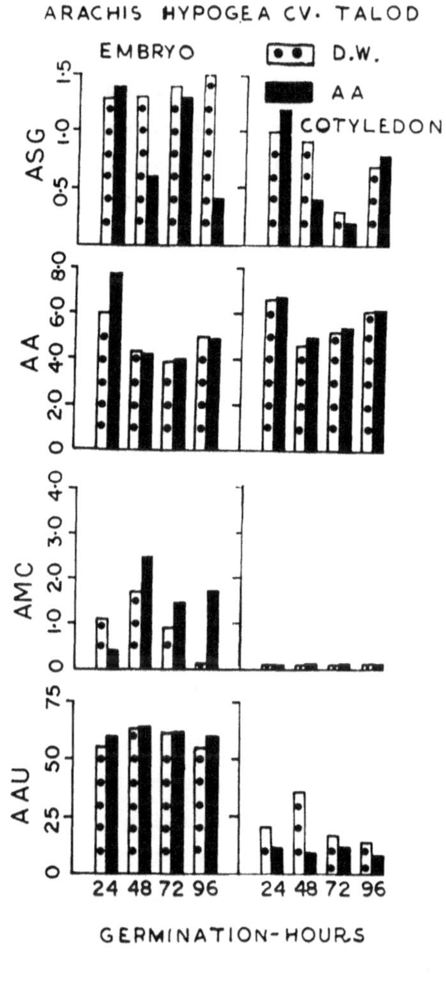

FIGURE 4.5. AA, ASG, AAU and AA - MM complexing in embryo axes and cotyledon of *Arachis hypogea* cv. Talod (Gurumurti, 1971).

corresponding decline in ASG content. An appreciable upsurge in the complexing of AA with macromolecules also occurred.

Enhancement in the water content of seedlings growing in AA + H_2O_2 as well as in H_2O_2 and to a lesser extent in AA alone, showed that these treatments led to a greater hydration of the protoplasm and of cell wall constituents

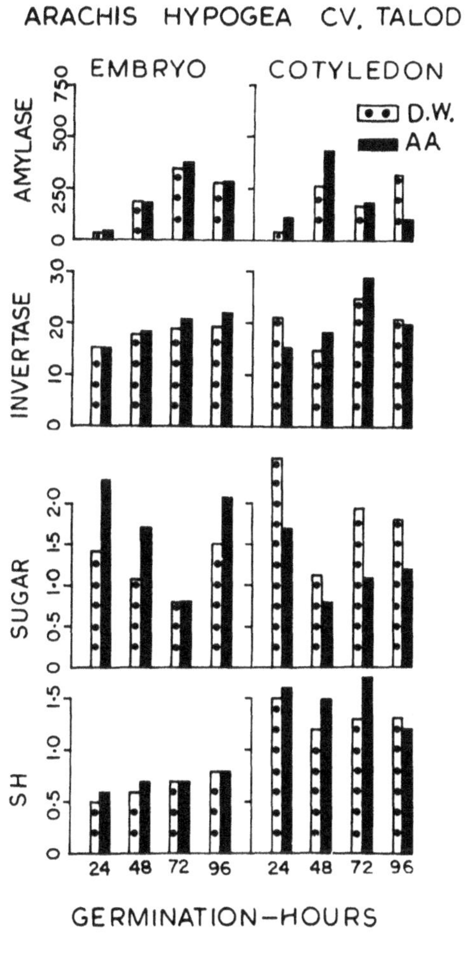

FIGURE 4.6. Activities of amylase, invertase and contents of sugar and -SH in embryo axis and cotyledon of Arachis hypogea cv. Talod (Gurumurti, 1971).

enabling the seedlings to grow faster in length.

Michniewicz and Rowicka (1961) determined the AA content of five winter and five spring varieties of wheat during germination and early growth. The plants were kept in darkness at 20 - 22° C or in natural light at a temperature of 5° C at night rising to 15° C during the day time. The

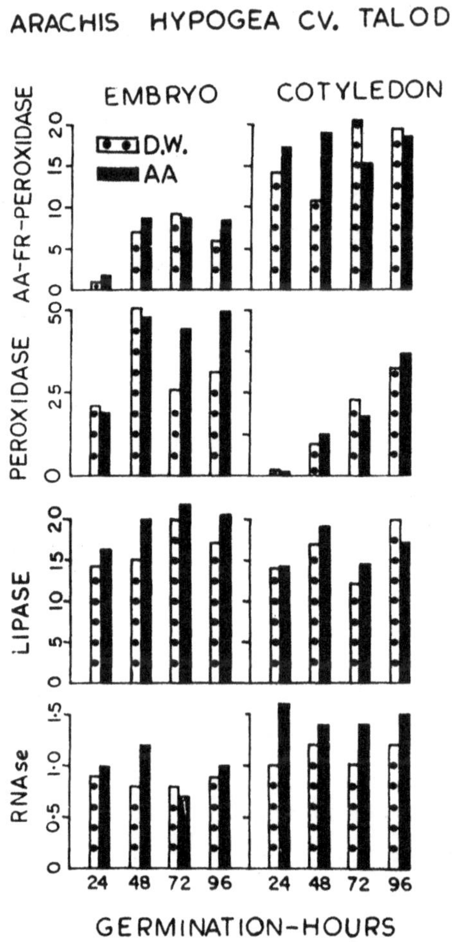

FIGURE 4.7. Activities of RNAse, lipase, general peroxidase and AA-FR-peroxidase in embryo axes and cotyledon of *Arachis hypogea* cv. Talod (Gurumurti, 1971).

amount of AA increased as the germination proceeded and was independent of the variety being a spring or a winter one. The difference in AA content depended chiefly on the physiological condition and on the stage of growth and development of the plant. Michniewicz (1961, 1961 a) reported AA contents of vernalized and control seed of three cultivars of winter wheat at different stages of vernalization. The three cultivars had different vernalization periods shown against each: (i) Leczezyanska Weresna - 20 days; (ii) "Eka" - 40 days; and (iii) Komorowska - 60 days. AA content of seeds continually increased during the course of vernalization right upto the end of the cold temperature treatment. There was a reduction in AA content if the seeds were kept at low temperature after the termination of the period of vernalization. The vernalized seeds had an invariably higher AA content as compared to the non-vernalized ones at the corresponding stages of growth. From this, the author concluded that the larger content of AA was responsible for the quicker passing of vernalized plants through the particular growth phase.

Chinoy et al. (1967) found that AA and ASG contents of the embryo-axis as well as the endosperm of the vernalized wheat were higher, indicating an accumulation of AA and ASG during low temperature treatment. The AAU on the other hand, was found to be higher at III and IV stages of germination in the non-vernalized embryo axis. It was only during the sixth stage of germination that AAU was found to be significantly higher in the vernalized set in both the embryo axis as well as the endosperm.

Patel (1967) studied the effect of vernalization of AA turnover in the late and early flowering cultivars of wheat, oat and barley. The embryo and endosperm were separately examined at different stages of germination for the control set (25 - 27° C) as well as for the vernalized set (3 - 5° C) for free ascorbic acid (AA), ascorbigen (ASG) and ascorbic acid utilization (AAU). Results for Triticum aestivum cv. Bledsoe and Avena sativa cv. Victory are reproduced here. These results clearly indicate that during germination at the inductive temperature (3 - 5° C), vernalization brings about an enhancement in the AA content of the embryo-axis of all three crop plants. During the initial stages of germination the AA content of the vernalized embryo axis of wheat cultivar Bledsoe was significantly higher compared to that of the control at a comparable stage of germination. In general, AA content of the embryo axis gradually declined with advance in growth (Fig. 4.8). In the case of the endosperm, on the other hand, the position was generally reversed, the AA content of the non-vernalized endosperm being higher at all stages compared to that of the vernalized one. Similarly, the ASG content of the embryo axis of non-vernalized germinating seed of Bledsoe was higher than that of the vernalized one during the initial stages of germination (Fig. 4.8). This result indicated that under vernalization, ASG was hydrolysed at a faster rate to release more free AA in the case of Bledsoe. The AAU was significantly higher both in the embryo axis as well as the endosperm of vernalized seed of Bledsoe upto IV and V stages of juvenile growth respectively (Fig. 4.9).

Vernalized embryo axis of Avena sativa cv. Victory (Fig. 4.10) showed a significantly higher AA content upto the Vth stage of germination. The non-vernalized endosperm on the other hand, registered higher values of AA, thereby suggesting that there was a faster mobilization of AA from the vernalized endosperm to its corresponding embryo axis (Patel, 1967).

During the first two stages of germination, ASG content of the embryo axis of non-vernalized seed was higher, but later on, the ASG content of vernalized embryo axis as well as that of the endosperm remained at a slightly higher level than that of the control. AAU in vernalized embryo axis and the endosperm of victory oat was significantly higher upto the VIIth stage in the former and upto the Vth stage in the latter respectively (Fig. 4.10).

A similar result was obtained by Patel (1967) in the case of Hordeum vulgare cv. Munsing. The AA level was generally higher in the vernalized embryo axis with the exception of the first reading and lower in the corresponding endosperm.

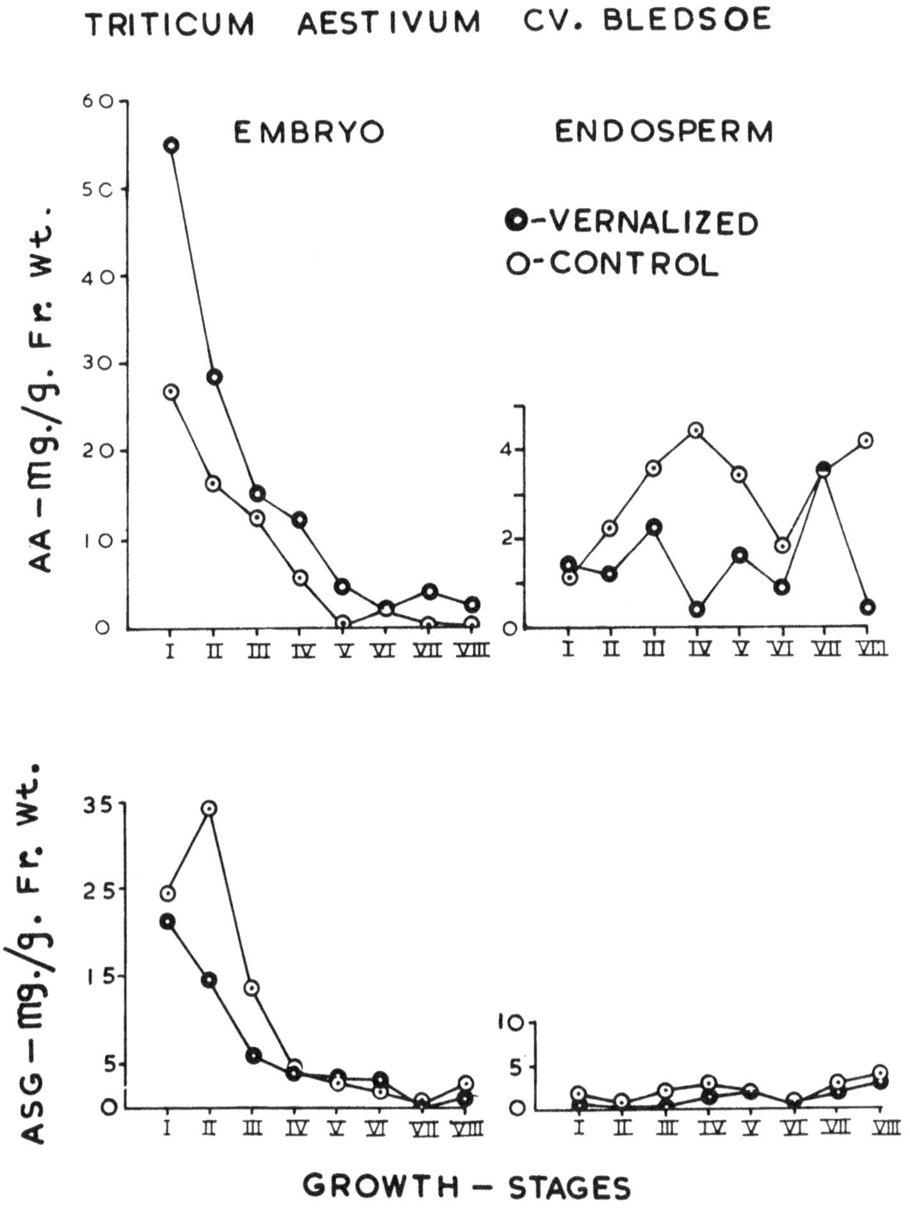

FIGURE 4.8. AA and ASG contents of vernalized and non-vernalized embryo axes and endosperm of *Triticum aestivum* L. cv. Bledsoe during juvenile phase of growth (Patel, 1967).

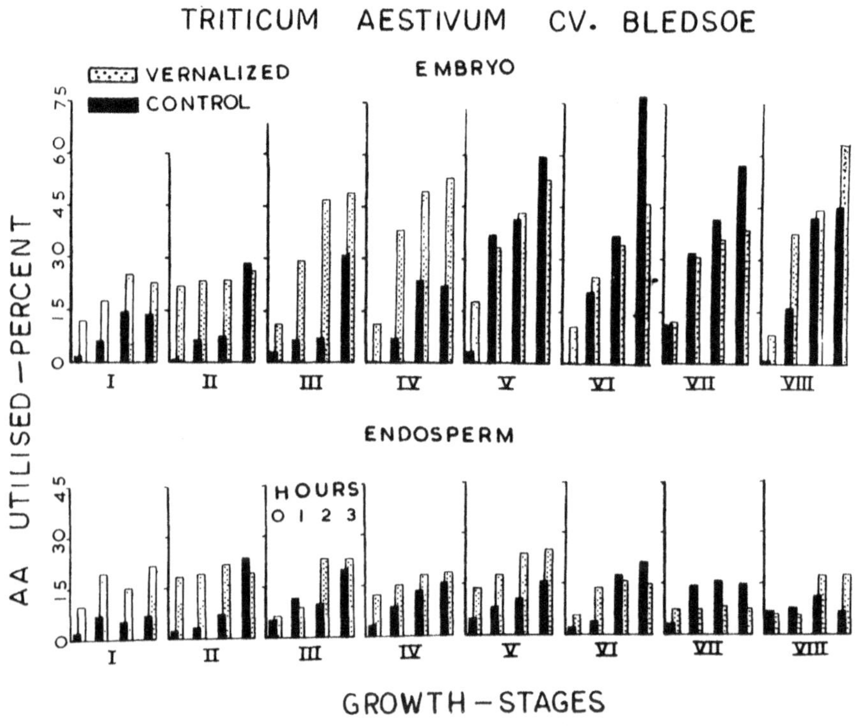

FIGURE 4.9. Percent AAU in vernalized and non-vernalized embryo axes and endosperm of *Triticum aestivum* L. cv. Bledsoe during juvenile phase of growth (Patel, 1967).

Considering the data of ASG content it became clear that there was a faster conversion of the bound form of AA into the free form in the vernalized embryo axis compared with that of the unvernalized one resulting into a quicker depletion of ASG in the former case. The same was more or less true in the case of ASG of vernalized endosperm except

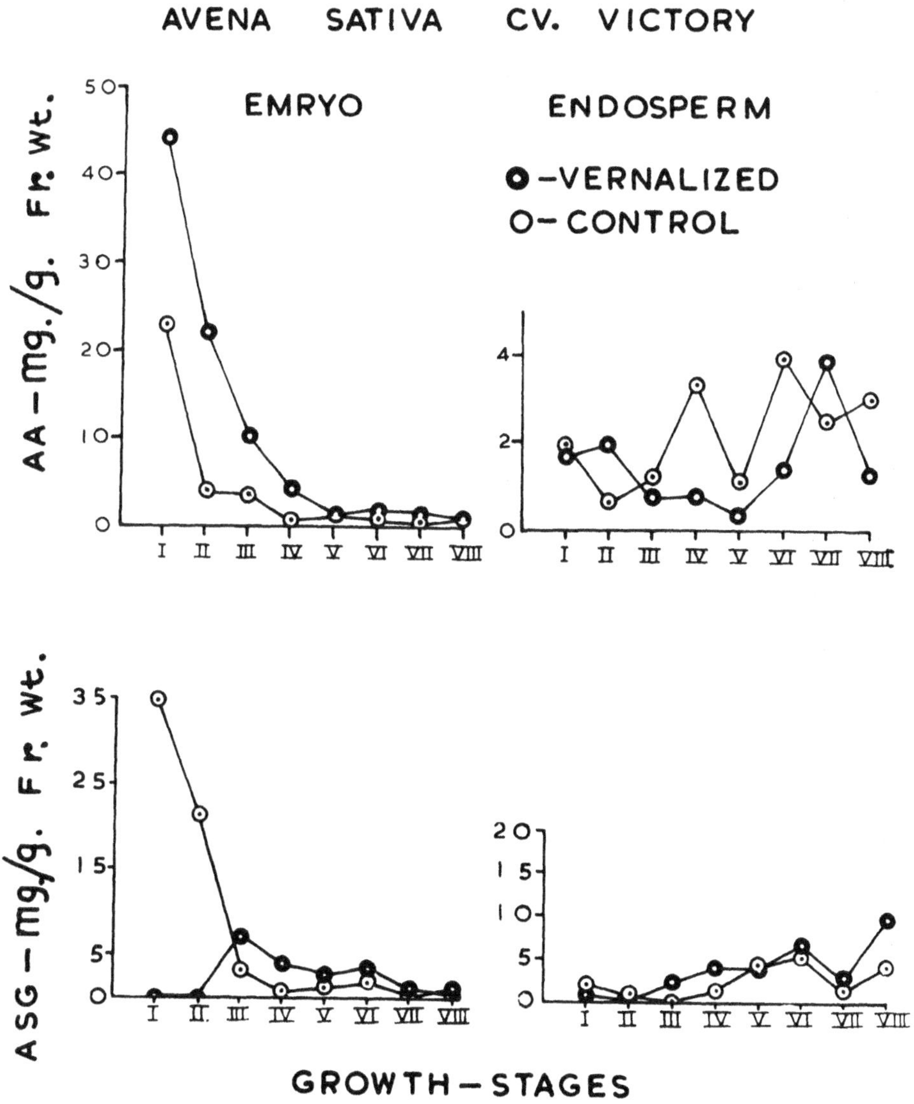

FIGURE 4.10. AA and ASG contents of vernalized and non-vernalized embryo axes as well as endosperm of <u>Avena sativa</u> L. cv. Victory during its juvenile phase of growth (Patel, 1967).

that the differences were not marked because of the low ASG content of the endosperm. Taking into consideration the data of enzymic utilization of AA (AAU), Patel (1967) presented conclusive evidence to show that there was significant acceleration in the AAU of vernalized embryo axis of Munsing as well as its corresponding endosperm. Only in the later stages of seedling growth the unvernalized set overtook and sometimes surpassed the vernalized one in AAU. For all the four periods of incubation, viz., 0, 1, 2 and 3 hours, AA utilization was significantly faster in the vernalized set (Fig. 4.11). Analysis of variance of the extensive data revealed that the turnover of AA was significantly enhanced by cold temperature treatment during germination (Patel, 1967).

Pandya et al. (1969) investigated AA turnover of germinating seeds of *Triticum aestivum* cv. N.P. 718 during vernalization in relation to catalase activity, which was significantly higher in the embryo-axis as well as in the endosperm of the non-vernalized set. In the case of the former organ, it decreased with advance in germination, whereas, in the latter (endosperm) it increased upto the fifth stage of germination after which it declined sharply. In this experiment too, AA and ASG contents of vernalized embryo-axis and endosperm were higher but not so markedly because N.P. 718 was an early flowering spring cultivar. In the case of AAU also the differences between the vernalized sets were not significant. In the endosperm however, it was higher in the high temperature set. This is in contrast to the significantly enhanced utilization of AA in Victory oat (Fig. 4.11), late flowering cultivar of barley Munsing and late cultivar of wheat Bledsoe (Fig. 4.9).

Chinoy et al. (1969 a, c) have reported further studies on change in total sugars, starch, amylase and invertase during the period of vernalization of the same cultivar of wheat (N.P. 718) used by Pandya et al. (1969) as well as for Bledsoe, a late American spring cultivar of wheat. As

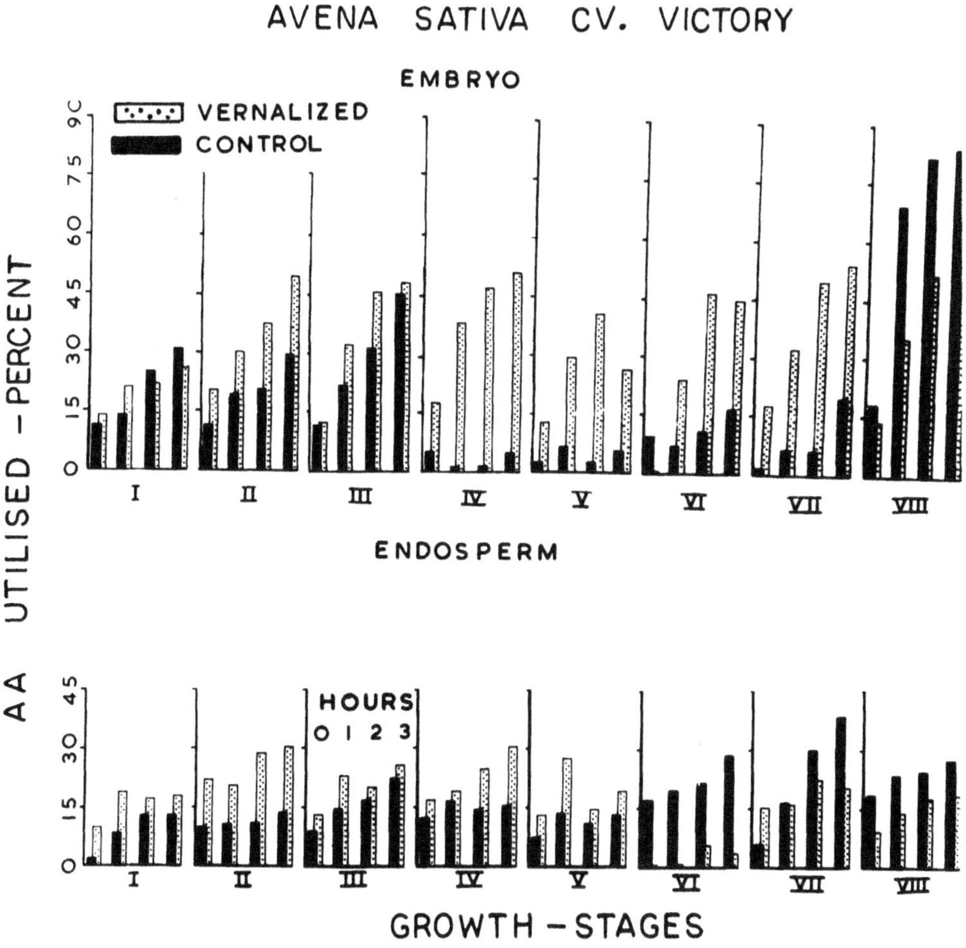

FIGURE 4.11. Percent AAU in vernalized and non-vernalized embryo axes as well as endosperm of Avena sativa L. cv. Victory during its juvenile phase of growth (Patel, 1967).

already reported, AA turnover was appreciably accelerated both in the embryo axis as well as in the endosperm of the vernalized set of N.P. 718 (Pandya et al., 1969), and also in those of Bledsoe which was significant in the light of the accelerated hydrolysis of starch and increased sugar content of the vernalized endosperm (Patel, 1967). Some of the biochemical data on AA turnover as well as other cell metabolites and enzymes presented by Pandya (1969) for the juvenile phase of wheat N.P. 718 and Bledsoe are summarized here: AA content of the embryo axis declined with advance of germination (Fig. 4.12). Generally it was more in the vernalized embryo axis of both the cultivars N.P. 718 and Bledsoe.

The AA content of the endosperm of unvernalized and vernalized seedlings of both the cultivars remained at more or less a constant high level. The AA content of vernalized N.P. 718 and Bledsoe endosperm was higher compared to that of unvernalized one throughout the course of the experiment (Fig. 4.12). The ASG content of vernalized embryo axis of both the cultivars was higher than that of the unvernalized one (Fig. 4.12). In the case of the endosperm, there were fluctuations and the ASG content of the vernalized endosperm showed an appreciably higher value at different stages of seedling growth in the two cultivars. Statistical analysis revealed highly significant effect of vernalization on the ASG content of the embryo axis. The interaction of vernalization with growth stage was also significant at the 5% point (Chinoy et al., 1969 a).

Ascorbic acid utilization in the embryo axis of N.P. 718 was not affected by vernalization (Fig. 4.12). In the embryo axis of Bledsoe, vernalization enhanced AAU to a significant extent. There was also a slight enhancement of AAU in vernalized endosperms of both the cultivars. The significant feature of the AA turnover is the determination of the degree of its complexing with macromolecules (AA - MM complexing) of the cell to produce a charge transfer complex (Fig. 4.12). AA - MM complex was significantly higher in the vernalized embryo axis of N.P. 718 and Bledsoe (Fig. 4.12). However, the general level of complexing was lower in Bledsoe compared with that in N.P. 718 and the beneficial effect of vernalization was seen more during the time of later growth stages (V and VII).

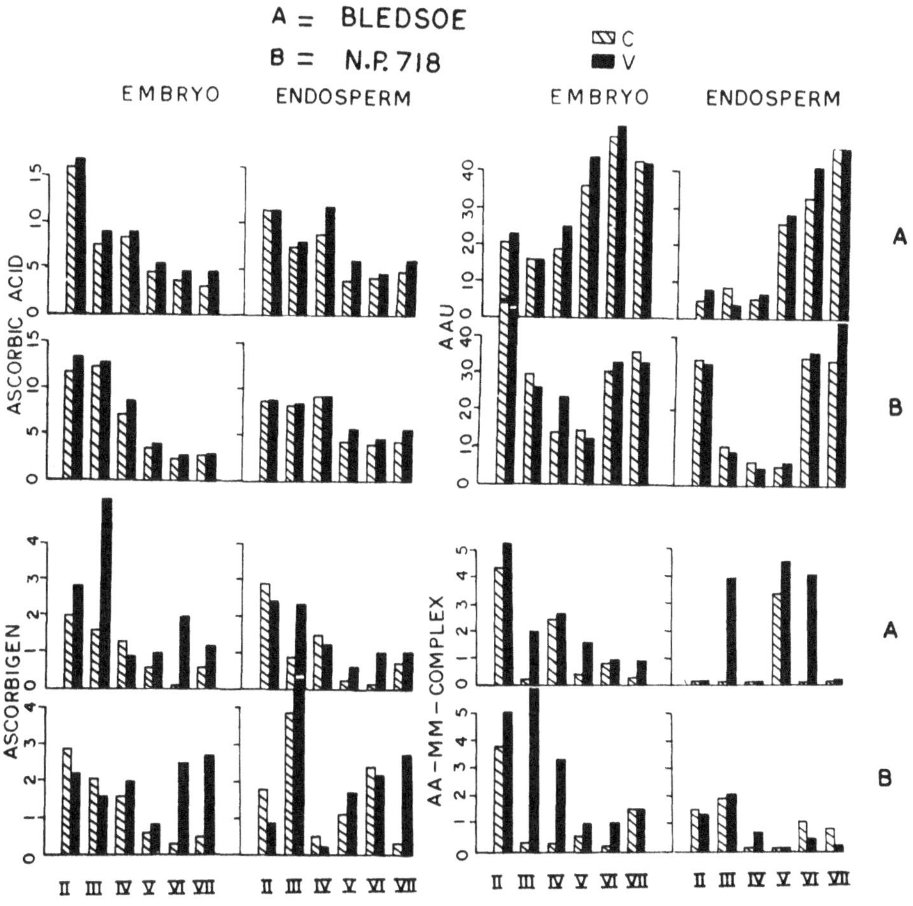

FIGURE 4.12. AA, ASG, AAU and AA - MM complexing in embryo axes and endosperm of Triticum aestivum L. cv. Bledsoe (A) and N.P. 718 (B) during their juvenile phase of growth (Pandya, 1969).

On the basis of the above evidence, together with the existing mass of data, it becomes evident that vernalization of seed brings about an acceleration in the AA turnover. It would be well at this stage to examine the experimental evidence pertaining to some enzymic activities and biosynthesis of some cell constituents during vernalization, which are correlated with greater AA turnover. While vernalization treatment considerably enhanced the activity of AA - FR peroxidase especially in the embryo axis of Bledsoe, the same was enhanced though to a lesser extent in the case of N.P. 718. The enzyme activity was appreciably lower in the endosperm compared to that in the embryo axis, especially in the case of Bledsoe (Fig. 4.13).

2. EFFECT OF VERNALIZATION ON ELECTRON FLOW AND ON METABOLISM DURING THE JUVENILE PHASE

Chouard (1960), Trione (1966) and Trione et al. (1967) have pointed out the difficulty in correlating the changes in germinative metabolism brought about by a low temperature with the vernalization response. This was due to the fact that most of the changes observed were found to be present in both spring as well as obligate winter varieties grown at a low temperature (Trione, 1966). Chouard (1960) has reviewed the studies on physico-chemical properties of vernalized tissues and is of the opinion that all these interesting results do not solve the problem of the biochemical nature of the process of vernalization.

A number of metabolic changes and alterations in rates of enzymatic activities have been observed as a result of low temperature treatment. The higher moisture content of the endosperm under low temperature suggests a greater hydration of the protoplasm which may be responsible for the enhanced activity of some of the hydrolytic enzymes, especially amylases. Faster metabolic activity was also reflected in a higher dry weight of the embryo axis under low temperature. This is corroborated by a steeper fall in the dry weight of the endosperm at low temperature indicating a faster mobilization of food reserves under this condition (Chinoy et al., 1967). Moreover, it was found that all the processes concerned with growth, synthesis and mobilization are at higher levels in this condition. A high level of AA and especially ASG and AA – MM complexing was observed in vernalized plants (Fig. 4.12). There was higher

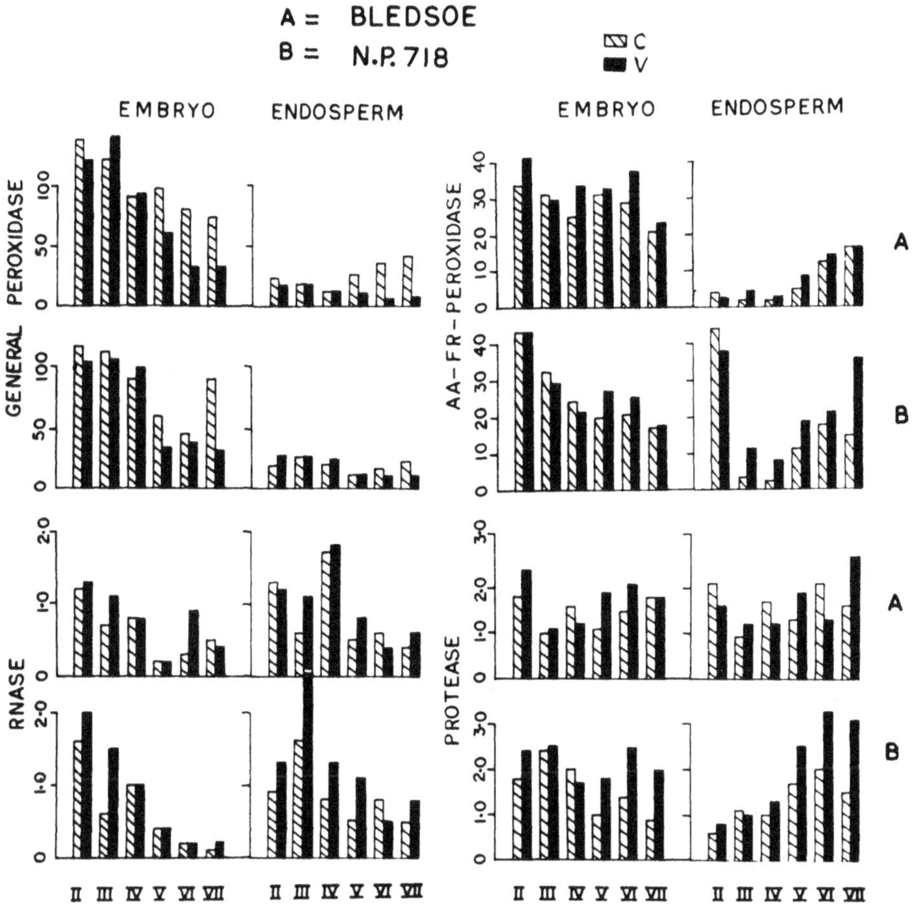

FIGURE 4.13. General peroxidase, AA - FR peroxidase, RNAse and protease activities of vernalized (V) and non-vernalized (C) embryo axes and endosperm of Triticum aestivum L. cv. Bledsoe (A) and N.P. 718 (B) during their juvenile phase of growth.

amylase activity at the low temperature resulting in faster hydrolysis of starch (Fig. 4.14) and consequently faster mobilization of the products as shown by the higher and

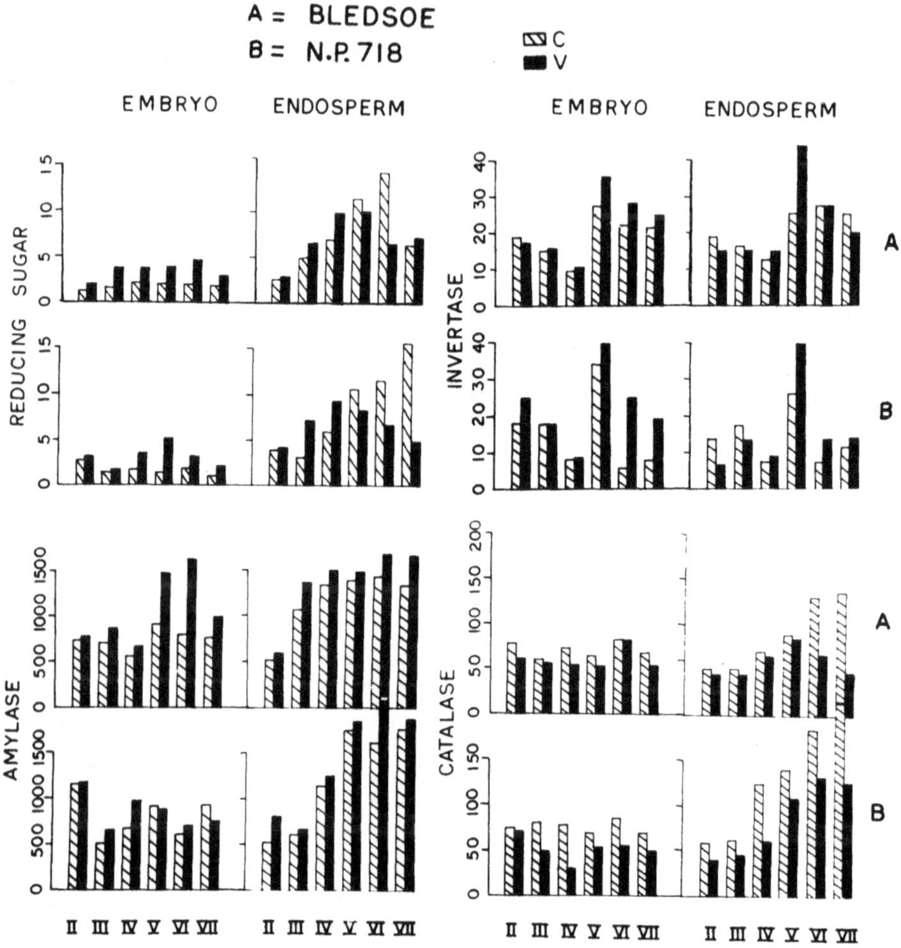

FIGURE 4.14.1. Reducing sugar content, invertase, amylase and catalase activities of vernalized (V) and non-vernalized (C) embryo axis and endosperm of *Triticum aestivum* L. cvs. Bledsoe (A) and N.P. 718 (B) during their juvenile phase of growth (Pandya, 1969).

increasing sugar content of the embryo axis under the temperature. The protein content was also higher in the embryo axis at the low temperature especially in the later stages of germination (Fig. 4.15). Correspondingly, the RNAse

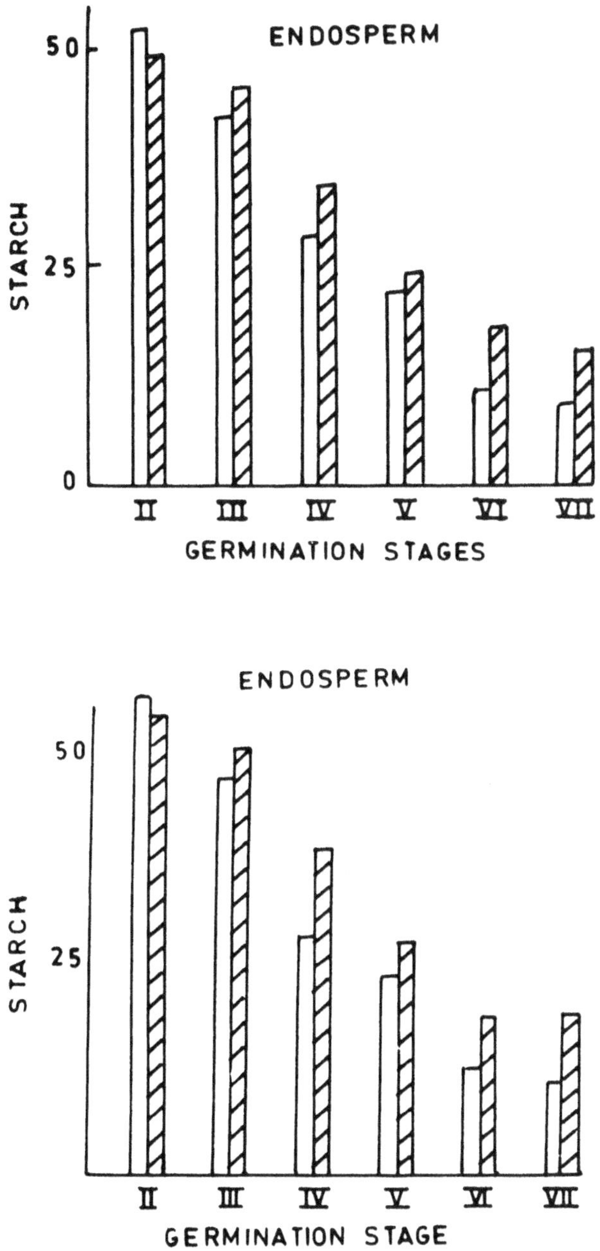

FIGURE 4.14.2. Starch content in the endosperm of wheat var. Bledsoe (A) and N.P. 718 (B)
Low temperature ▭ and room temperature ▨

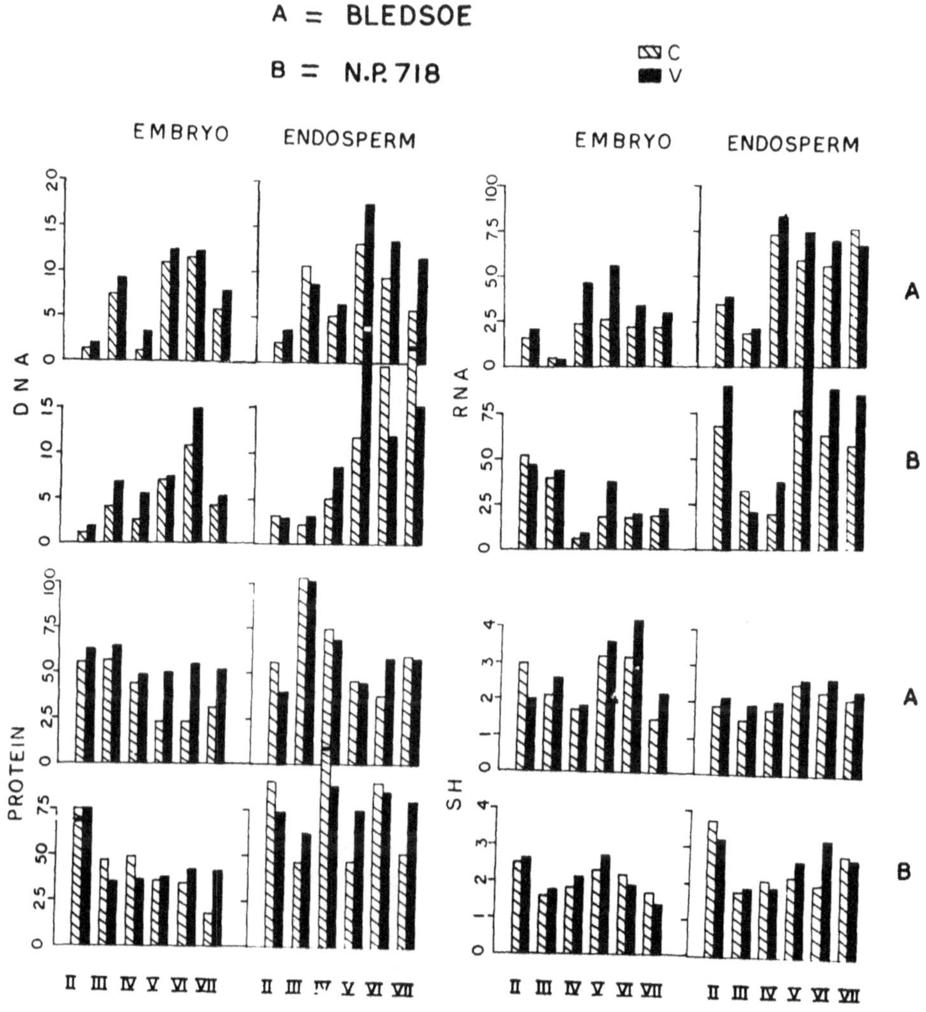

FIGURE 4.15. DNA, RNA, protein and -SH contents of vernalized (V) and non-vernalized (C) embryo axis and endosperm of <u>Triticum aestivum</u> L. cvs. Bledsoe (A) and N.P. 718 (B) during their juvenile phase of growth (Pandya, 1969).

activity was higher at the low temperature (Fig. 4.13), and indicated a faster turnover and transport of RNA. It has been shown that the biochemical mechanism of vernalization was partially connected with nucleic acid metabolism (Devay, 1965 a, b) and was probably followed by the synthesis of specific proteins (Devay, 1965 a, b; 1966). A claim has also been made that a specific RNA is synthesized during germination, at the beginning of cold induction (Devay, 1966). The importance of nucleic acids in thermal induction has been shown by Suge and Yamada (1965). Trione et al. (1967) found the concentrations of sugars, nitrogen, -SH and a number of amino acids at higher levels in vernalized as compared to unvernalized controls. Salcheva et al. (1966) also observed large amounts of soluble sugars and free amino acid accumulation during vernalization of winter wheat. Similar results have been reported by a number of workers (Kirillova, 1958; Pauli and Mitchell, 1960; Moskov and Bozova, 1962; Grzesiuk and Kulka, 1963; Shvedskaya and Kruzhilin, 1964; Kruzhilin, 1963). Semenenko and Krasilnikova (1963) reported higher incorporation of ^{32}P in nucleic acids during the vernalization of winter wheat. The possible connection of vernalization with oxidative phosphorylation was brought out by Krekule (1961). This again points to the importance of metabolic changes as a result of vernalization. It may be added here that the activity of mobilizing enzymes at low temperature may probably be due to enhanced synthesis of these enzymes under the inductive effect of enhanced AA turnover on m-RNA synthesis (Chinoy and Saxena, 1972; Schopfer, 1967; Price, 1966).

The enzymic activities concerned with oxidative processes like catalase, peroxidase and AAU were higher at higher temperature (Figs. 4.12 - 4.14). A lower catalase activity in the vernalized seeds after full germination has been observed by Parthasarthy (1940). Stanislawskii (1966) reported that the effect of raised temperature on catalase activity depended upon the growth period in which the plants were subjected. Yokoyama (1956) has pointed out that catalase activity directly affects the oxidation and reduction of cytochrome - c- oxidase system. The effect of temperature on -SH in the embryo appears to be dependent on the stage of germination since the earlier and later stages show distinctly different trends for the temperature effect. Identical results were obtained with free radical peroxidase activity of the embryo. The activity was distinctly higher at low temperature in the earlier stages, while in the later stages the trend was reversed. Moreover, changes in the -SH content of the two cultivars showed distinctly different trends in temperature effect during the course of germination (4.15). Biosynthesis of RNA and DNA were enhanced by vernalization especially in late spring and winter varieties (Fig. 4.15).

From the foregoing data it appears that low temperature enhances the anabolic processes concerned with mobilization and synthesis whereas, the higher temperature accelerates oxidative and catabolic utilization of substances.

Weinberger and Godin (1966) rightly pointed out that, the primary effect of cold treatment was to get growth and development off to a more rapid start. This conclusion also emerges from the work carried out by other investigators (Chinoy, 1962; Lee, 1955; Mansuri, 1965; Patel, 1967). It may be added here that the immediate immobile products of vernalization could be named as the accumulation of synthetic products like RNA, proteins and especially ascorbigen - the immobile storage form of AA.

The point worthy of note is the enhanced accumulation of ASG in vernalized seedlings. It is also clear from the data already presented, that the complexing of AA with macromolecules was enhanced in vernalized embryo axis, thus forming a charge transfer complex (CTC). Further, the activity of AA - FR forming special peroxidase was greater in vernalized seedlings compared with that in nonvernalized seedlings. All these facts suggest a faster flow of electrons from the free radical of AA and from the break up of the AA - macromolecule CTC complex. The enhanced energy flow was evidently utilized for accelerating the biosynthesis of DNA and m-RNA which in their turn produce enzymes and other proteins at enhanced rates leading to faster cell division and differentiation. The essential difference between spring and winter type of plants is not due to the very profound qualitative difference in the metabolites but to the enhanced rates of biosynthesis of macromolecules and other cell constituents which are the result of an interaction between the heredity quantum and the environment (Chinoy, 1970).

Reports published during recent years confirm some of the observations stated earlier (Cheltsova, 1978; Michniewicz et al., 1978; Reda et al., 1978; Barabas and Csepley, 1978). Based on the electrophoretic analysis of proteins formed in the cell free system (Fukushi et al., 1977) and DNA-RNA hybridization experiments (Ishikawa and Midori, 1977), it was concluded that vernalization induced the formation of new species of RNA especially m-RNA in the embryos of Triticum aestivum. The rate of incorporation of labelled thymidine into DNA was higher in rib meristem in the sub-apical region of winter plant Arabidopsis thalliana L. Heynh subsequent to vernalization (Basnard-Wibont, 1977). Similarly, vernalized shoot tips of Hyoscyamus niger was characterized by a decrease in the vegetative organogenesis and an increase in apical meristem activity pronounced by an increase in nuclear and nucleolar volume as well as RNA content of cytoplasm and nucleolus (Diomainto-Bonnand et al., 1980). Salisbury et al. (1979) are of the opinion, that for breeding purposes, adapting the wheat varieties to a particular environment making use of vernalization genes should be evaluated for their action under appropriate field conditions as the correlation values for various growth parameters and days to anthesis in four near-isogenic lines of wheat post vernalization differed significantly.

The study of the paramagnetic behaviour of excised embryo axis and endosperm revealed that the free radical concentration was maintained higher throughout in the endosperm (Chinoy et al., 1969 d; Saxena, 1969; Gurumurti, 1971), indicating the participation of free radicals in the mobilization of reserve nutrients. This finding further supports earlier observations from this laboratory that free radicals actively participate by providing electron energy to the metabolic reactions in the germinating seeds culminating in the growth of the embryo (Chinoy et al., 1969 d). Zholkevich and Chetverikov (1967) have also shown that free radicals of plant tissues are positively correlated with the energy efficiency.

The paramagnetic behaviour of excised embryo axis and cotyledon (Gurumurti, 1971; Chinoy et al., 1972 a, b) clearly showed that the initial free radical level in the cotyledon was much higher compared to that of embryo axis. The enzyme activity catalysing the production of MDHA was higher initially in the embryo axis. It can thus be said that free radicals participate in mobilization of reserve nutrients in the cotyledon, and synthesis of new macromolecules in the embryo axis. Gurumurti (1971) had shown that free radicals sharply rose again in the embryo axis after 96 hours. This coincided with the protrusion of the first pair of embryonal leaves from the cotyledon. During this period, there was a high level of AA, total sugar, proteins and -SH alongwith a high AA - free radical peroxidase. This suggests an initiation of enhanced synthetic activity in young seedlings to help maintain itself independently of the cotyledon during further periods of juvenile differentiation. It has already been shown that free radicals are being formed during active phases of juvenile differentiation as well as during periods of vegetative and reproductive differentiation (Chinoy, 1962; Chinoy et al., 1969 a, 1970 a, b; 1971 a, b; 1972 a; Fellenberg, 1969; Saxena, 1969; 1979 a, b; Saxena and Singh, 1983; Singh, 1971). Enhanced AA turnover synchronized with active cell division and elongation (Chinoy, 1962). In this investigation, exogenous supply of H_2O_2 as well as AA not only increased the biosynthesis of AA, but also its utilization. Concomitant with increase in free AA content there was a corresponding greater fall in ASG content in H_2O_2 and AA treated seedlings and greater mobilization of the bound AA under the influence of these media. There was also a very significant upsurge in the AA - MM complexing and increase in AA - FR - peroxidase activity, suggesting a more rapid formation of free radical of AA. EPR studies have revealed that the formation of the free radical of AA by the addition of H_2O_2 to a solution of AA can be considerably enhanced if DNA or RNA was added to the system. The rates of formation of FR as well as its decay were controlled by the relative concentrations of AA and macromolecules in the system (ref. Chapter 1). EPR studies as well as spectrometric and biochemical studies have also revealed that complexing of AA with the above mentioned macromolecules took place; the rate of complexing being governed by the relative concentrations of these metabolites.

Taking into consideration the above mentioned evidence it is postulated that the enhancement in the extention growth of <u>Sesamum</u> seedlings brought about by treatment with H_2O_2 alone or in combination with AA is the resultant of acceleration in the formation of the free radical, MDHA, which synergizes the hydration of cell constituents and also the biosynthesis of DNA, RNA as well as other cell constituents. Faster synthesis of the macromolecules in its turn accelerates the peroxidative mechanism leading to the faster formation of AA free radical as well as an accelerated rate of its decay (Chinoy et al., 1972 b). Presumably, this integrated mechanism of electron flow and biosynthesis of cell constituents (including the macromolecules) is responsible for acceleration in growth.

3. EFFECT OF CHEMICAL FERTILIZERS ON ASCORBIC ACID CONTENT

Aberg and Ekdahl (1948) reviewed the work done on the effect of nitrogen fertilization of AA content in plants. They reported a regular decrease of the AA content of full grown leaves in the range of increasing growth. A very high dose of fertilizer decreased growth while it increased the

AA content. They suggested that structural and metabolic factors were at work in causing the observed phenomena. Carbohydrates are rapidly consumed due to protein synthesis which might compete with the AA synthesis and thus the AA content might decrease initially. The phenomenon could be compared with the effect of light intensity of AA content. With increasing light intensity, AA disappeared. A high nitrogen supply might also depress the AA content of fruits through the increased shading caused by a richer foliage (Somers et al., 1951), however, this factor may only be partly responsible for the nitrogen effect (Murneek et al., 1954).

An increase in nitrogen content reduced the concentration of sugars which in turn led to a reduced AA content of both leaves and fruits of tomatoes (Popovskaya, 1952).

The content of crude protein, fat, carotene and AA in the dry matter markedly increased on application of nitrogen (N) especially in quantities of 100 and 150 mg/kg soil (Filimonov, 1969). Largskii (1971) showed that doses of N above 60 mg/kg soil maintained the amount of total sugar and AA in the cucumber. Phosphate fertilizer increased the concentration of these substances. Sugar content and dry matter were also increased by the fertilizer. An increase in the phosphorus level also enhanced the sugar and AA concentrations and decreased the nitrogen content in plant tissues, especially under water deficit (Popovskaya, 1957). The AA content was found to be inversely related to the nitrogen concentration and the amount of water in the plant tissue.

Carotene and AA contents were considerably enhanced in pea plants grown from seeds treated with silicate and phosphate fertilizers. AA content was higher in leaves than in stem or pods and showed the greatest response to seed treatment (Martovitskaya, 1967).

Potassium sulphate and super-phosphate had no effect on the ontogenetic drift of AA in the roots of berseem. AA content of the tops was reduced by potassium sulphate and increased by super phosphate three weeks after fertilizer application. No significant effect on AA content was observed after a two month growth period (Kamal, 1957). A drop in the level of sulphur nutrition intensified the growth of young pea plants, increased the AA content and reduced the amount of glutathione in the above ground organs and roots (Bugakova et al., 1969).

Watson and Noggle (1947) studied the effect of mineral deficiencies on AA content of oat plants at the jointing stage. Generally, AA was maximum at this stage. A deficiency of magnesium increased AA concentration in leaves and stems, whereas, potassium deficiency increased the AA content in leaves only. On the contrary, nitrogen deficiency reduced AA in stem and leaves. Negative correlation was found between the accumulation of magnesium and potassium in leaves and its AA content.

Experiments were carried out in winter and spring to study the effect of nitrogen, potassium and phosphorus on the AA content of lettuce leaves (Harding and David, 1954). The level of mineral nutrients had a slight effect on AA concentration. Of the three elements studied, nitrogen was most effective in causing variation in AA content. Highly significant variations were found in winter and non-significant results were reported in spring. This can be explained on the basis of different environmental conditions. Generally, increasing the nitrogen and phosphorus increased the AA content however increasing the potassium caused a decrease in AA of lettuce leaf.

Ishevskaya (1965) investigated the relationship between prolonged use of fertilizer and AA in potato tubers. AA content increased considerably during continued application of fertilizers. Prolonged use of the fertilizer increased vitamin C content by 32 - 60%.

The effects of nitrogen, phosphorus and potassium on carotenoids, AA and protein fractions of sugar beet and potato have been reported by Klemm and Ansorge (1967). The stimulation of the formation of chlorophyll as well as the outflow of photosynthetic products like carbohydrates and AA from leaves into other organs by an improvement in the water supply, application of fertilizer and deeper ploughing of winter wheat has been reported by Kirichenko (1972).

4. EFFECT OF TRACE ELEMENTS ON ASCORBIC ACID CONTENT OF PLANTS

Lucas (1948) studied the effect of copper fertilization on carotene, AA, protein and copper content of plants raised on organic soils and reported beneficial effect of copper on AA content. Hewitt and coworkers (1950) also noted positive effect of molybdenum on AA content of plants in sand culture. Lebrec (1951) studied the effect of various concentrations of copper and iron on AA content in wheat, tomato and tobacco plants. A greater stability of AA in the drying plant tissue was associated with a greater concentration of Cu and Fe in the nutrient solution. In fresh tissue such relationship was not found. A positive relationship was more apparent between the various amounts of Fe and stability of AA compared with various amounts of Cu in the nutrient solution (Lebrec, 1951). In leaves and fruits of spinach, tomato and black currants, 12 - 30% increase in vitamin C content was found after application of Mn or a dry seed treatment with finely pulverized MnO_2 (Bronsart, 1950). Kokin (1957) studied the effect of microelements on physiological processes in cereal crops. The redox system in plants is greatly affected by microelements and especially by manganese. AA and its oxidation were also greatly affected.

Absence of molybdenum from the nutrient solution caused an increase in bound AA and nitrate, while the total amount of AA was not affected. Under optimum concentration of nitrate nutrition, an increase in bound AA was observed. Bound AA may accumulate in the deficiency of molybdenum because of the high content of nitrate in plants (Chernavian and Kupke, 1959). Turkevich (1963) found enhanced sugar content, AA and dry weight in tomato fruits after spraying with Mn and molybdenum.

El-Fouly (1964) studied the effect of manganese nutrition on AA content of leaves of *Phaseolus vulgaris*, *Helianthus annuus* and *Zea mays*. Plants were grown in water culture with addition of 2 mg of $MnCl_2$, 4 H_2O per plant. Leaves in $MnCl_2$ treatment contained higher AA content compared with that in leaves grown without Mn. Okhrimenko (1965) also showed that the Mn treatment of corn plants lowered DHA and enhanced AA content. Mn treatment increased AA, dry matter, chlorophyll and protein. Lupinovich and Getsevich (1967) reported increased chlorophyll content, intensified photosynthesis and increased vitamin C in barley, oat and sugar-beet plant by the application of Mn. Foliar spray with Mn of *Psidium guava* increased the AA content, number of flowers/shoot, fruit set, fruit retention as well as carbohydrate contents (Singh and Rajput, 1977). A similar treatment of red currant berries with 0.04 - 0.05% $MnSO_4$. 5 H_2O twice during flower and fruit setting increased the dry matter and AA content (Tolpekina, 1975). In leaves of tea plants, a positive correlation between Mn and AA content has been reported (Takayanagi, 1977).

The addition of 300 ppm of molybdenum (Mo) to the soil or soaking of seed in 70 ppm of Mo for 5 - 10 minutes increased carotene and AA levels in alfalfa. The production of green mass was also increased by Mo (Nadzhakova et al., 1970). Presowing treatment of red clover seeds with Mo increased chlorophyll, carotene, AA, number of root nodules, seeds and fresh weight (Orlova and Gulyaeva, 1970). One year old lemon trees when grown for three years with Mo and Mn added as a side dressing to soil or applied as foliar feeding, significantly increased the AA content of leaves (Pachuliya, 1965). The second addition of Mo further increased AA synthesis in leaves. Slight increase was observed by other trace elements, except Cu which caused a decrease. Application of trace elements also increased growth. AA synthesis was the highest in young developing leaves. On the other hand, Hivon et al. (1951) did not find any consistent relationship between Mn nutrition and AA content of Soyabean leaves.

Dobroliubskii and Slawo (1958) studied the effect of Vanadium on AA, sugar contents and yield of grape plants. When grape plants were sprayed with vanadium sulphate at the time of massive flowering and with grapes about the size of peas, AA concentration in leaves and berries increased appreciably. Sugars accumulated in greater concentrations, ripening was accelerated and yield was increased.

Girko and Repetun (1965) reported the effects of trace elements on some biochemical processes in potato plants. Trace elements boron, Mn, Mo and Cu increased AA content in potato leaves. These elements also increased the formation of chlorophyll and carotene in the leaves especially at the beginning of flowering. The yield and starch contents also increased. The beneficial effect of boron, Mn, and Cu on AA, and some physiological and biochemical processes of potato plants was reported by Gedz (1965). Potato plants were sprayed with H_3BO_3, $MnSO_4$ and $CuSO_4$. AA content increased by 23% in boron and Mn treated plants. This enhancement was twice as much in the stem as in the leaves. The starch content of roots also increased fourfold. The yield increased by 22% in Mn and 16% in boron plants.

Boron increased AA in tobacco plants. Participation of AA in respiration of boron treated plants was also indicated (Gupta, 1966). Corn plants were sprinkled with 0.025% H_3BO_3 at the development of the 7th leaf and at the start of flowering. Control plants were sprayed with water. AA content was found to be 80 mg %, reducing sugar 1.09% and crude protein 1.14% in leaves of boron treated plants compared to water sprinkled ones which had 60 mg %, 1.19% and 0.74% respectively (Popovich, 1963). Izergina et al. (1969) found that H_3BO_3, $ZnSO_4$, $CuSO_4$ increased AA and protein contents in cabbage. Cu was more effective than Zn.

Treatment of the loamy, sandy soil with boron (borax) in combination with nitrogen and phosphorus with nitrogen - 120, phosphorus - 40 and boron - 15 Kg/ha increased the AA content as well as the yield of cauliflower (Randhawa and Bhail, 1976).

The micronutrient, boron plays an important role in a number of biochemical processes and recently its deficiency has been reported to cause changes in nucleic acid and protein metabolism in plants (Kannan, 1980). Further, premature flower-fall and floral sterility also occur. Brown (1979) has shown that when boron efficient (Rutgers) and boron inefficient (T3238FFR) varieties of tomato were grown in nutrient solutions receiving various levels of boron, there was an upsurge in the ascorbic acid oxidase (AAO) activity in plants receiving boron stress. It was significantly higher in T3238FFR variety than in Rutgers. According to the author, AAO activity is considered as an index for the effect of born on copper activity and in his opinion, the results provide an evidence that the activity level may be responsible for more effective rise of boron by Rutgers than by T3238FFR tomato plants.

The elements, especially aluminium and iodine promoted AA synthesis in the leaves and roots of cotton plants under condition of nonsalinization as well as under salinization (Kustova, 1965). Romanchuk and Prilutskii (1968) reported that trace elements Mn, Mg, cobalt, nickel and boron had a stabilizing effect on AA in alkaline medium.

Godfrey - Sam-Aggrey et al. (1979) found significant correlation between the copper content of leaves and AA content in the fruit juice of grapes. Similarly, leaf sulphur content was closely related to the AA content in fruits of tangarins. Mozaleva and Ganyushkina (1969) did not find any correlation between Cu and Mo content accumulation in the needles of pine.

5. ROLE OF AA IN MOBILIZATION OF BOUND IRON

The availability of iron to the plants from the growing medium is influenced by a number of factors (Kannan, 1982). It has been suggested that the intracellular ferritins release the iron without undergoing degradation and with a concomitant reduction of iron. In the plants, the ferritins are localized in the plastids of non-photosynthesizing tissues as electron microscopic evidences show that green leaves generally lack ferritins. The transfer of dark grown leaves is marked by a rapid disappearance of ferritin-iron accompanied by a rise in pH, high oxygen and AA content in the stroma of illuminated chloroplasts. Under *in vitro* conditions, Bienfait and Van den Briel (1980) found almost a 100% release of iron from ferritin when the reaction mixture containing 475 μm horse-spleen ferritin, 50 mM AA, 25 mM Tris and 25 mM phosphate (pH 8.0) was incubated in the presence of Cu^{2+} (1.10^{-5} M $CuSO_4$) strictly under aerobic conditions. The lag period before the iron release was 0.5 - 1 hour or 3 - 4 hours in the presence or absence of Cu^{2+}, denoting thereby formation of some intermediate compound between AA and DHA. Oxalate, threionate and DHA could release only < 5 of iron from ferritin. The intensity of EPR signals of free radical from the incubation mixture was 4 - 10 times more in presence of Cu^{2+}. The authors strongly contend that MDHA mobilizes the ferritin iron by reduction. This radical is known to reduce free or liganded iron.

Sugawara studied (1941) the AA content of leaves of spinach, barley, wheat, broad bean and Chinese cabbage and found that it was high in fair weather and much less in rainy weather with almost no change during the entire day. The maximum concentration of AA in leaves correspond with the accumulation of carbohydrates. Increase of CO_2 in the air was also found to lead to an almost proportional increase in AA content.

Bregetova (1952) found diurnal fluctuations in AA content of the leaves of a number of species of cotton with a rise during the day and a fall during the night. Mechanic (1952) has also carried out extensive studies to show that the presence of chlorophyll and sugar greatly increased the concentration of AA in leaves upto a temperature of 25° C. During summer the maximum AA content was found in the morning and in spring and autumn at noon. At night the AA content dropped to a minimum. Haagen (1953) also found the maximum AA- DHA content in various plants at about mid-day or in the afternoon and the minimum during the morning. The change in the vitamin content took place only during the light phase of the day and not during the night. The AA content was found to be dependent upon weather and light. Fränke (1954) considered the synthesis of AA as oxygen dependent and concluded that it was tied up with respiratory activity.

AA increased with advance in germination. There was no specific difference in the pattern of AA levels of both winter and spring varieties. The difference in AA content were due mainly to the physiological condition of the plant and its stage of growth and development (Michniewicz, 1961).

A 200 ppm AA treatment given through cotton wad feeding of the apical region had a stimulatory effect on the growth of Trigonella foenum graecum and Brassica chinensis. This is brought out from a higher rate of stem elongation and leaf length as also by an increase in leaves which emerged in plants treated by AA as compared with control plants (Chinoy et al., 1957 a) (Table 4.1). The stimulatory action of AA on growth as well as flowering observed in this work clearly indicates that AA not only acts as a growth promoting substance but also accelerates the developmental process of these plants (Fig. 4.16). Chinoy, Grover and Sirohi (1957 b) studied the interaction between IAA and AA at varying concentrations and showed that extension growth of Avena coleoptile sections in low concentration of IAA was greatly accelerated by the addition of AA (Fig. 4.17). This growth was found to surpass even that obtained by the exogenous application of IAA (0.5 mg/l) alone to coleoptile sections (Fig. 4.18).

Table 4.1. Mean values of different growth characters along with the standard errors of the differences of means and the values of F (treatment variance/error variance).

Treatment	Height (cm)	Leaf No. after 58 days	Leaf No. after 73 days	Leaf length after 58 days	Leaf Length after 73 days	Days to flowering	No. of flowers per plant	No. of fruits per plant	Seeds per fruit	Thousand seed weight (g)
Control	56.9	7.3	8.6	7.6	9.8	101.2	141.8	1292.5	303.8	10.0
Col	48.4	7.1	7.7	6.1	7.9	107.0	125.5	413.4	136.9	8.5
AA	81.6	8.9	11.2	9.6	13.0	93.3	418.6	2492.3	367.0	12.5
Col + AA	52.2	7.4	8.8	6.8	8.1	101.5	143.3	906.2	245.0	9.0
S.E.	1.476	0.159	0.182	0.145	0.173	0.549	55.83	139.45	1.794	0.062
Value of F	205.69	43.89	161.62	200.38	362.2	210.44	12.77	60.7	53.96	65.94

Degrees of freedom in all cases is 3.

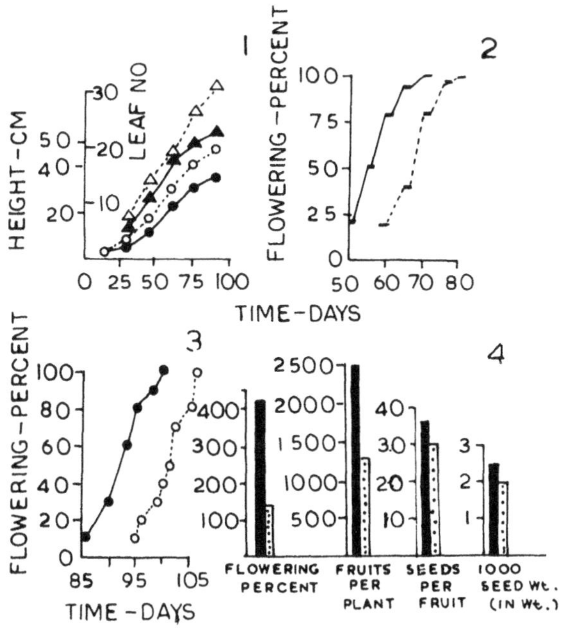

FIGURE 4.16. Stimulatory effect of exogenous application of AA on growth and flowering of <u>Trigonella foenum graecum</u>, L. and <u>Brassica chinensis</u> (Chinoy, Nanda and Garg, 1957 a).

The striking stimulatory effect of exogenous application of 50 mg/l of AA together with 0.03 mg/l of IAA in the presence of one percent sucrose in the medium (soln. 5 AB - Fig. 4.17) and an equally striking increase in the section growth by the exogenous use of a high concentration of IAA (0.5 mg/l) in the presence of a relatively low concentration of AA (5 mg/l), renewed every three hours in the medium containing 1% sucrose (soln. 3 AC - Fig. 4.17) are suggestive of important physiological roles of these vital plant constituents. The stimulatory action of AA observed in this work suggests that the enhanced coleoptile section growth may partly be due to the lowering of structural viscosity of the cytoplasm. It is well known that dispersion and hydration of plasmic colloids are affected by changes in viscosity.

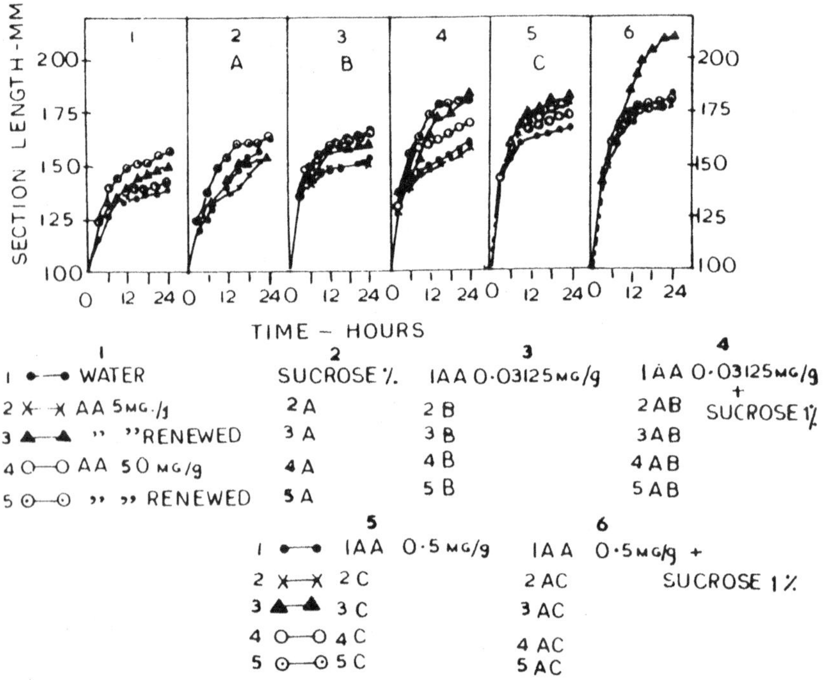

FIGURE 4.17. Coleoptile section lengths as per cent of the original length, for 30 combinations of AA, IAA and sucrose (1%). For details see key to Figs. 4.17, 1-6 (Chinoy, Grover and Sirohi, 1957 b).

6. SUMMARY

Parallelism between growth rate and ascorbic acid content in higher plants has led a number of workers to consider AA as an essential growth factor or even as a phytohormone. However, it has been shown that such a parallelism is by no means a general one. During germination in the dark, AA

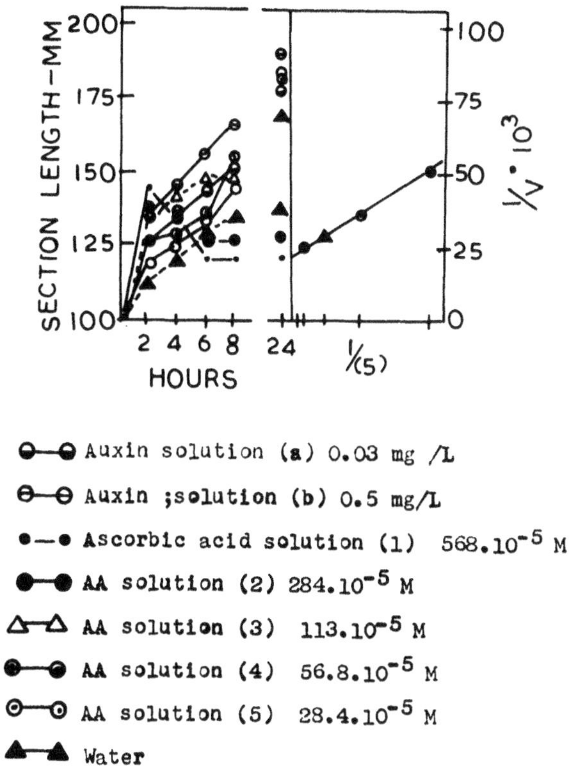

FIGURE 4.18. Elongation of Avena coleoptile sections in varying concentrations of AA (Grover et al., 1958).

content increased. Addition of a hexose to the medium still further enhanced the biosynthesis of AA. A comparative study of AA content of a number of winter and spring cultivars of wheat during germination and early growth did not reveal any specific difference in the pattern of AA levels in both the types of cultivars. Some workers were able to show an appreciable amount of stimulation of growth due to exogeneous application of AA. It was also shown that stimulation or retardation of growth by exogenous application of AA depended upon its concentration. This treatment of AA not only accelerated growth but also stimulated earlier flowering in Trigonella foenum graecum and Brassica chinensis. The retarding effect of colchicine on growth and flowering of the latter plant was considerably overcome when applied in combination with AA. The Milan group of workers have, on the other hand, shown an inhibitory effect of AA on growth.

A study of the paramagnetic behaviour of excised embryo axis and endosperm using the EPR technique revealed the sustained high level of free radical concentration in the endosperm throughout the period of juvenile differentiation. These studies have also revealed that the formation of the free radical of AA was considerably enhanced as well as its decay significantly accelerated by the addition of DNA and RNA to the medium containing AA and a little H_2O_2. The rates of formation of free radical of AA and its decay are controlled by the relative concentrations of AA and the macromolecules in the system.

AA turnover plays a vital role during germination and juvenile differentiation. AA serves as an enzyme mobilizing hormone by stimulating the activity of hydrolyzing and oxidative enzymes resulting in rapid breakdown of reserve nutrients. AA undergoes transformation into its free radical form by the catalytic action of a special isozyme of the peroxidase, called AA - FR - peroxidase, thus creating a reductive atmosphere which helps in the activation of synthesis of macromolecules and other cell constituents by providing electron energy. The hydrolytic products are translocated to the growing points of the embryo axis leading it to rapid growth and differentiation.

Early and more rapid biosynthesis of AA by wheat embryo axis from sucrose in the presence of IAA at a low temperature throws some light on the mechanism of vernalization. The difference in AA content depends chiefly on the physiological condition and on the stage of growth and development and not on the spring or winter habit of the plant. AA content of seeds continuously increased during the entire course of vernalization treatment. Vernalized seed had invariably a higher content compared with that in the non-vernalized ones at the corresponding stage of growth. From these data it was concluded that the larger content of vitamin C was responsible for the quicker passing of vernalized plants through the particular growth phase. The primary effect of vernalization appears to be to get growth and development off to a more rapid start.

REFERENCES

1. Aberg B 1961. Vitamins as Growth Factors in Higher Plants. In: Encycl. Plant Physiol. Ruhland W (ed) Berlin: Springer Verlag, 14: 418.
2. Aberg B, Ekdahl I 1948. Physiol Plantarum (Copenh) 1: 290.
3. Abraham PG 1969. Ph.D. Thesis, Botany Department, Gujarat University, Ahmedabad - 380009, India.
4. Barabas Z, Csepely IT 1978. Euphytica 27: 831.
5. Basnard-Wibaut C 1977. Plant Cell Physiol 18(5): 949.
6. Bienfait HF, Van den Briel ML 1980. Biochim Biophys Acta 631: 507.
7. Bitancourt AA 1955. Rev Gen Bot 62: 498.
8. Bonner WD Jr 1957. Ann Rev Plant Physiol 8: 427.
9. Bregetova LG 1952. Biol Abstr 7202 26: 679.
10. Bronsart HV 1950. Zeitschr Pflanzenernar Dung V Bodenk 51(2): 153.
11. Brown JC 1979. J Plant Nutrition 1(1): 39.
12. Bugakova AN, Beleva VI, Tulunina AK, Topchieva VT 1969. Agrokhimiya 11: 128.
13. Chel'tsova LP 1978. Fiziol Biokhim Kul't Rast 10(5): 481.
14. Chernavian IA, Kupke G 1959. Nauch Dokl Vysshei Shokoly Biol Nauki 2: 149.
15. Chinoy JJ 1962. Indian J Plant Physiol 5: 172 (F.G. Gregory Memorial volume).
16. Chinoy JJ 1970. Indian J Plant Physiol 13: 115.
17. Chinoy JJ 1977. Indian J Plant Physiol 20: 1.
18. Chinoy JJ, Saxena OP 1972. Proc 16th Int Seed Test Assoc (ISTA) Congr Washington DC, June 7-12, 1971, 37: 903.
19. Chinoy JJ, Nanda KK, Garg OP 1957 a. Physiol Plant 10: 869.
20. Chinoy JJ, Grover R, Sirohi GS 1957 b. Physiol Plant 10: 92.
21. Chinoy JJ, Nanda KK, Garg OP 1958. Proc Sem "Modern Develop Plant Physiol", Delhi Univ Aug 26-30, 1957 p. 43.
22. Chinoy JJ, Grover R, Nanda KK 1961. Flora 150: 1.
23. Chinoy JJ, Pandya RB, Saxena OP, Dave IC, Abraham PG 1967. J Indian Bot Soc 46: 344.
24. Chinoy JJ, Saxena OP, Abraham PG, Pandya RB, Dave IC,

Gurumurti K, Sastri K 1969 a. Report PL 480 Research Project FG - In- 182 (A7-CR-87) p. 1 - 350.
25.Chinoy JJ, Patel Hemlata D, Jani BM, John D, Vyas AV, Raj KPS 1969 b. Report PL 480 Research Project FG-In-208 (A7-CR-108) p. 1 - 450.
26.Chinoy JJ, Sastry K, Pandya RB, Gurumurti K, Abraham PG, Saxena OP, Shah PN 1969 c. J Biol Sci. 12: 1.
27.Chinoy JJ, Saxena OP, Gurumurti K, Abraham PG, Pandya RB 1969 d. Science and Culture 35: 690.
28.Chinoy JJ, Saxena OP, Gurumurti K 1970 a. ISPP Symp "Plant Growth Regulators" Kharagpur Jan 3-8, Abstr 1.
29.Chinoy JJ, Saxena OP, Gurumurti K 1970 b. Ibid Abstr 2.
30.Chinoy JJ, Shah CK, Madhavan Unni NV, Neelkandan N 1971 a. Internl Sem "Physiology of Differentiation in Plants" May 17-22, Simla India.
31.Chinoy JJ, Singh YD, Gurumurti K 1971 b. Indian Agriculturist 15: 33.
32.Chinoy JJ, Gurumurti K, Saxena OP, Singh YD 1972 a. UGC All India Symp "Biology of Land Plants" Meerut Univ, Meerut June 18-20, p. 188.
33.Chinoy JJ, Singh YD, Laloraya MM, Saxena OP 1972 b. Sem "Current trends in plant sciences" Delhi University, Delhi Oct 3-10 p. 47.
34.Chouard P 1960. Ann Rev Plant Physiol 11: 191.
35.Davies W, Atkins GA, Hudson PCB 1937. Ann Bot (NS) I: 329.
36.Dennison RA 1940. Science 92: 70.
37.Devay M 1965 a. Acta Agronomica Hung 14: 93.
38.Devay M 1965 b. Ibid 14: 275.
39.Devay M 1967. Acta Agronomica Hung 16: 289.
40.Diomainto-Bonnand J, Houivet J, Picard C 1980. Bull Soc Bot France Lett Bot 127: 427.
41.Dobroliubskii OK, Slawo AV 1958. Dokl Akad Nauk SSSR 118: 36.
42.El-Fouly Mohamed M 1964. Plant Soil 24: 473.
43.Emmerich H 1941. Ib Wiss Bot 90: 99 See Aberg B 1961.
44.Fellenberg G 1969. Planta 84: 324.
45.Filimonov DA 1969. Khimiya Sel'Khoz 7: Herb Abstr 39: 1990, 1969.
46.Franke W 1954. Planta 44: 437.
47.Fukushi S, Koh I, Kimike S 1977. Plant Cell Physiol 18: 969.
48.Garg OP, Chinoy JJ, Nanda KK 1958. Proc Sem "Modern Develop Plant Physiol" Delhi University, Delhi Aug 26-30, 1957, p. 20-21.
49.Gedz SM 1965. Primenenie Mikroelementov V Sel'sk Khoz Acad Nauk Ukr SSR 73.
50.Girko PA, Repetun SI 1965. Primenenie Mikroelementov V Sel'sk Khoz Akad Nauk UKR SSR 82:
51.Godfrey-Sam-Aggrey W, Hague I, Garber MJ 1979. J Plant Nutrition 1: 185.
52.Goh C 1950. Biol Abstr 26, 7203 679, 1952 (in Japanese).
53.Grzesiuk S, Kulka K 1963. Acta Soc Botan Polon 32: 313.
54.Gupta US 1966. Proc Indian Acad Sci 64 (B): 199.
55.Gurevich AA 1963. Dokl Biol Sci 145: 674.

56. Gurumurti K 1971. Ph.D. Thesis, Botany Department, University School of Sciences, Gujarat University, Ahmedabad - 380009, India.
57. Haagen U 1953. Biol Abstr 29: 945, 2754 (1955).
58. Harding CF, David JJ 1954. Food Res 19: 138.
59. Havas L, Gal I 1936. Nature (Lond.) 138: 586.
60. Hewitt EJ, Agarwala SC, Jones EW 1950. Nature (Lond) 166: 1119.
61. Hivon KJ, Doty DM, Quackenbush FW 1951. Soil Sci 71: 353.
62. Ishevskaya IM 1965. Dokl Mosk Sel'Khoz Akad KA Timiryazeva No. 108, 161.
63. Ishikawa K, Midori T 1977. Plant Cell Physiol 18: 875.
64. Izergina MM, Pochashev MS, Pogodina TN 1969. Nauka Sel'Khoz 132.
65. Kamal MAM 1957. Ann Agri Sci Ain Shams Univ Cairo 2: 65.
66. Kannan S 1980. J Plant Nutrition 2: 717.
67. Kannan S 1982. J Plant Nutrition 5: 395.
68. Kirichenko VP 1972. Fiziol Biokhim Kult Rast 3: 646.
69. Kirillova GA 1958. Fiziol Rast 5: 175.
70. Klemm KH, Ansorge H 1967. Albrecht-Tmaer Arch 11: 449.
71. Knox WH 1960. Proc IVth Internl Congr Biochem Vienna Umbreit W, Molitor H (eds) New York: Pergamon Press p. 307.
72. Kokin AY 1957. Fiziol Rast 4: 333.
73. Krekule J 1961. Biol Plant (Prague) 3: 107.
74. Kruzhilin AS (1963) Fiziol Rast 10: 374.
75. Kustova AKH 1965. Izv Akad Nauk Turkm SSR ser Biol Nauk 5: 3.
76. Largskii YN 1971. Agrokhimiya 2: 138.
77. Lebrec RL 1951. Dissertation Publ No 3568 Michigan State Univ p. 450.
78. Lee AE 1955. Bull Torrey Bot Club 82: 1.
79. Lucas RE 1948. Soil Sci 65: 461.
80. Lupinovich IS, Getsevich NA 1967. Vestsi Akad Nauk Belorus SSR SER Sel'ska Gaspad Nauk 1967: 42.
81. Mansuri AD 1965. Ph.D. Thesis, Botany Department, University School of Sciences, Gujarat University, Ahmedabad - 380009, India.
82. Mapson LW 1958. Ann Rev Plant Physiol 9: 119.
83. Mapson LW 1960. In: Vitamin Metabolism Umbreit W, Molitor H (eds) London: Pergamon.
84. Mapson LW, Cruickshank EM 1947. Biochem J 41: 197.
85. Mapson LW, Moustafa EM 1956. Biochem J 62: 248.
86. Martovitskaya AM 1967. Vest Sel'khoz Nauki Alma-Ata No 5 30. Herb Abstr 38: 193 1968.
87. Mechanic FY 1952. Biol Abstr No 6912 28: 636.
88. Michniewicz M 1961. Biol Abstr 36 (14): 45935.
89. Michniewicz M 1961 a. Acta Agrobotanica 10(2): 119.
90. Michniewicz M, Rowicka K 1961. Acta Agrobotanica 10(2): 19.
91. Michniewicz M, Rozej B, Chrominski A, Porazinski Z 1978. Acta Soc Bot Pol 47 (1-2): 115.

92. Mitra J, Datta C 1951. Sci Culture 16: 428.
93. Moskov I, Bozova L 1962. Compt Rend Acad Bulgare Sci 15: 559.
94. Mozaleva LD, Ganyushikina LG 1969. Uch Zap Pertozavodsk Gos Univ 17(6): 74.
95. Murneek AE, Maharg L. Wittwer SH 1954. Univ Missouri Agricult Expt Station Res Bull No 568, 1.
96. Nadzhakova ZA, Ragimov AV, Shakuri BK 1970. Khim Sel's Khoz 8: 580.
97. Okhrimenko MF 1965. Primenenie Mikroelementov v Selśk Khoz Akad Nauk Ukr SSR 63.
98. Orlova IF, Gulyaeva VE 1970. Nauch Trud Orlov Sel's Khoz Optyn Stan No. 5, 105.
99. Pachuliya KF 1965. Primenenie Mikroelementov v Sel's Khoz Akad Ukr SSR 190.
100. Pandya RB 1969. Ph.D. Thesis, Botany Department, University School of Sciences, Gujarat University, Ahmedabad - 380009, India.
101. Pandya RB, Dave IC, Saxena OP, Abraham PG 1969. Vidya Gujarat University School of Sciences Journal 12(1): 181.
102. Parthasarthy SV 1940. Madras Agri J 28: 133.
103. Patel HD 1967. Ph.D. Thesis, Botany Department, University School of Sciences, Gujarat University, Ahmedabad-380009, India.
104. Pauli AW, Mitchell HL 1960. Plant Physiol 35: 539.
105. Popovich IA 1963. Mikroelementov Zhizni Rastenii Zhivotn i Cheloveka Akad Nauk Ukr SSR Inst Fiziol Rast Tr Koordinats Soveshch p 192.
106. Popovskaya EM 1952. Biokhimiya 17: 145.
107. Popovskaya EM 1957. Fiziol Rastenii 4(4): 326.
108. Price CE 1966. Nature (Lond.) 212: 1481.
109. Raadts E 1948. Planta (Berl) 36: 103. See Aberg B 1961.
110. Randhawa KS, Bhail AS 1976. Indian J Hortic 33(1): 83.
111. Reda F, Poul L, Ole SR 1978. Physiol Plant 42 (1): 109.
112. Romanchuk PS, Prilutskii GM 1968. Plodorodie Pochv Eff Udobr Polese p 267.
113. Rowsell EW, Goad LJ 1962. Biochem J 84: 73.
114. Ruge U 1947. Planta (Berl) 35: 297.
115. Salcheva G, Vaklinova St, Gramatikova Khr (1966) Fiziol Rast 13(6): 1004.
116. Salisbury PA, Berry GJ, Halloran GM 1979. Can J Gene Cytol 21(3): 429.
117. Saxena OP 1969. Ph.D. Thesis, Botany Department, University School of Sciences, Gujarat University, Ahmedabad - 380009, India.
118. Saxena OP 1979. Varian Winter School in EPR, IIT, Bombay (Nov 19-30) p. 1-3.
119. Saxena OP 1979 a. Current Adv in Pl Reproductive Biol (Ed) Malik CP p. 323. New Delhi: Kalyani Press.
120. Saxena OP, Ghesani PI, Shukla SP, Chinoy JJ 1969. Indian J Plant Physiol 12: 1.
121. Saxena OP, Singh G 1983. The widening horizons of plant sciences (Ed) Malik CP, p. 41, New Delhi: Kalyani Press.

122. Scheuermann R 1952. Planta (Berl) 40: 265, See Aberg B 1961,
123. Schöpfer P 1967. Planta (Berl) 212: 1481.
124. Semenenko GL, Krasilnikova LA 1963. Uch Zap Kharkovsk Univ 137: Tr Nauch-Issled Inst Biol, Biol Fak 35, 98. Raferat Zhur Biol 1963, No 24 G 36 (Transl).
125. Shvedskaya ZM, Kruzhilin AS 1964. Fiziol Rast 11(2): 235.
126. Singh YD 1971. Ph.D. Thesis, Botany Department, University School of Sciences, Gujarat University, Ahmedabad - 380009, India.
127. Singh NP, Rajput CBS 1977. Indian J Hortic 34(2):120.
128. Somers GF, Kelly WC 1951. Plant Physiol 26: 90.
129. Somers GF, Kelly WC, Hamner KC 1951. Amer J Bot 38: 472.
130. Sprent JI 1968. Planta (Berl) 82: 299.
131. Sreenivasan A, Wandrekar SD 1950 a. Proc Indian Acad Sci 32 B: 143.
132. Sreenivasan A, Wandrekar SD 1950 b. Nature (Lond.) 165: 765.
133. Stanislawskii JJ 1966. Proc II Symp Plant Growth Reg Zesz Nauk 12, Biologia VIII: 113.
134. Sugawara T 1941. III - IV Jap J Bot 11: 343.
135. Suge H, Yamada N 1965. Plant and Cell Physiol 6: 147.
136. Takayanagi H 1977. Chagyo Gijutsu Kenkyu 52: 50.
137. Tolpeikina GI 1975. Vopr Khim Biokhim Sist Soderzh Marganets Polifenoly 3: 45.
138. Tonzig S, Marre H 1955. Instituto Lombardo Di Scienze e Lettere (Milano) 89: 243.
139. Trione EJ 1966. Plant Physiol 41(2): 277.
140. Trione EJ, Yound JL, Yamamoto M 1967. Phytochemistry 6: 85.
141. Watson SA, Noggle GR 1947. Plant Physiol 22: 228.
142. Weinberger P, Godin C 1966. Can J Biochem 44: 1035.
143. Yokoyama K 1956. Proc Plant Physiol Meetings Respiratory Metabol P XXVI.
144. Yurkevich IV 1963. Soviet Plant Physiol 10: 70.
145. Zholkevich VN, Chetverikov AG 1967. Soviet Plant Physiol 12(2): 167.

5. REPRODUCTIVE GROWTH

Studies have been undertaken on the relationship between AA turnover and metabolism, growth and flowering of numerous crop plants over a number of years (Chinoy, 1962; 1977; Chinoy et al., 1957 a, b; 1958 a, b; 1965; 1966; 1968; 1969 a–d; 1970 a, b; 1971 a–d; 1972 a, b; 1973 a, b; 1974; 1975; 1976; Chinoy and Saxena, 1972; Patel, 1967; Acharya, 1968; Jani, 1969; Abraham et al., 1968 a, b; 1969; John and Jani, 1969; Raj, 1971; Vyas, 1971; Gurumurti, 1971; Ghesani, 1972; Shah, P. 1972; Laloraya et al., 1972; Shah, 1974).

The special feature of this work is the periodic determination of AA and ASG contents as well as enzymic utilization of AA and the degree of complexing of AA with macromolecules, both in the shoot apex and the leaf.

Colchicine (Col) - 1000 ppm inhibited growth. AA (400 ppm) on the other hand, had a stimulatory effect. The retarding effect of Col on growth was considerably overcome when applied in combination with AA. Flowering was delayed in Col treated plants and accelerated in AA treated ones (Fig. 5.1). The delaying effect of Col on flowering was eliminated by its combination with AA. The number of flowers and fruits per plant, the number of seeds per fruit and the thousand seed weight were reduced in Col treated plants and increased in AA treated ones. Weights were reduced in Col treated plants and increased in AA treated ones. This reduction in weight as a result of Col treatment was considerably overcome by combined treatment with Col and AA (Garg, Chinoy and Nanda, 1968; 1958 a) (Fig. 5.2). A part of the retardation of growth by Col may be due to the destruction (or inactivation) of the endogenous AA. The fact that the retarding influence of Col on growth is considerably removed when it is applied in combination with AA strongly suggests that Col treatment brings about a decrease in the endogenous AA. Garg, Chinoy and Nanda (1958) postulated that an initial decrease in the endogenous AA brought about by Col treatment adversely affects the reproductive processes of

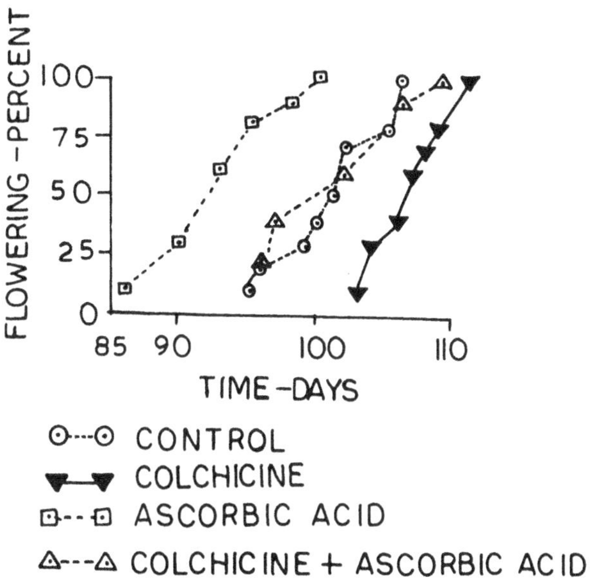

FIGURE 5.1. Antagonism between colchicine and AA.

the plant in so far as it takes the plant much longer to reach that critical AA concentration which by interacting with endogenous IAA may bring about the initiation of flowering. When an exogenous application of Col in combination with AA is made, the retarding influence of Col is removed to a considerable extent, thus giving support to the role of AA in flowering.

Chailakhyan (1956) obtained slight stimulatory effects of AA, thiamine, and nicotinic acid on growth and flowering of plants. He however, considered these effects to be non-specific.

Chinoy, Singh and Sirohi (1958 a) grew both the vernalized (V) and non-vernalized (C) seedlings of Triticum aestivum cv. Hansa (Swalof, Sweden) under normal day (11 hour sunlight - ND) and long day (11 hours of sunlight + 13 hours of artificial light - LD) conditions. The growing species were periodically analyzed(at 7 day intervals)

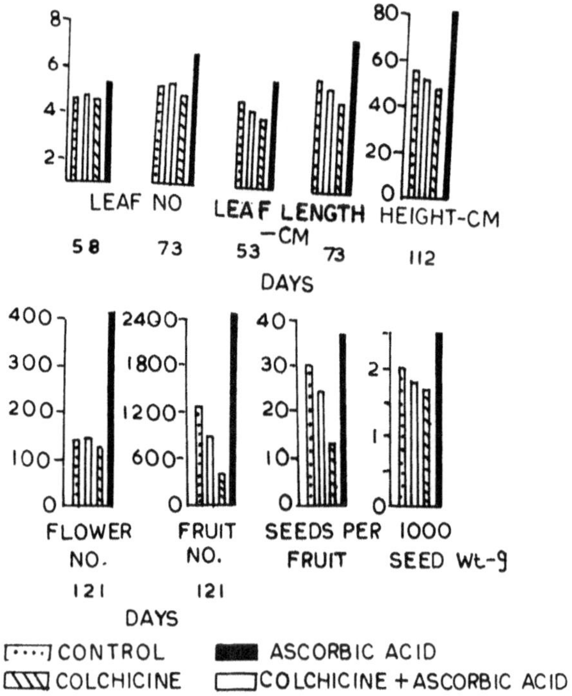

FIGURE 5.2. Elimination of the retarding influence of colchicine treatment by the simultaneous exogenous application of AA on various growth and yield characters of *Brassica chinensis* L.

for AA and auxin contents. Determinations of AA and auxin for only two extreme treatment combinations, unvernalized under ND (NDC) and vernalized under LD (LDV) are given here (Fig. 5.3). At first the AA content of the shoot apex varied within a narrow range. But at the time of the first detectable morphological change in the growing point from the vegetative to the reproductive state, the AA content increased significantly. It remained at a high level for some weeks after which it declined. Auxin concentration on the other hand, showed a decline at the time of the change in the growing point from vegetative to reproductive state.

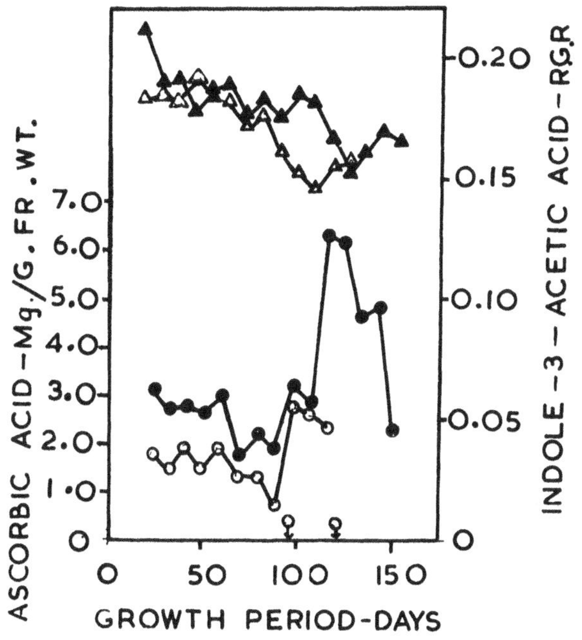

FIGURE 5.3. Ascorbic acid and auxin contents of apical organs at seven day interval throughout the growth period of <u>Triticum</u> <u>aestivum</u> L. cv. Hansa (Swalof, Sweden).

AA-Long day-vernalized (LDV): 0 —— 0; and AA- normal day non-vernalized (NDC) : ● —— ●
Auxin-Long day-vernalized (LDV): Δ —— Δ; Auxin-normal day-nonvernalized (NDC): ▲ —— ▲
(Chinoy, Singh and Sirohi, 1958).

This decline occurred earlier in LDV compared with that in NDC ones. In the light of the above mentioned findings Chinoy et al. (1958 a) postulated that the endogenous AA of a plant, interacting with its IAA, had an important role to perform in the transformation of the growing point of the plant from the vegetative to the reproductive state.

Chinoy (1962) carried out weekly determinations of AA and AAU in the shoot apices and reproductive organs of vernalized and non-vernalized plants of three cultivars of *Triticum aestivum* (Fig. 5.4). AA content as well as its utilization increased and reached a significantly higher level at the time of the change in the shoot apex from the vegetative to the reproductive state. This enhanced formation and utilization of AA continued throughout the period of reproductive differentiation (from double ridges to the first anthesis).

Vernalization and long photoperiod brought about an earlier enhancement in AA-concentration and its utilization (AAU) which produced an earlier upsurge in laying down of growth centres and cell elongation. As a consequence of this, all the cultivars flowered earlier. Metabolic activity of the plant was found to depend to a considerable extent upon AA - concentration and its utilization. A metabolic index was worked out by taking the product of AA - concentration and AA - utilization with development (number of days from sowing to anthesis). Linear regressions of growth centres and cell elongation as well as organizer centres and cell elongation on the metabolic index were found to be highly significant (Fig. 5.5.1,2). These investigations revealed that AA turnover was accelerated by vernalization and long photoperiod.

Chinoy and Mansuri (1966) have provided further evidence on AA metabolism of barley under normal and long photoperiod with and without vernalization. AA content and its utilization were at higher levels compared with those obtained during the period of vegetative differentiation (Fig. 5.6). These enhanced levels were maintained throughout the period of reproductive differentiation. Vernalization and long photoperiod, both brought about an earlier upsurge in AA content and AAU (Fig. 5.6). Concomitant with faster AA turnover, nucleic acids and protein contents rose earlier and remained high in long day vernalized (LDV) plants (Fig. 5.7).

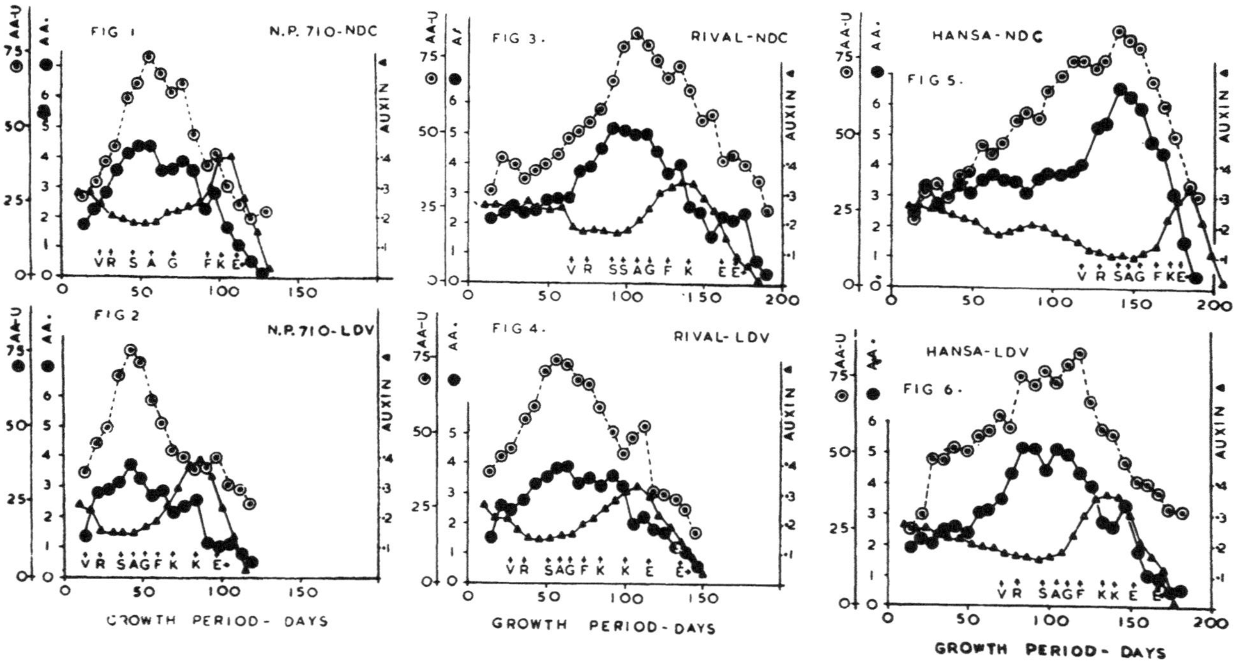

FIGURE 5.4. AA and auxin contents as well as AA-U in the apical organs of three cultivars of <u>Triticum aestivum</u> under LDV and NDC conditions (Chinoy, 1962).

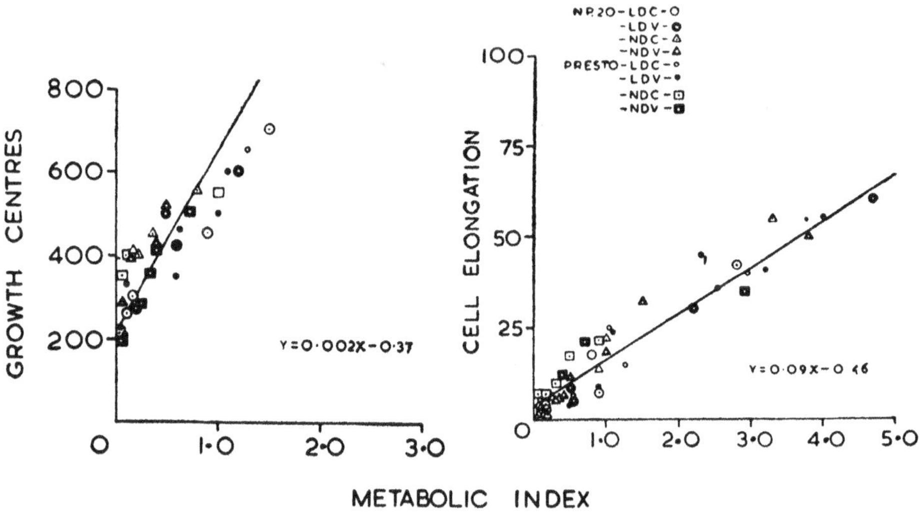

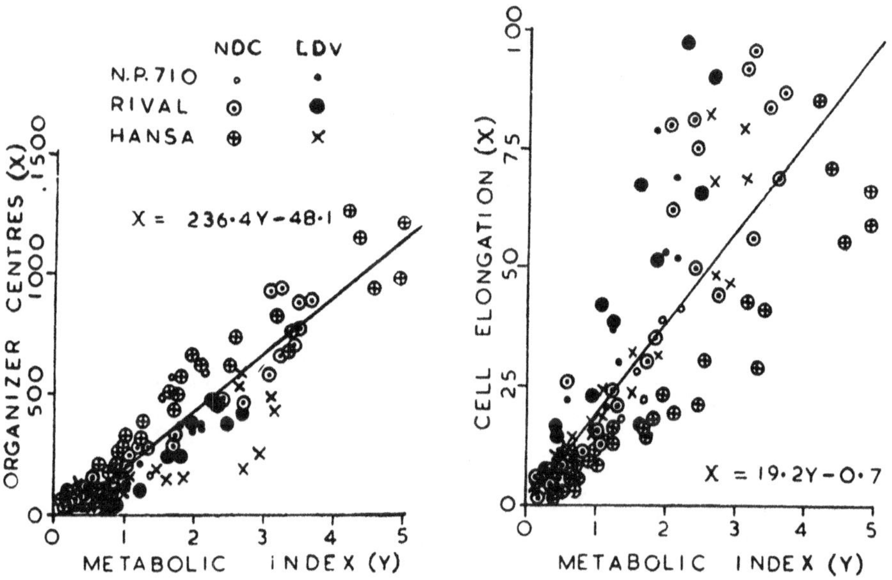

FIGURE 5.5.1, Regression of growth and organizer centres
FIGURE 5.5.2. and of cell elongation on metabolic index
(Chinoy, 1962).

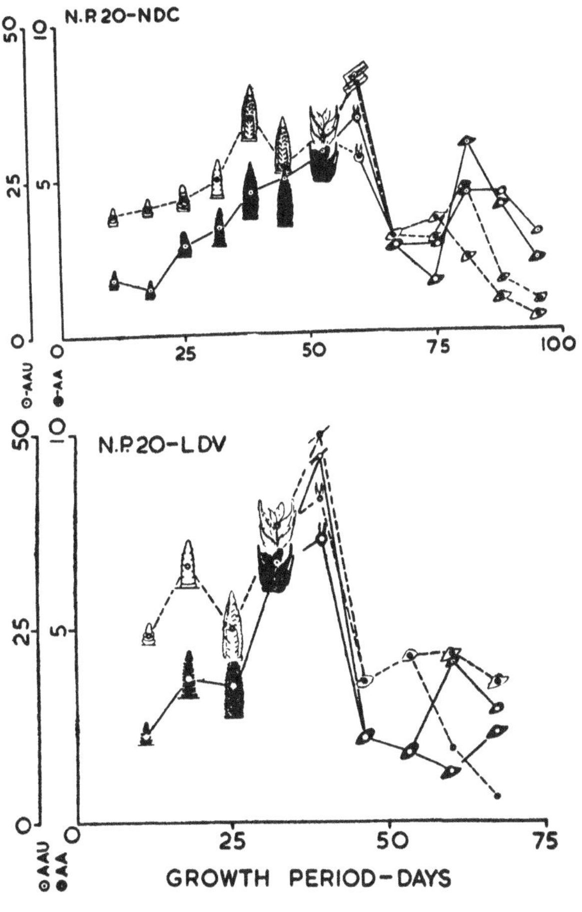

FIGURE 5.6. AA content and AAU in apical organs of <u>Hordeum vulgare</u> L. cv. N.P. 20 under NDC and LDV conditions (Chinoy and Mansuri, 1966).

Lamparska (1966) made periodic determinations of the AA and DHA concentrations which continued to increase in a short day plant <u>Perilla cymoides</u> L. and that of <u>Calendula officinalis</u>, a long day plant, both grown under a 16-hour and a 10-hour day. This enhancement continued upto the 35th day after which it declined. DHA was generally in greater concentration compared to AA. Total AA content was invariably

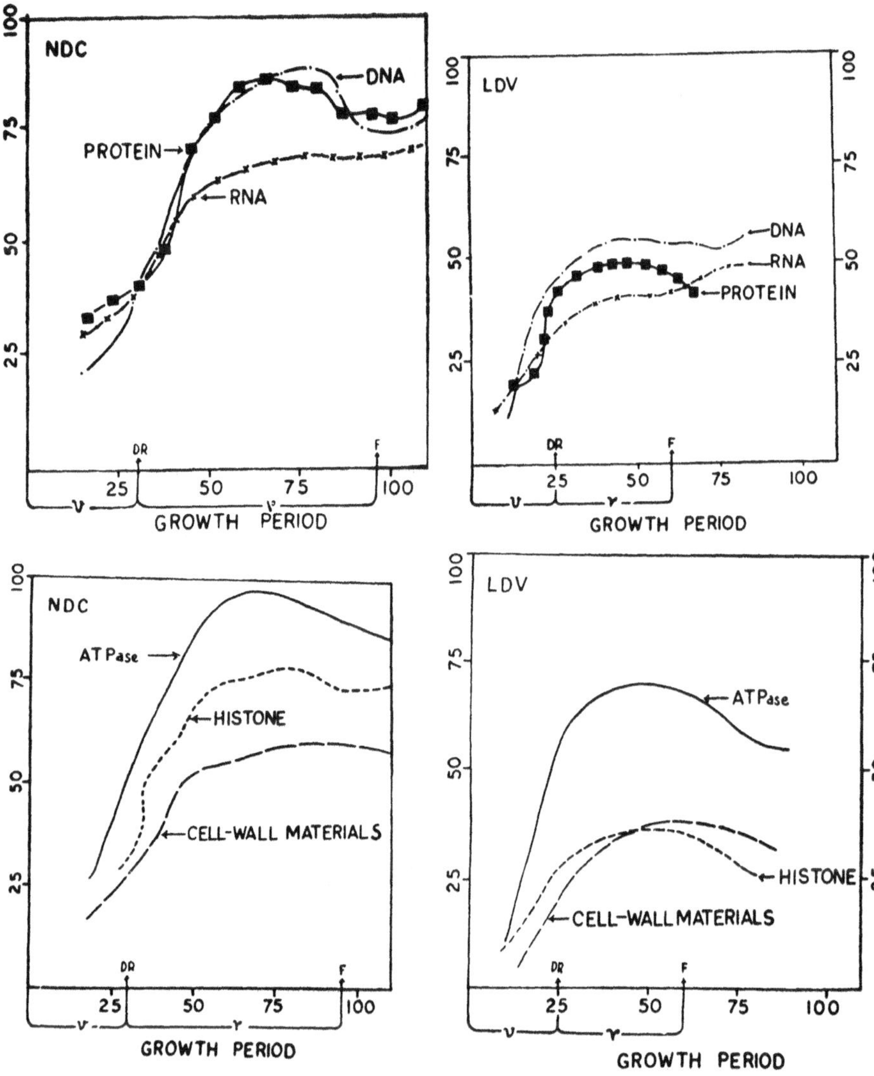

FIGURE 5.7. DNA, RNA, protein, histone and cell-wall material as well as ATPase activity in apical organs or Hordeum vulgare L. cv. N.P. 20 grown under NDC and LDV conditions.

higher under long day conditions. However, no correlation was established between AA leval and the process leading to blossoming.

The dynamics of AA in wheat under long day differed markedly from that in etiolated plants germinated at the same temperature (Stanislawski, 1966 b). Devernalization also caused profound changes in the AA content. Total AA (AA + DHA) fluctuated in a different manner under all light and thermic conditions compared to AA alone. The level of AA in plants kept under darkness and devernalized by raising the temperature, was the lowest compared to that found in plants under other treatments. The process of devernalization, which was carried out by raising the temperature from the 10th to the 20th day, lowered the AA content in plants grown both in light and darkness compared to the controls. Photoperiodic and vernalization treatments differently affected the AA content of winter and spring varieties of wheat.

A comparative study of the effect of vernalization and photoperiod on AA turnover and its correlation with major metabolic events during the adult phase of wheat was carried out (Pandya, 1969). The results are summarized for photoperiodic and vernalization effects for four major stages, viz., vegetative apex (v), initiation of reproductive apex (r), mean of four different stages of reproductive differentiation (rd) and mean of four different stages of grain development.

AA content rose at the initiation of reproductive apex (r) as well as during reproductive differentiation (rd) in ND as well as LD plants of both the cultivars, with the exception of (r) of N.P. 718. During the (rd) stage there was an appreciable enhancement in AA content of Bledsoe (Fig. 5.8).

The specialized feature of LD and vernalized plants (V) is the overall higher ASG content in comparison with that of ND and non-vernalized (C) plants respectively. In this case also, the ASG content was higher during (r) and (rd) stages in both the cultivars compared with that in the vegetative apex (Fig. 5.8).

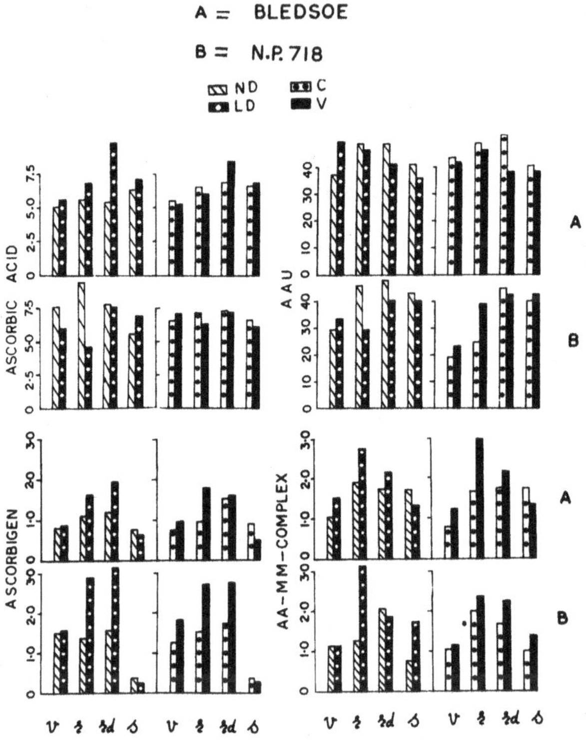

FIGURE 5.8.1. AA and ASG contents, AAU, AA - MM complexing, in the shoot apices and reproductive organs of wheat cultivars Bledsoe (A), N.P. 718 (B) under long (LD) and normal (ND) photoperiod and vernalization (V) and without vernalization (C) during the period of vegetative differentiation (v), induction of reproductive phase (r), period of reproductive differentiation (rd) and period of maturation (s).

Ascorbic acid utilization (AAU) was generally at a higher level in LD during (V) stage in Bledsoe as well as in N.P. 718. During the (r) and (rd) states there was an upsurge in AAU compared to that in the (r) stage in almost all cases (Fig. 5.8).

The AA - MM complexing was generally at a higher level in LD and (V) plants compared to that in ND and (C) plants respectively. In (r) as well as during reproductive

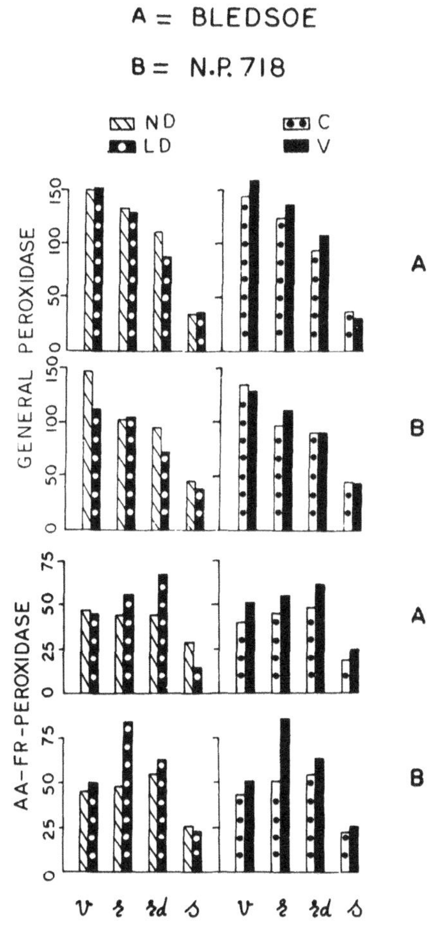

FIGURE 5.8.2. AA - FR and general peroxidase in the shoot apices in reproductive organs of wheat cultivars Bledsoe (A), N.P. 718 (B) under long (LD) and normal (ND) photoperiod and vernalization (V) and without vernalization (C) during the period of vegetative differentiation (v), induction of reproductive phase (r), period reproductive differentiation (rd) and period of maturation (s).

differentiation AA - MM complexing was, as a rule, taking place at an enhanced rate compared to that in the vegetative apex of both the cultivars (Fig. 5.8).

1. EFFECT OF PHOTOPERIOD AND VERNALIZATION ON ASCORBIGEN CONTENT OF CROP PLANTS

The ASG content of embryo axis and endosperm of *Triticum aestivum* L. cv. Bledsoe was determined during juvenile differentiation. There was faster conversion of ASG to AA in the vernalized set, thereby lowering the ASG concentration especially during the earlier phase of germination (Chinoy et al., 1969 a) (Fig. 5.9). In both oat cv. Victory and barley cv. Munsing, the ASG content was depleted more rapidly especially in the vernalized embryo axis during the early stages of germination. This result is in conformity with that obtained for wheat - Bledsoe (Patel, 1967) (Fig. 5.9).

2. EFFECT OF VERNALIZATION AND PHOTOPERIOD ON ASCORBIGEN CONTENT OF TWO CULTIVARS OF RYE

The apex and leaf of control and vernalized plants of *Secale Cereale*, cv. Petkus from Swalof, Sweden and cv. Gatore from U.S.A., were analysed for ASG content during vegetative (v), reproductive (r) and senescence (s) phases (Patel, 1967). ASG content in apex of Gatore and Petkus was high in the vegetative phase under short day and normal day (SD and ND), afterwards it declined and remained at a low level. Under long day (LD) condition, ASG content was low during the vegetative phase. As the plant entered reproductive phase, it increased and then remained more or less at the same level till completion of maturation (Fig. 5.10). ASG content was lower in the leaf compared with that in the apex.

ASG content of Petkus rye was higher during vegetative phase and then declined, and very little ASG was found at the time of senescence. In all the photoperiodic treatments except in the leaf of long day vernalized (LDV) plants, ASG content was lowest at the time of vegetative differentiation. In this cultivar, vernalization treatment and LD condition were found to be beneficial for increasing ASG content in the leaf during reproductive differentiation

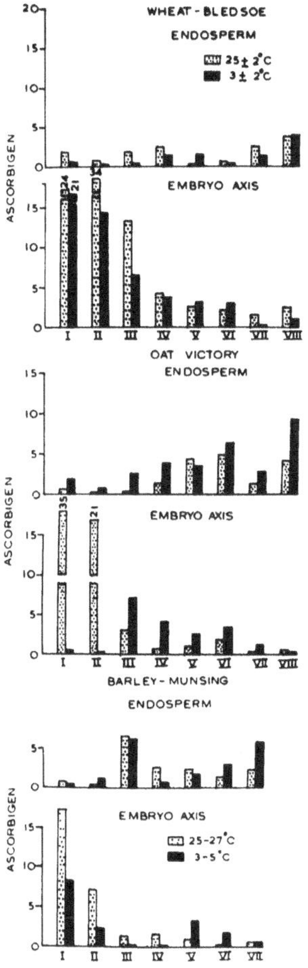

FIGURE 5.9. ASG content of the embryo axis and endosperm of vernalized (3 ± 2° C) and nonvernalized (25 ± 2° C) seeds of wheat, oat and barley at different stages of juvenile growth (Chinoy et al., 1969 a).

till maturation. This experiment also indicates that AA accumulates in a bound state during vegetative phase and is released in free condition in the shoot apex at the time of reproductive differentiation for sustaining active metabolism (Fig. 5.10).

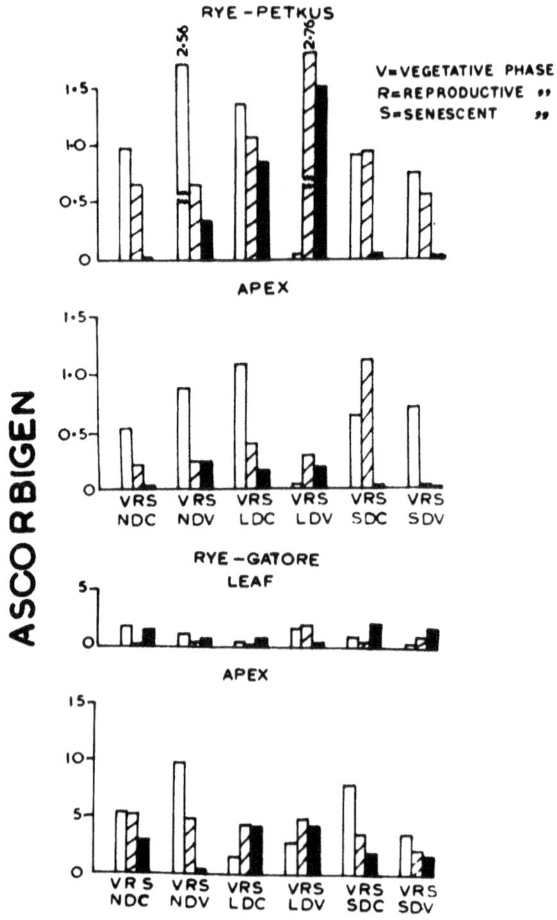

FIGURE 5.10. Effect of vernalization (v) and photoperiod (SD, ND, LD) on ASG content of two cultivars of *Secale cereale* during vegetative, reproductive and senescent phases (Patel, 1967).

Michniewicz (1961) also studied the effect of vernalization on the AA content of winter wheat - during its ontogenetical development and the results are very interesting. AA content of leaves increased during the course of its ontogenetic development and reached its maximum just before blossoming. The content of this vitamin then declined during fruiting stage and was the lowest at harvest time. Plants raised from vernalized seeds had higher AA content throughout the growth period.

Histochemical studies correlating AA content of an organ with its macromolecular content have been carried out on Zea mays, a thermaphyte (Suthar, 1972; Chinoy et al., 1973). In all the cells of the vegetative apex (except the middle apical region cells), the cell wall, cytoplasm, nuclei, and nucleoli showed silver deposits. The cell wall and nucleoli registered very high amount of AA (Fig. 5.11). There was a marked changed in the AA localization during the transformation of the apex from vegetative to reproductive stage (Fig. 5.11). The apex of the cells along with the tunica and sub-apical regions were heavily loaded with AA. The developing inflorescence, branch primordia, spikelet primordia along with glumes contained high amount of AA (Fig. 5.12), which was confined mainly to the nucleus especially in the central mother cell zone and peripheral zone of the shoot apex (Fig. 5.12). The shoot apices at this stage and the succession of new growth centres also possessed higher concentrations of AA compared with those of the growth centres of the preceding stage. This was maintained until florets were formed (Fig. 5.12). Considerable amount of AA was present in the nucleus as well as the cytoplasm.

Upper and lower floret meristems were then derived from the spikelet meristem. The lower floret meristem with young glumes registered higher AA content than the upper floret meristem, which produced two flowers. The flowers were protandrous and the stamen had more AA than the carpel initials. In tassel both flowers were functional, but at the time of differentiation into lemma, palea and stamen, there was a slight decline in the AA content. Younger flowers contained higher AA than the upper one. There was a two-fold increase in the AA content at the evocation and it remained high at the time of formation of branch primordia, spikelet primordia and even the floret meristem but declined slightly at the time of flower formation. In the case of maize, regression of DNA, RNA, basic protein and -SH contents

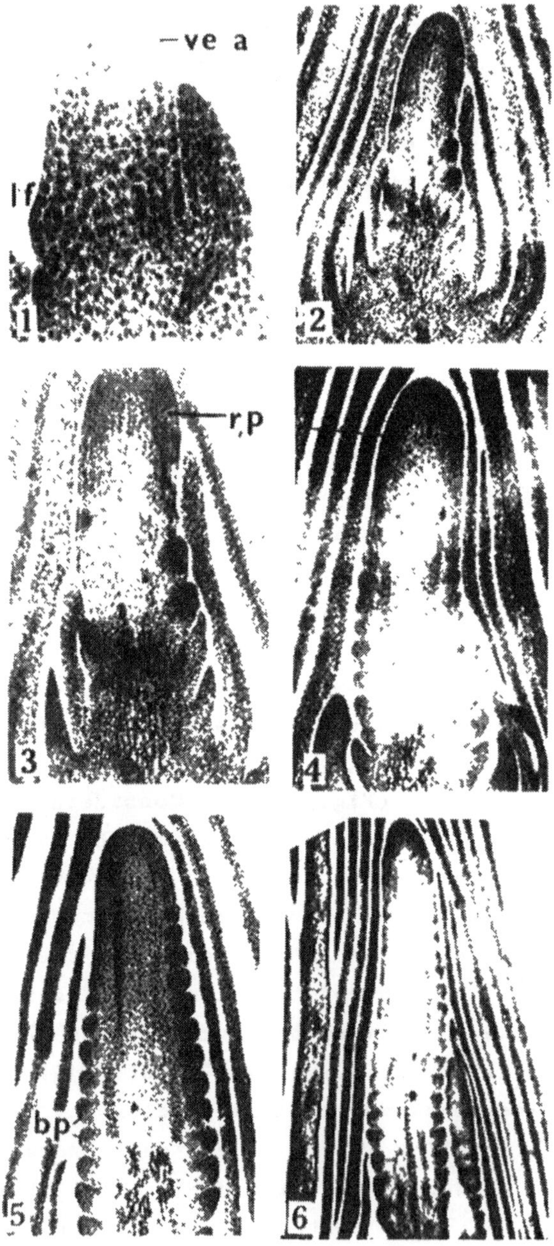

FIGURE 5.11. Histochemical localization of AA during the vegetative and reproductive differentiation of the shoot apex of Zea mays L. 1. Vegetative apex x 270; 2, 3. Transforming apex. 2 : x 60, 3 : x 270; 4,5. Transforming apex

with branch primordia 4 : x 100; 5 : x 60. 6 : Elongated apex with many branch primordia x 35.
Ve. a = vegetative apex; lf = leaf buttress; rp = reproductive apex; b_p = branch primordia.

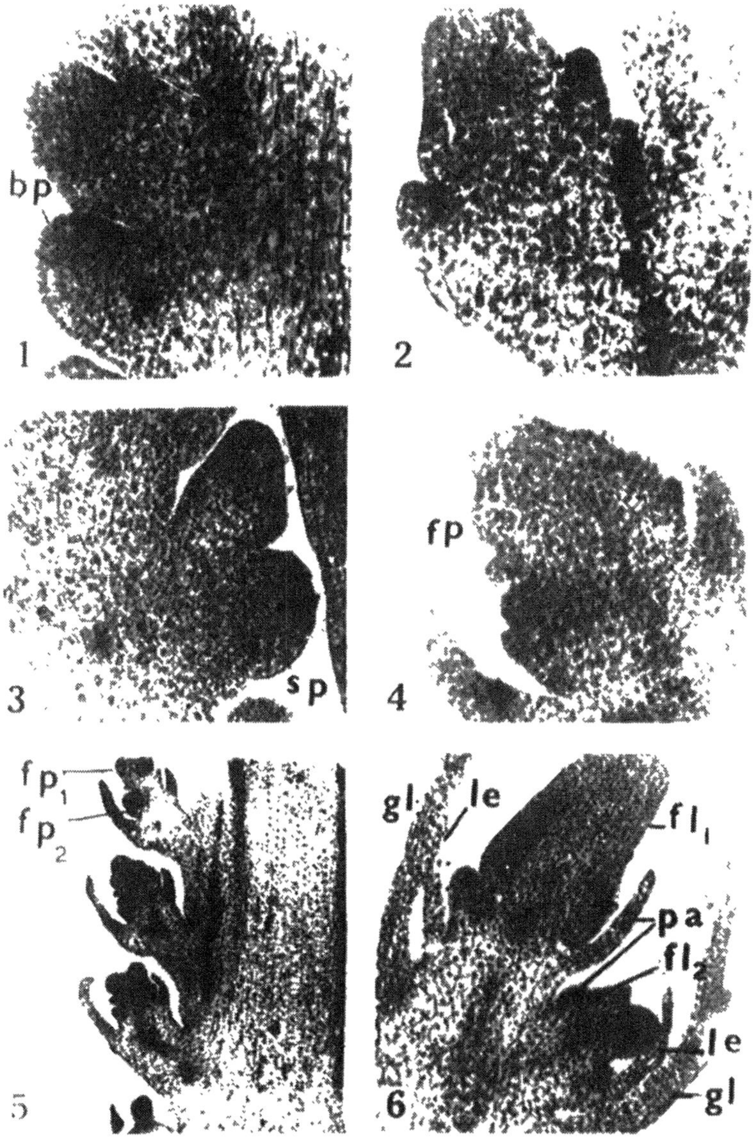

FIGURE 5.12. Histochemical localization of AA during the reproductive differentiation of shoot apex of Zea mays L.
1. Branch primordia x 270; 2. Spikelet meristem x 270;
3. Spikelet primordia x 270; 4. Floret primordia x 270;
5. Floret primordium 1 & 2 x 60; 6. Floret differentiation x 90;

bp = branch primordium; sp = spikelet primordium; fp = floret primordium; fp_1 = floret $primordium_1$; fp_2 = floret $primordium_2$; gl = glumes; le = lemma; pa = palea; fl_1 & fl_2 = $floret_1$ and $floret_2$.

on that of AA were determined (Fig. 5.13). This analysis clearly brings out the fact that there was a positive linear correlation between the content of macromolecules and that of AA.

In a thermophobe (wheat), the transition began with elongation of the vegetative apex which assumed an elongated conical shape (Pathan, 1974) (Fig. 5.14), and there was a clear change in the AA localization. The apical cells along with the tunica and subapical regions showed a high amount of AA but central mother cells had lower amount (Fig. 5.14).

Histochemical determinations of DNA, RNA and basic protein at the corresponding stages of differentiation were also carried out (Pathan, 1974). A parallelism between AA, DNA, RNA and basic protein contents was found which was further confirmed by studying the regression of DNA, RNA and basic protein contents on that of AA (Fig. 5.15). Prasad et al. (1981) studied the metabolic correlations during ontogeny of axillary bud in Murraya koenigii L. Spreng by histochemical techniques and concluded that the axillary bud was metabolically very active throughout the course of development as was evident from the rich distribution of the metabolites and AA. Ontogenic development was marked throughout the developmental course by a sustained high content of these metabolites, except in the shell zone. A close correlation could be established between the capability of the bud retaining the tunica corpus organization throughout its development and the high content of these metabolites.

SHOOT APEX

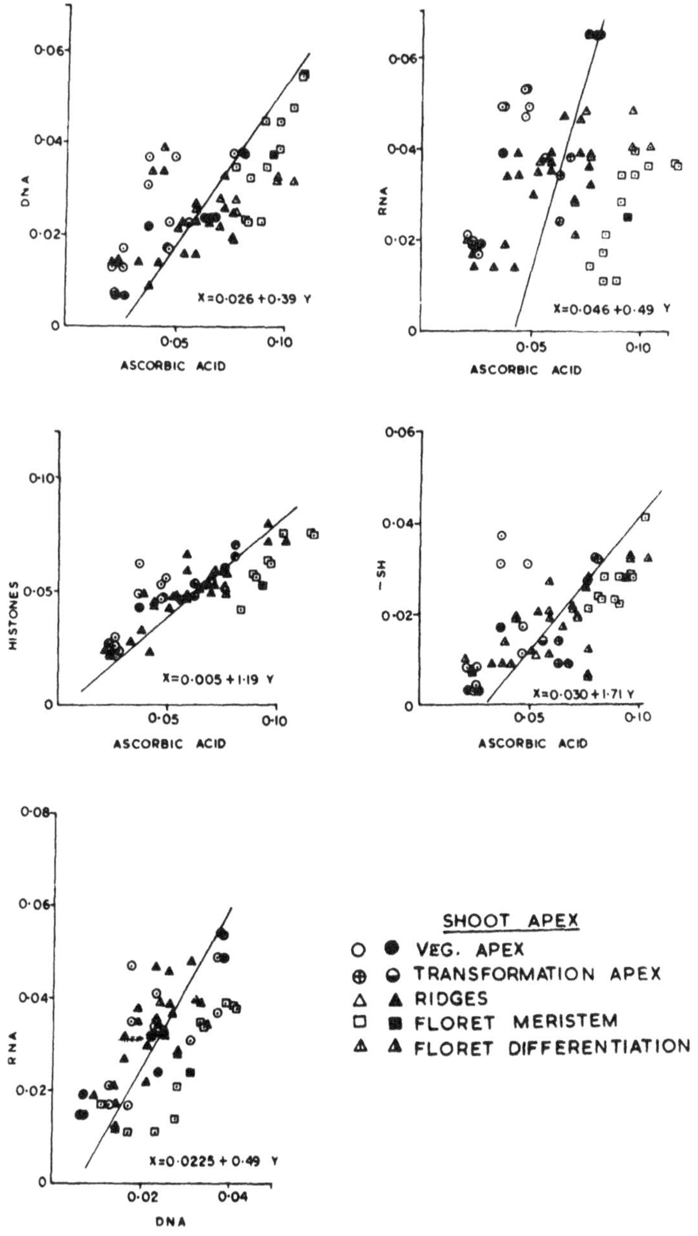

FIGURE 5.13. Regressions of DNA, RNA, histones and -SH on AA and that of RNA on DNA during the reproductive differentia-

tion of shoot apex of Zea mays L.
0 ——— total values for the regression
θ ——— mean values for the regression

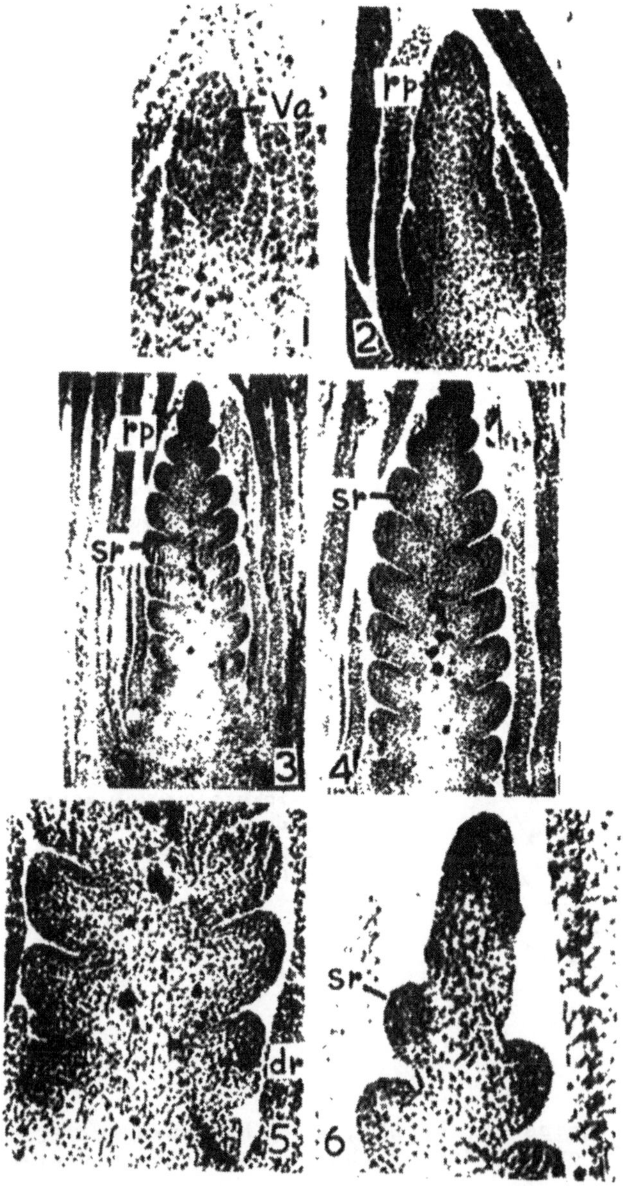

FIGURE 5.14. Histochemical localization of AA during vegetative and reproductive differentiation of shoot apex of Triticum aestivum.
1. Vegetative apex x 60; 2. Transforming apex x 60;
3,4. Shoot apex with single ridges, 3 x 36; 4 x 60;
5. Shoot apex with initiation of double ridges x 100;
6. Shoot apex with single ridges x 100.
va = vegetative apex; rp = reproductive apex; sr = single ridge; dr = double ridge.

In Cyperus rotundus (Madhavan Unni, 1971) and Withania somnifera (Neelkandan, 1971) during vegetative and reproductive differentiation, there was a parallel increase of AA, nucleic acids, especially RNA, basic protein and -SH in the organogenetic centres.

The DNA, RNA and protein content of the apical organ of both the cultivars of wheat under LD and (V) treatments was higher as compared to that in the plants subjected to ND and (C) treatments respectively. The activity of the AA - FR peroxidase, on the other hand, increased appreciably at the initiation of reproductive apex as well as in the differentiating reproductive organs. Long photoperiod and vernalization enhanced the AA - FR peroxidase activity both during periods of vegetative and reproductive differentiation. Catalase activity was higher during vegetative differentiation. Vernalization appeared to have a slight enhancing effect only in the case of Bledsoe. Long photoperiod and vernalization treatment significantly enhanced invertase, protease and amylase activities in N.P. 718 as well as in Bledsoe. Onset of reproductive phase also increased amylase activity in the apical organs of both the cultivars.

The far reaching inductive effects of vernalization and that of the following photoperiod on the growth and flowering response of long day plants observed by Pandya (1969) is supported by the findings of a number of earlier workers (Chinoy, 1962; Chinoy and Nanda, 1951, 1951 a, b; 1952; Chinoy et al., 1959; Chinoy and Mansuri, 1966; Mansuri, 1965; Nanda et al., 1959; Sirohi et al., 1959; Sawhney et al., 1959; Raj, 1971; Vyas, 1971; Patel, 1967; Shah, 1974;

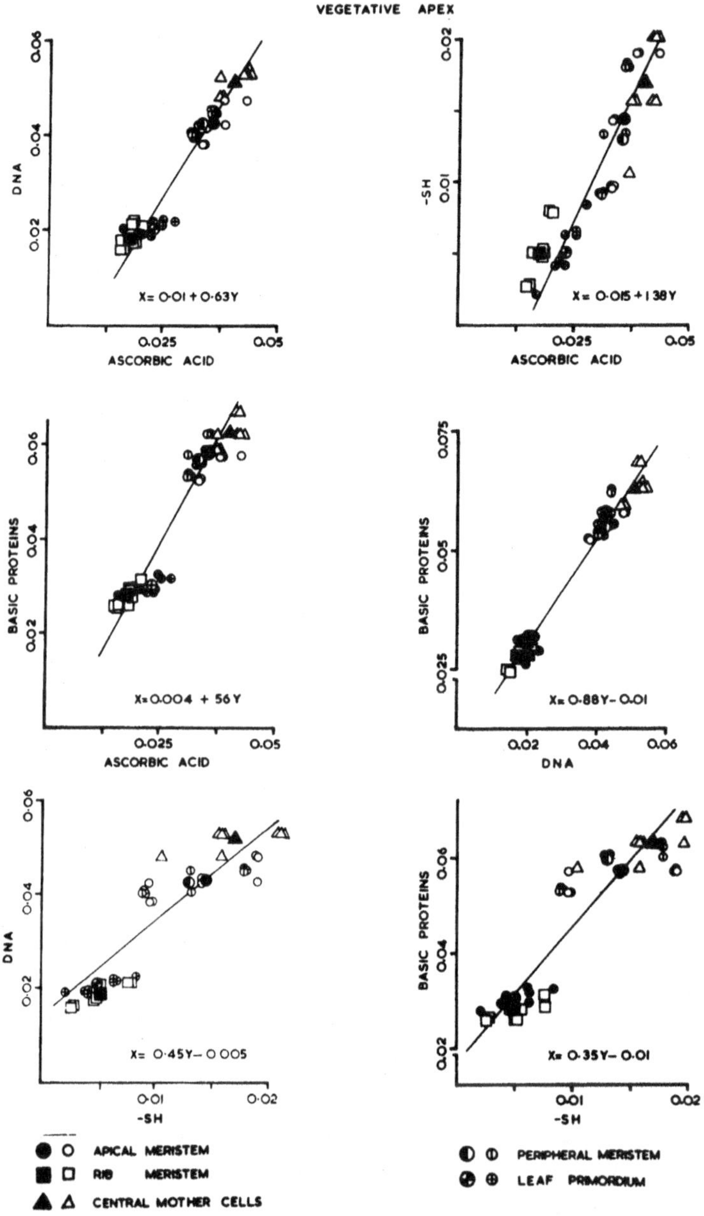
VEGETATIVE APEX
DNA
0.02
0.04
0.06
X= 0·01 + 0·63Y
0.025
0.05
ASCORBIC ACID
-SH
0.01
0.02
X= 0·015 + 1·38Y
ASCORBIC ACID
BASIC PROTEINS
X=0.004 + 56Y
ASCORBIC ACID
BASIC PROTEINS
0.025
0.05
0.075
X= 0·88Y− 0·01
DNA
X= 0·45Y− 0·005
-SH
BASIC PROTEINS
X= 0·35Y− 0·01
-SH
APICAL MERISTEM
RIB MERISTEM
CENTRAL MOTHER CELLS
PERIPHERAL MERISTEM
LEAF PRIMORDIUM

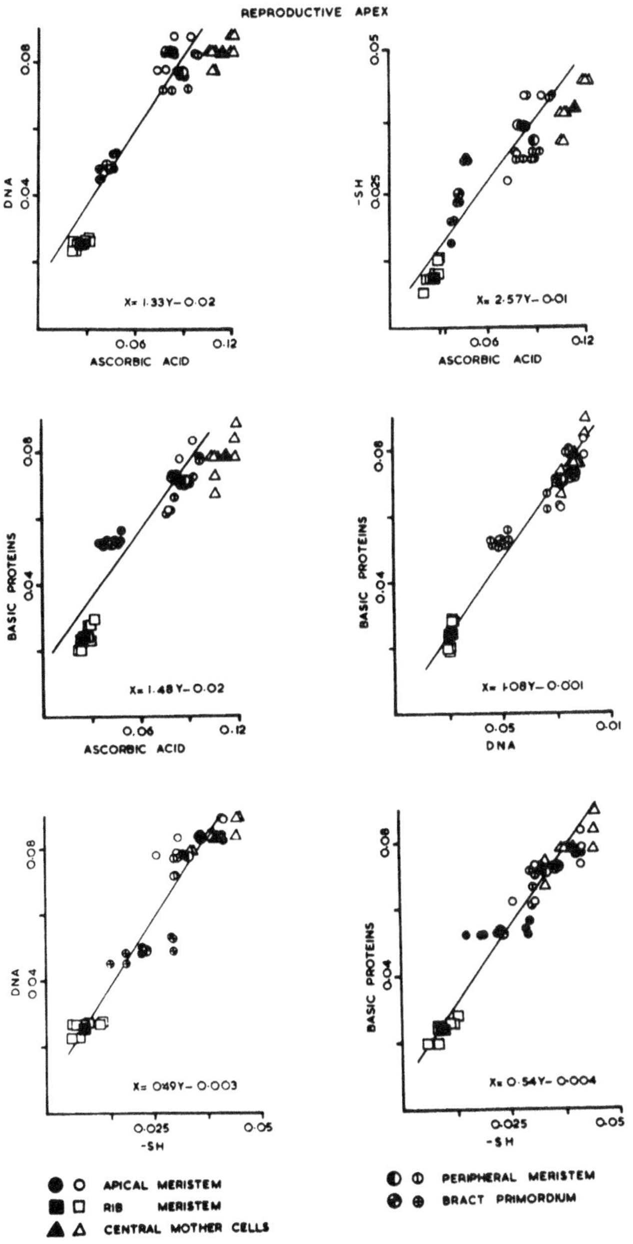

FIGURE 5.15.1. Regression of DNA, -SH and basic proteins on
FIGURE 5.15.2 ascorbic acid, those of DNA and basic proteins

on -SH and that of basic proteins on DNA in the vegetative and reproductive apices of *Cyperus rotundus* L. (Madhavan Unni, 1971).

Devay, 1965; Lamparska, 1966; Stanislawskii, 1966 a; Michniewicz, 1966; Michniewicz and Lamparska, 1965). Vernalization and long photoperiod brought about earlier shooting followed by earlier spike emergence and anthesis (flowering). As a consequence of all this, the vegetative growth in terms of leaf production, tiller number, and dry weights of the vegetative parts was considerably reduced. The correlation between metabolism, growth and development is thus an established fact.

The importance of metabolism in flowering has been emphasized by Sen (1964) and Chinoy (1962) and is provided by a number of correlations established by various workers between the metabolic changes and the flowering process. Studies of dynamics of regulatory substances such as enzymes, vitamins, auxins, nutritional substances both organic and inorganic as well as changes in shoot apices have been made (Chinoy, 1962; 1975; Patel, 1967; Chinoy and Sharma, 1957; 1958; Chinoy et al., 1958 a, b; Garg, 1960; Nanda and Tayal, 1976; Nanda et al., 1958 a, b: Sawhney, 1958; Garg et al., 1958 a; Garg and Chinoy, 1964; Bonner and Zeevaart, 1962). All these extensive studies have established that most of the physiological processes of a plant are not only dependent upon each other but are also fully integrated with growth and development into a harmonious system leading to a specific growth pattern. It is also concluded from these studies that although the rate of utilization of these various substances differed in different varieties these rates were profoundly influenced by photoperiodic and temperature treatments and that the pattern of their production and utilization were linked with the flowering process. Of all the regulatory substances, production and utilization of AA were found to be consistently correlated with flowering (Chinoy, 1962; Chinoy and Mansuri, 1966). The rates of cell division and cell elongation as well as the laying down of new growth centres increase tremendously with the advance of the flowering processes in a plant and this was paralleled in most of the cases with a considerable enhancement in the formation and utilization of AA.

3. ASCORBIC ACID - NUCLEIC ACID - PROTEIN METABOLISM CONCEPT OF GROWTH AND DEVELOPMENT IN PLANTS

The dynamics of AA production and utilization as well as the metabolic drifts of nucleic acids, proteins and other cell constituents have been studied in the shoot apex and differentiating floral organs of a number of species under different photoperiodic and vernalization treatments and based upon these findings the ascorbic acid - nucleic acid - protein metabolism concept of growth and development was advanced (Chinoy, 1962) and subsequently elaborated (Chinoy and Mansuri, 1966; Chinoy, 1969). A brief resume of this concept is reproduced here.

During the juvenile phase, the AA content rose significantly soon after germination and declined rapidly thereafter attaining a low level during the later stages of germination. The concentration of ascorbigen (ASG), showed a rhythmic fluctuation, whereas, on the other hand, ascorbic acid utilization steadily increased during the juvenile-phase. The activity of AA - FR - peroxidase was also found to fluctuate during germination.

During the period of vegetative differentiation ascorbic acid and ascorbigen contents as well as ascorbic acid utilization (AAU) registered an upward trend. At the time of transformation of the apex from vegetative to reproductive stage (double ridges) there was a substantial increase in the concentration of free AA as well as AAU. The AA - FR - peroxidase activity also increased during floral induction. Further, during the entire period of reproductive differentiation, AA - FR - peroxidase activity as well as AA content and its utilization remained at substantially higher levels.

Complexing of AA with DNA represented by black circle with white star (CTC) was the highest during the period of reproductive differentiation (Fig. 5.16). During the period of senescence all the above mentioned metabolites as well as enzymic activity declined with the exception of auxin which registered a steep rise during the early stages of grain development. The CTC was also at a very low level. The analysis of the shoot apex revealed that DNA, RNA and protein contents increased significantly during the period of reproductive differentiation (Figs. 5.16 to 5.18), together with the level of complexing and declined during the period of senescence. In the case of leaf, AA and ASG contents were the highest during the vegetative phase and decreased during the reproductive phase presumably due to their enhanced transport to the transformed apex. AAU and AA - FR - peroxidase activities in the leaf were the highest during the reproductive phase. The most pronounced increase in the leaf during the reproductive phase was that of catalase.

In both apical organs and leaf, the free radical (FR) signal increased sharply in the reproductive apex as well as in fertilized carpel, whereas at other stages of apical organs free radical signal was maintained at a lower level. The leaf always maintained a much higher level of free radicals throughout the course of reproductive development as compared to vegetative apex.

4. FR OF AA AND CTC OF AA - DNA / RNA AS SOURCES OF ELECTRON ENERGY

In the ascorbic acid - nucleic acid - protein metabolism concept of growth and development, the redox energy derived from ascorbic acid, the universal participant in biological oxidation - reduction processes, is considered as the motive force behind the sequence of events culminating in floral induction and differentiation (Chinoy, 1962). Accelerated rates at which metabolic turnover is taking place especially during reproductive differentiation suggests that the increased demand for electronic energy cannot be met solely

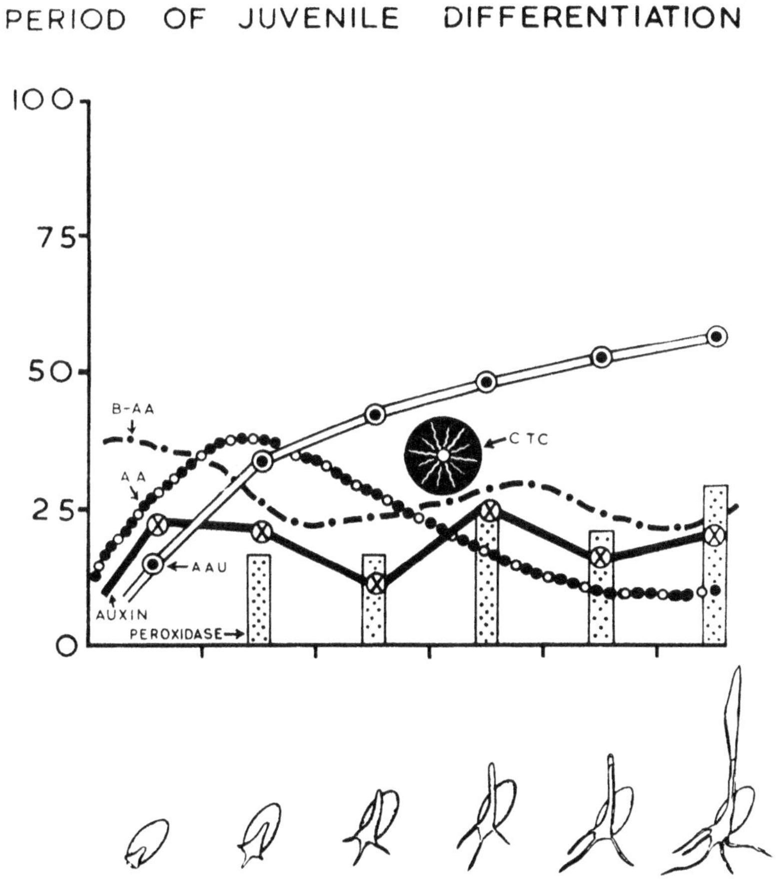

FIGURE 5.16. Free AA and bound (BAA) ascorbic acid contents as well as its utilization (AAU), peroxidase activity, auxin content and charge transfer complex (CTC) during the period of juvenile differentiation.

from ~ p. This necessitates a study of free radicals (FR) and charge transfer complexes (CTC) in which AA is involved.

An appreciable upsurge in the complexing ability of AA with macromolecules (CTC) alongwith sustained high level of AA and AAU during the period of reproductive differentiation (Fig. 5.17) shows that the potent system of energy flow for biological processes participates in the increased energy

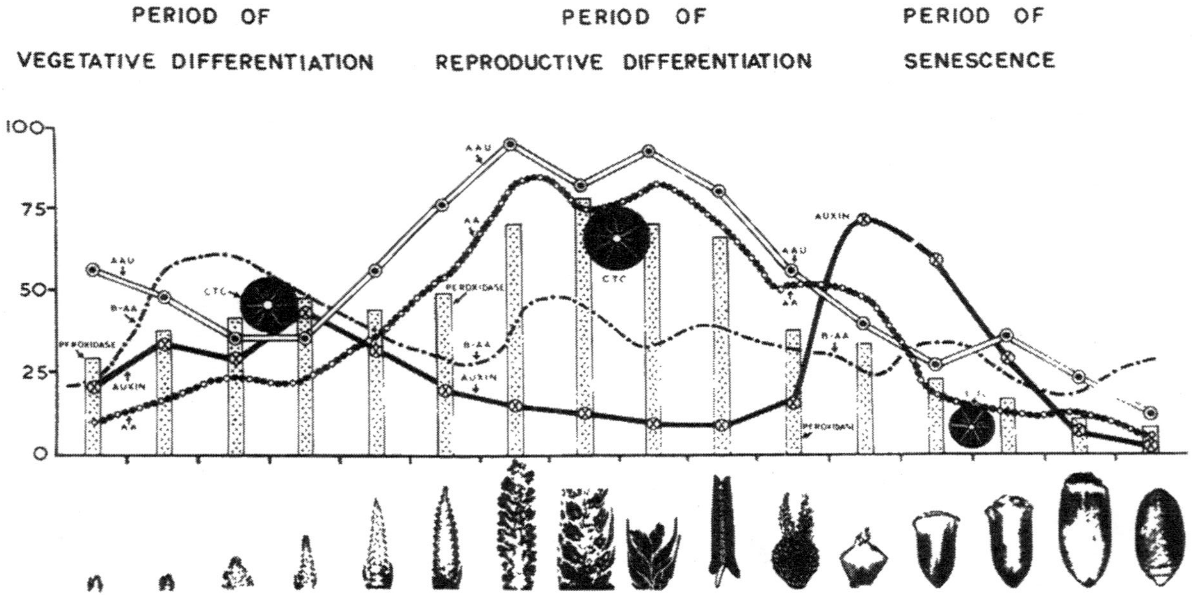

FIGURE 5.17. AA, BAA, peroxidase, auxin and CTC during the periods of vegetative and reproductive differentiation and during senescence.

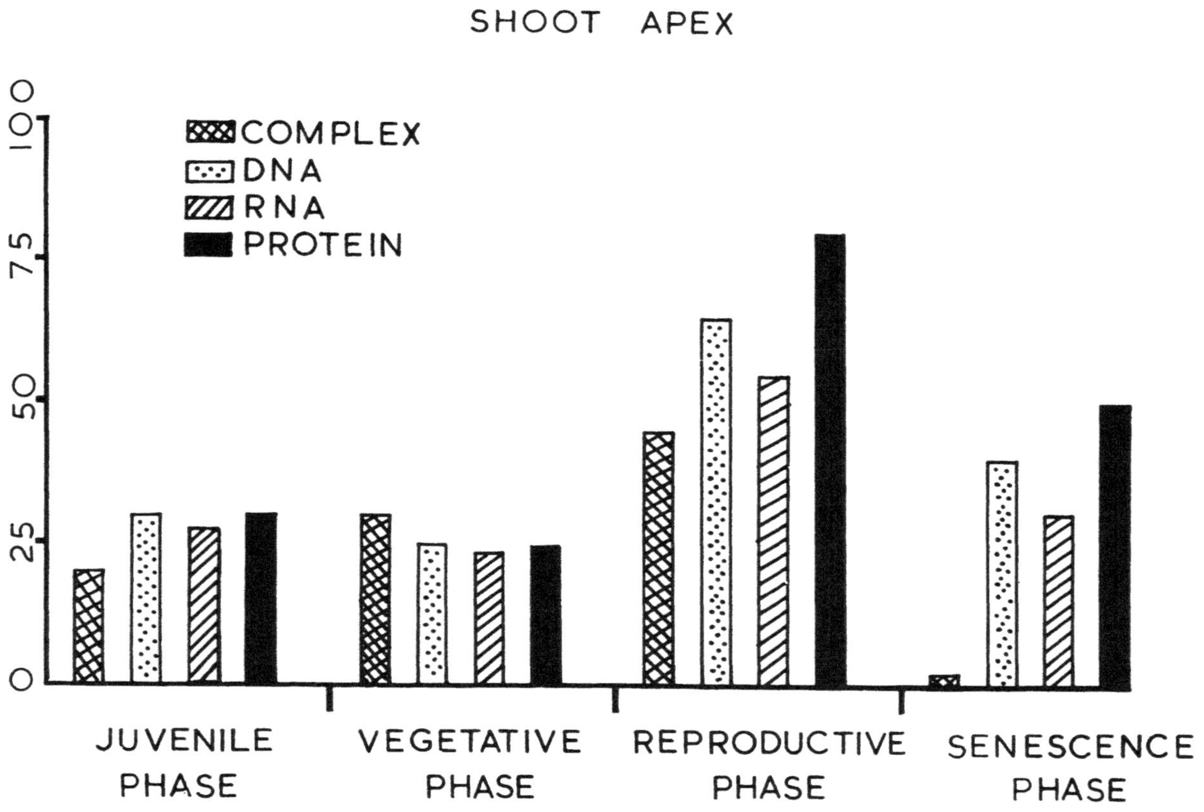

FIGURE 5.18. Mean DNA, RNA, and protein content as well as CTC during the four phases of development of plants.

flow during floral induction and differentiation. In this system of energy transfer, the complexing of two molecules takes place as a result of the overlapping of electron clouds of the two units of the complex. This facilitates the transfer of an electron of one molecule from its own orbital to that of its partner (ref. Chapter 1).

These results further demonstrate and confirm the earlier findings that the concentration of AA - free radical increased, as shown by the enhanced activity of AA - FR - peroxidase during the reproductive phase, as well as by the fact that EPR signals were more intense in the reproductive apex (double ridges), fertilized carpel as well as in the corresponding leaf at all stages of reproductive differentiation compared with those obtained from the vegetative shoot apex (Figs. 5.19 to 5.21) (Chinoy et al., 1971 a, Gurumurti and Chinoy, 1974). Free radicals are also reported in the endosperm during the early stage of germination suggesting that they are involved in energy transfer mechanism required for the 'active' transport of solubilized nutrients to the embryo axis (Chinoy et al., 1969 d; 1971 b; Saxena, 1979; 1983; Singh and Saxena, 1983).

On the basis of the foregoing the following sequence of molecular and submolecular events is visualized in floral induction and differentiation (Chinoy, 1969). AA undergoes monovalent oxidation in aerobic condition with the help of a special peroxidase, whose activity increases considerably during the period of reproductive differentiation, and which produces a powerful reducing agent-monodehydroascorbic acid (MDHA), which being a free radical serves as an electron donor in the biosynthesis of macromolecules. Photo-phosphorylation as well as oxidative-phosphorylation are also energized by the flow of electrons from MDHA, thus, augmenting the ATP pool. The increased production of free radicals during the period of reproductive differentiation is confirmed by considerable enhancement in the magnitude of EPR signals. The formation of a charge transfer complex (CTC)

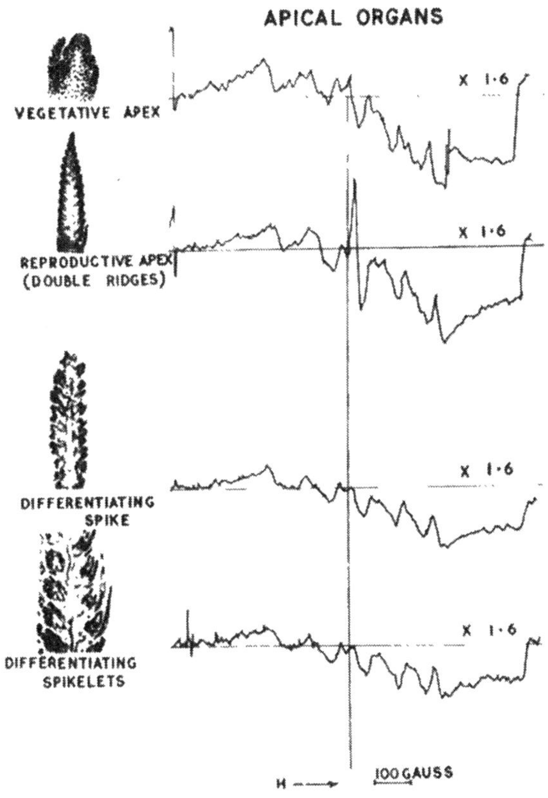

FIGURE 5.19. EPR spectra of wheat cv. S. 22 vegetative and differentiating reproductive apex recor ıt 77°K; magnetic field of 3300 gauss and scanning f of 1000 gauss. x denotes the relative amplifier ga each sample.

between AA and DNA in the shoot apex helps in establishing a direct flow of electron energy for biosynthesis of cell constituents. This charge transfer paves the way for the production of different types of RNA and subsequently of structural proteins, enzymes and other cell constituents at greatly accelerated rates.

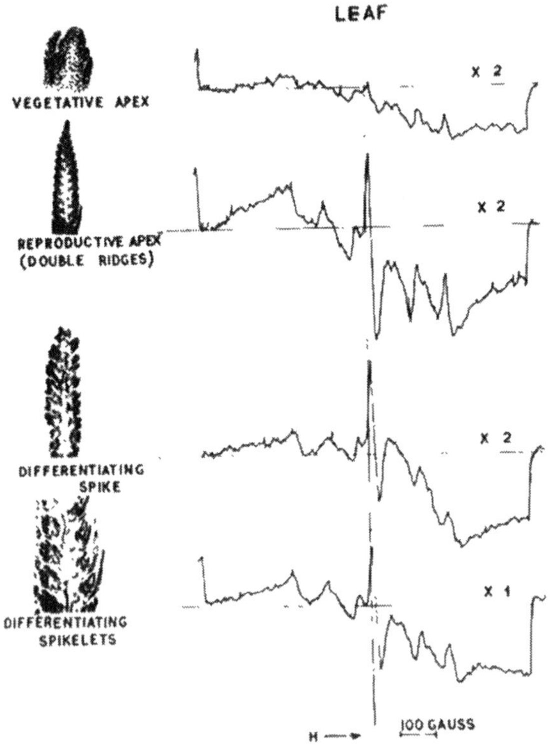

FIGURE 5.20. EPR spectra of wheat cv. S. 227 in top most unfolded leaf recorded at 77°K; magnetic field at 3300 gauss and scanning field of 1000 gauss. x denotes the relative amplifier gain of each sample.

It is a significant fact that Chinoy, Singh and Laloraya (1974) have shown *in vitro* that addition of DNA and RNA to an ascorbic acid solution with a trace of H_2O_2 accelerates both the formation of the free radical of AA-monodehydroascorbic acid and its subsequent faster decay which ultimately leads to the formation of a complex between AA and the macromolecules (ref. Chapter 1).

The production of CTC and free radicals under photo- and thermo-inductive conditions is further accelerated during the

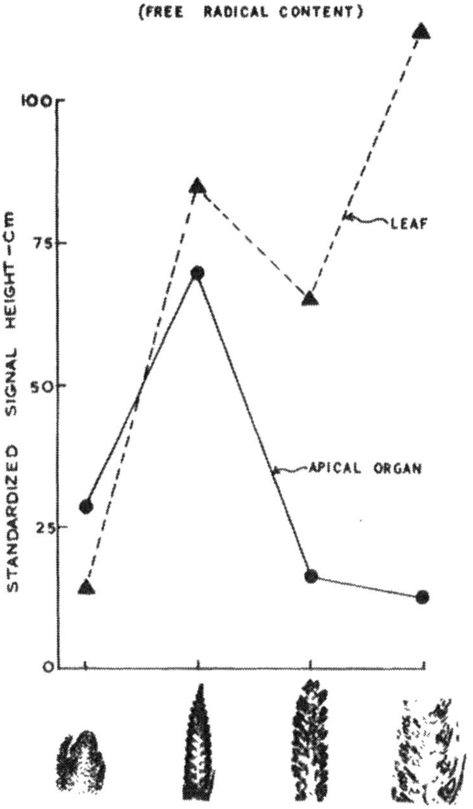

FIGURE 5.21. Free radical contents as signal heights (cm) per gram fresh weight of the sample.

period of reproductive differentiation thus bringing about an upsurge in the biosynthetic activity noted above. Consequently rates of cell division and formation of growth centres are increased considerably, resulting in floral induction and differentiation. In fact the onset of meiosis is the resultant of a change in the rhythm of nuclear and cytoplasmic processes participating in energy regulation and consequently in the formation of new cells and growth centres at enormously intensified rates.

5. SUMMARY

At the time of change in apex from vegetative to reproductive state, the AA content began to increase and IAA content decrease and this trend continued throughout the period of reproductive differentiation. A postulation has been advanced, stating that the endogenous AA of plant interacts with its IAA concentration to bring about the transformation of the growing point from the vegetative to the reproductive state. Throughout the period of reproductive differentiation, AA content and its utilization increased and reached a significantly higher level at the time of the transformation of the shoot apex from vegetative to the reproductive state. Vernalization and long photoperiod brought about an earlier enhancement in AA content and its utilization. A marked change in AA localization during transformation of the vegetative to the reproductive stage occurred. The summit of the apex along with the tunical and sub-apical regions were heavily loaded with AA. There are linear regressions of DNA, RNA, basic proteins and -SH contents on that of AA.

Experiments with Triticum aestivum, Cyperus rotundus and Withania somnifera also yielded similar regressions of macromolecular concentrations on that of AA for a given stage of development signifying the important role of AA in the biosynthesis of macromolecules. Further it clearly showed that the concentration of AA as well as those of the macromolecules are considerably enhanced during early reproductive differentiation.

A new concept called: The Ascorbic acid - Nucleic acid - protein metabolism concept of growth and development in plants was advanced. This concept is based on the fact that an upsurge in AA turnover brings about faster biosynthesis of nucleic acids which in their turn enhance synthesis of proteins, enzyme proteins and other cell constituents leading to a general upsurge in metabolism. There is an upsurge in ATPase activity also, indicating faster utilization of ATP. All these intense metabolic activities result in rapid formation of growth centres and elongation which leads to the differentiation of various organs of the spike in quick succession.

An appreciable enhancement in the complexing ability of AA with macromolecules (CTC) along with sustained high level of AA and AAU during the period of reproductive differentiation shows that the potent sources of electron flow for biological processes are free radicals and charge transfer complexes especially during the period of reproductive differentiation. This postulation is further supported by the enhanced activity of AA - FR - peroxidase as well as by the intensified EPR signals during reproductive differentiation compared with those obtained during the period of vegetative differentiation.

REFERENCES

1. Abraham PG, Dave IC, Saxena OP, Pandya RB 1968 a. Abstr. 37th Ann Session Nat Acad Sci India p. 45.
2. Abraham PG, Dave IC, Pandya RB, Saxena OP 1968 b. Abstr. Ibid p. 48.
3. Abraham PG, Gurumurti K, Dave IC, Pandya RB, Saxena OP 1969. Vidya Gujarat University Journal 12: 194.
4. Acharya UH 1968. Ph.D. Thesis, Botany Department, University School of Science, Gujarat University, Ahmedabad - 380009, India, p. 208.
5. Bonner J, Zeevaart JAD 1962. Plant Physiol 37: 43.
6. Chailakhyan MKh 1956. Dokl Akad Nauk SSSR 111: 894.
7. Chinoy JJ 1962. Indian J Plant Physiol 5 (1,2): 172, FG Gregory Memorial Volume.
8. Chinoy JJ 1969. Ibid 12 (1,2): 67.
9. Chinoy JJ 1975. Proc. XII Internl Bot Cong Leningrad July 3-10 Vol II: Abstr p 283.
10. Chinoy JJ 1977. Indian J Plant Physiol XX (1): 1.
11. Chinoy JJ, Mansuri AD 1966. Proc Seminar on "Recent Advances in growth and development of plants" Bot Dept, Punjab University, Chandigarh India Dec 28-30, p 68-83, 1965.
12. Chinoy JJ, Nanda KK 1951. Physiol Plant 4: 209.
13. Chinoy JJ, Nanda KK 1951 a. Physiol Plant 4: 427.
14. Chinoy JJ, Nanda KK 1951 b. Ibid 4: 575.
15. Chinoy JJ, Nanda KK 1952. Ibid 5: 11.
16. Chinoy JJ, Saxena OP 1972. Proc 16th Int Seed Test Assoc (ISTA) Congr Washington DC June 7-12, 1971. 37(3): 903.
17. Chinoy JJ, Sharma SN 1957. Agrobiologia No. 3 (105): 32 (In Russian)

18. Chinoy JJ, Sharma SN 1958. Ibid No 2 (110): 53 (In Russian).
19. Chinoy JJ, Nanda KK, Sirohi GS, Sawhney KL 1959. Indian J Plant Physiol 2(1): 29.
20. Chinoy JJ, Jani BM, John D 1966. Proc Sem "Growth and development of plants", Bot Dept, Punjab Univ Chandigarh, India, Dec 28-30, 1965 p 169.
21. Chinoy JJ, Grover R, Sirohi GS 1957 a. Physiol Plant 10: 92.
22. Chinoy JJ, Nanda KK, Garg OP 1957 b. Ibid 10: 869.
23. Chinoy JJ, Grover R, Nanda KK 1958 a. Proc Sem "Modern developments in plant physiology" Uni Delhi, Delhi, India, Aug 26-30, 1957 p 12.
24. Chinoy JJ, Singh N, Sirohi GS 1958 b. Proc Ibid. 1957, p 45.
25. Chinoy JJ, Shah CK, Patel HD, Suthar HK 1965. Proc Internl Seminar "Tissue Culture" Baroda University, Baroda, India Jan 21-28, p 377, Ramakrishnan CV (ed).
26. Chinoy JJ, Jani BM, Raj KPS 1968. Sym "Natural Resources of Rajasthan" Jodhpur University, Jodhpur & UGC, Oct 23-26, Abstr No. A.19, p 17.
27. Chinoy JJ, Saxena OP, Abraham PG, Pandya RB, Dave IC, Gurumurti K, Sastri K 1969 a. Report PL 480 Research Project, FG-In-182 (A7-CR-87) p 1-350, Botany Department, University School of Sciences, Gujarat University, Ahmedabad - 380009, India.
28. Chinoy JJ, Saxena OP, Dave IC, Gurumurti K, Pandya RB, Abraham PG 1969b. Current Sci 38: 153.
29. Chinoy JJ, Patel HD, Jani BM, John D, Vyas AV, Raj KPS 1969 c. Report PL 480 Research Project FG-In-208 (A7-CR-108) p 1-456, Botany Department, Gujarat University, Ahmedabad - 380009, India.
30. Chinoy JJ, Saxena OP, Gurumurti K, Abraham PG, Pandya RB 1969 d. Science & Culture 35: 690.
31. Chinoy JJ, Saxena OP, Gurumurti K 1970 a. ISPP Symp "Plant Growth Regulators", Kharagpur, India, Jan 3-8, Abst 1.
32. Chinoy JJ, Saxena OP, Gurumurti K 1970 b, ISPP Symp Ibid Abst 2.
33. Chinoy JJ, Gurumurti K, Singh YD, Dave IC, Saxena OP 1971 a. Biophysik 7: 157.
34. Chinoy JJ, Gurumurti K, Singh YD, Saxena OP 1971 b. Indian J Plant Physiol. XIV: 232.
35. Chinoy JJ, Shah CK, Madhavan Unni NV, Neelkandan N 1971 c. Internl Sem "Physiology of Differentiation in plants" May 17-22, Simla, India.
36. Chinoy JJ, Singh YD, Gurumurti K 1971 d. Indian Agriculturist 15: 33.
37. Chinoy JJ, Gurumurti K, Saxena OP, Singh YD 1972 a. UGC All India Symp "Biology of Land Plants" Meerut Univ, Meerut, India, June 18-20, p 188.
38. Chinoy JJ, Singh YD, Laloraya MM, Saxena OP 1972 b. Sem "Current Trends in Plant Sciences" Delhi Univ, Delhi, Oct 3-10 p 47.

39. Chinoy JJ, Gurumurti K, Singh YD 1973 a. Ind. J Plant Physiol 16: 42.
40. Chinoy JJ, Vyas AV, Raj KPS, Jani BM 1973 b. Ind J Plant Physiol XVI: 35.
41. Chinoy JJ, Singh YD, Laloraya MM 1974. Vth Int Symp Mag Resonance Jan 14-18, Bombay, India, Abstr 17 A 64.
42. Chinoy JJ, Patel HD, Patel JN 1975. Vidya Gujarat University Journal 18: 69.
43. Chinoy JJ, Singh YD, Gurumurti K 1976. Ind J Plant Physiol XIX: 122.
44. Devay M 1965. Acta Agronomica Hung 14: 93.
45. Garg OP 1960. Ph.D. Thesis, Botany Department, University of Delhi, Delhi, India.
46. Garg OP, Chinoy JJ 1964. Indian J Plant Physiol 7 (1,2): 71.
47. Garg OP, Chinoy JJ, Nanda KK 1958. Fyton 10 (1): 17, III.
48. Garg OP, Chinoy JJ, Nanda KK 1958 a. Proc Sem "Modern Developments in Plant Physiol" Delhi Univ, Delhi, Aug 26-30, 1957 p 22.
49. Ghesani PI 1972. Ph.D. Thesis, Botany Department, Gujarat University, Ahmedabad - 380009, India.
50. Gurumurti K 1971. Ph.D. Thesis, Botany Department, Gujarat University, Ahmedabad - 380009, India.
51. Gurumurti K, Chinoy JJ 1974. Biochem Physiol Pflanzen (BPP) 165: 137.
52. Jani BM 1969. Ph.D. Thesis, Botany Department, Gujarat University, Ahmedabad - 380009, India.
53. John D, Jani BM 1969. Proc Natl Acad Sci India, 39 b, p 217, Abstr 132, 1968.
54. Lamparska K 1966. Proc II Symp Plant Gr Reg Zesz Nauk UMK 12 Biol VIII: 69.
55. Laloraya MM, Srivastava HN, Bakshi N, Shah N, Pandya KJ, Singh YD 1972. All India Symp "Biol Land Plants" Meerut, India.
56. Madhavan Unni NV 1971. Ph.D. Thesis, Botany Department, Gujarat University, Ahmedabad - 380009, India.
57. Mansuri AD 1965. Ph.D. Thesis, Botany Department, Gujarat University, Ahmedabad - 380009, India.
58. Michniewicz M 1961. Acta Agrobotanica 10 (2): 119.
59. Michniewicz M 1966. Proc II Symp Plant Gr Reg Zesz Nauk UMK 12, Biol VIII; 247.
60. Michniewicz M, Lamparska K 1965. Acta Soc Bot Pol 34: 375.
61. Nanda KK, Tayal MS 1976. Plant Physiol 57: 227.
62. Nanda KK, Grover R, Chinoy JJ 1958 a. Proc Sym "Modern Developments Plant Physiol" Univ Delhi Aug 26-30, 1957, p 63.
63. Nanda KK, Chinoy JJ, Sawhney KL 1958 b. Proc Sym "Recent Advances in the Study of Plant Metabolism", Univ Allahabad, Jan 1958, p 41.
64. Nanda KK, Sirohi GS, Sawhney KL, Chinoy JJ 1959. Ind J Plant Physiol 2: 52.
65. Neelkandan N 1971. Ph.D. Thesis, Botany Department, Gujarat University, Ahmedabad - 380009, India.

66. Pandya RB 1969. Ph.D. Thesis, Botany Department, Gujarat University, Ahmedabad - 380 009, India.
67. Patel HD 1967. Ph.D. Thesis, Botany Department, Gujarat University, Ahmedabad - 380009, India.
68. Pathan TS 1974. Ph.D. Thesis, Botany Department, Gujarat University, Ahmedabad - 380009, India.
69. Prasad TK, Dave YS, Mehta PM 1981. Biol Plant 23: 91.
70. Raj KPS 1971. Ph.D. Thesis, Botany Department, Gujarat University, Ahmedabad - 380009, India.
71. Sawhney KL 1958. Proc Sem "Modern Development in Plant Physiology" Delhi Univ, Delhi, Aug 26-30, p 81.
72. Sawhney KL, Chinoy JJ, Nanda KK, Sirohi GS 1959. Ind J Plant Physiol 2 (2): 116.
73. Saxena OP 1979. Varian Winter School in EPR, IIT, Bombay, Nov 19-30, p 1.
74. Saxena OP 1983. Adv in Pl. Reprod. Biol (Ed) Malik CP p 1, New Delhi: Kalyani Publishers.
75. Sen SP 1964. Ind J Plant Physiol 7 (1): 1.
76. Shah P 1972. Ph.D. Thesis, Botany Department, Gujarat University, Ahmedabad - 380009, India.
77. Shah H 1974. Ph.D. Thesis, Botany Department; Gujarat University, Ahmedabad - 380009, India.
78. Singh G, Saxena OP 1983. Silver Jubilee Seminar ISPP Jan 20-22, VII 7: 73 Abstr.
79. Sirohi GS, Sawhney KL, Chinoy JJ, Nanda KK 1959. Ind J Plant Physiol 2: 83.
80. Stanislawski JJ 1966 a. Proc II Symp Plant Growth Reg Zesz Nauk 12, Biologia VIII: 113.
81. Stanislawski JJ 1966 b. Ibid VIII: 131.
82. Suthar HK 1972. Ph.D. Thesis, Botany Department, Gujarat University, Ahmedabad - 380009, India.
83. Vyas AV 1971. Ph.D. Thesis, Botany Department, Gujarat University, Ahmedabad - 380009, India.

6. SENESCENCE

Senescence is the phase in the life of a plant wherein the termination of functional life occurs and the oxidative/catabolic processes are at a higher level than the reductive/anabolic ones. From this point of view, it is interesting to study the role of redox regulating substances such as AA in senescence.

Franke (1965, 1965 a) studied AA and DHA contents in autumnal leaves of deciduous trees and shrubs during and upto defoliation. AA decreased when leaves turned yellow and increased in those leaves turning red by anthocyanin formation. The AA/DHA ratio also decreased. At the very onset of defoliation, and thereafter, the oxidation of AA to DHA increased rapidly, and exceeded that of AA. From this experiment it was concluded that AA participated in the senescence phenomenon. Similarly, the trend in the concentration of AA observed in young and maturing tissues of Coffea, Phyllanthus and cotton plants revealed a close correlation between the content of AA in the tissue and its morphological status during the senescence or ripening phase. AA content was maintained at a higher level in the young floral buds of Coffea arabica. A progressive decline in its content coupled with a low AA oxidase activity was observed as the buds advanced towards the floral differentiation stages (Vasudev and Gopal, 1977). The concentration of AA in the young and maturing floral buds of Phyllanthus simplex Retz declined gradually with maturity (Dogra and Sinha, 1979). In comparison to cotton bolls retained on the plant, AA concentration was markedly lower in the abscissing bolls. The former were characterized by a higher AA content in the abscission zone in comparison with the proximal or distal regions (Varma, 1977). Thus tissues entering senescence or ripening phase have a low biosynthesis of AA as well as its decreased utilization.

Kinetin/cytokinin is assigned the function of retarding the degradation of chlorophyll and hence senescence (Lang, 1957; Mothes, 1961; Wong et al., 1961; Osborne, 1962; Srivastava, 1963; 1965; Waygood, 1965). During senescence, protein was rapidly lost, which in turn, disrupted the chloroplast resulting in the yellowing of leaf. A number of workers have reported that benzimydazole (Bz) also behaved like kinetin (Pearson et al., 1957; Mothes and Engelbracht, 1961). The metabolism of AA during senescence in the presence of kinetin and Bz in Datura has also been studied in this laboratory (Patel, K.R., 1973), using chlorophyll degradation as an indicator of senescence. It was observed that the kinetin floated leaf discs maintained higher level of chlorophyll for a longer period i.e. upto 8 days, as compared to those floated in distilled water (2 days). Bz maintained chlorophyll content upto the 5th day and then a sudden fall was observed.

The level of AA in kinetin treated discs were invariably higher as compared to that in the distilled water (D.W.) control. In both control as well as kinetin treated leaf discs, the trend was similar, as AA declined sharply upto second day and was then maintained practically at the same level upto the end of the experiment, whereas, a sharp decline occurred in the control. A similar trend was observed in Bz, but the difference was not so marked. The higher AA content found in kinetin treated discs appeared to be beneficial in retarding chlorophyll degradation. However, the utilization of AA (AAU) decreased slightly during the initial period of kinetin treatment and then subsequently increased and was maintained at the same level. On the 10th day, AAU was found to be as high as 60%. There was no difference in AAU of control and Bz. In control leaf discs the AAU was slightly increased in the initial stage and then declined sharply. On the 10th day, there was practically no utilization of AA.

Peroxidase activity increased in the initial period (upto 4th day) and then declined. On the 10th day very little activity was found in the control. In kinetin treated discs, peroxidase activity increased during the incubation period with a peak level on the 8th day and then declined, although the decline was not as much as was observed in the control. A similar trend of peroxidase activity was found in Bz, as that in the control, although the level of activity remained higher in the latter (Bz). Thus in Datura, Bz may not act as a senescence retardant. The high AA content, peroxidase activity, and AA - FR peroxidase in the kinetin treated discs, suggest that AA is being oxidized through the peroxidative mechanism to produce its free radical at a faster rate than in the control. Consequently, more rapid synthesis of nucleic acids and protein occur which retard senescence (Patel, K.R., 1973).

Sarkar and Chaudhari (1981) reported depletion in the content of chlorophyll, protein and glycolate with a parallel inhibition in the enzymic activity of glycolate oxidase and catalase in leaf discs of sunflower as a result of a 6 day dark incubation. Senescence was hastened by the addition of glycolate or H_2O_2 which further depleted the chlorophyll and protein contents, coupled with stimulation of glycolate oxidase activity. However, a significant decline in the depletion of chlorophyll and protein content was observed when the leaf discs were treated with ascorbate or 2-mercaptoethanol. The treatment also markedly deferred the onset of senescence. Siegenthaler (1969) suggested that ageing of photosynthetic apparatus may be considered as an attack on the energy transducing mechanism in chloroplast. The addition of ascorbate-dichlorophenol-indophenol (DCPIP) electron donor system restored the loss in NADP photoreduction, which suggests that ageing interferes with the integrity of photosystem II.

Garg and Kapoor (1972) demonstrated the effect of AA on retardation of leaf senescence in <u>Solanum melongena</u> Linn. The chlorophyll retention percentage was higher in the presence of growth regulators (GA_3, kinetin and AA). The effect of GA_3 was enhanced with increasing concentrations (1 to 500 ppm), while, kinetin and AA were effective at lower concentrations. The decrease in RNA and DNA contents were slower at low concentrations of AA compared with those in kinetin and GA_3. The total and protein nitrogen retention were found maximum at lower concentration of AA while kinetin and GA_3 helped in slowing down a fall in nitrogen content at higher concentration. Generally AA (20 ppm) was best for retarding leaf senescence while 50 ppm of kinetin also produced good results, whereas, AA was not effective at that contentration. Generally, small difference in the content of RNA, DNA, protein were found by using lower to higher concentration of GA_3. The mechanism of chlorophyll degradation and action of AA in the inhibition of yellowing in harvested parsley leaves was investigated (Yamauchi et al., 1980). AA contents and activities of AAO and chlorophyllase decreased markedly with increasing yellowing.

Studies on the effects of pretreatment with AA and sucrose on flowering and ripening (senescence) of wheat (Chinoy et al., 1969; Dave, 1969) revealed that pretreatment with AA (25 ppm) and sucrose (1.0%); early flowering occurred and senescence was delayed as manifested by the prolongation of the ripening period. This effect was very beneficial for grain filling processes. The grain weight of the main spike, extra spikes and total grain weight showed higher values in AA and sucrose treated plants as compared with those in the control or DW treated ones.

SUMMARY

AA content decreased in yellowing leaves and increased in those turning red by anthocyanin. AA/DHA ratio also decreased. At the very onset of defoliation, oxidation of AA to DHA increased rapidly, but after defoliation, DHA content exceeded that of AA. It was, therefore, concluded that, AA participated in the senescence phenomenon. The high AA content, its enzymic utilization, peroxidase and AA - FR peroxidase in the kinetin treated discs suggest that AA is being oxidized through the peroxidative mechanism to produce its free radical at a faster rate than the control and consequently more rapid synthesis of nucleic acids and protein occur which retards senescence. Total and protein nitrogen retention was at its best in treatment of leaf discs with lower concentration (20 ppm) of AA. On the other hand, GA_3, and kinetin were effective at higher concentration (50 ppm). Prolongation of the ripening period of wheat grain as a result of pretreatment with AA (25 ppm) and sucrose (1.0%) also helps in retarding senescence.

REFERENCES

1. Chinoy JJ, Saxena OP, Abraham PG, Pandya RB, Dave IC, Gurumurti K, Sastry K 1969. PL 480 Research Project Report, Project FG-In-182 (A7-CR-87) p. 1-350, Botany Department, University School of Sciences, Gujarat University, Ahmedabad - 380009, India.
2. Dave IC 1969. Ph.D. Thesis, Botany Department, Gujarat University, Ahmedabad - 380009, India.
3. Dogra JVV, Sinha SKP 1979. Comp Physiol Ecol 4 (2): 98.
4. Franke W 1965. Z Pflanzenphysiologie 53 (4): 289.
5. Franke W 1965 a. Z Pflanzenphysiologie 53 (4): 311.
6. Garg OP, Kapoor V 1972. J Exptl Bot 23 (76): 699.
7. Lang A 1957. Proc. Natl Acad Sci (Wash) 43: 709.
8. Mothes K 1961. Ber Deutsch Bot Ges X 74: 24.
9. Mothes K, Engelbracht L 1961. Phytochemistry 1: 58.
10. Osborne DJ 1962. Plant Physiol 37: 595.
11. Patel KR 1973. M.Sc. Dissertation, Botany Department, Gujarat University, Ahmedabad - 380009, India.
12. Pearson C, Samborski DJ, Forsyth FR 1957. Nature (Lond.) 180: 1294.
13. Sarkar U, Chaudhari MA 1981. Can J Bot 59 (3): 392.
14. Siegenthaler Paul-Andre 1969. Plant and Cell Physiol 10 (4): 811.
15. Srivastava BIS 1963. Arch Biochem Biophys 103: 200.
16. Srivastava BIS 1965. Arch Biochem Biophys 110: 97.
17. Varma SK 1977. Indian J Plant Physiol 20 (1): 85.
18. Vasudev N, Gopal NH 1977. Turrialba 27 (4): 355.
19. Waygood ER 1965. Plant Physiol 40: 1242.

20. Wong D, Hao MSH, Waygood ER 1961. Canad J Bot 39: 1029.
21. Yamauchi N, Hamaguchi S, Ogata K 1980. Engei Gakkai Zasshi 49 (3): 414.

7. ROOT NODULES

Nitrogen fixation mainly involves the reduction of N_2 to NH_3; it is worthwhile considering the role of AA in root nodulation and nitrogen fixation. Intensive work has been undertaken by Garg and his associates (Swaraj and Garg, 1970) on the role of AA in symbiotic nitrogen fixation and nodulation in gram, Cicer arietinum cultivar G 24, when applied through the rooting medium it considerably enhanced nodule initiation. The fresh and dry weights as well as nodule numbers were also increased. The nodules were more pink. Nitrogen content was also more in AA treated plants, and thus all plant organs had more weight. Earlier, pod formation and higher seed weight were noticed in AA treated plants compared with that in controls. AA plays an important role in nodulation by promoting nucleic acid and protein synthesis and by virtue of its action as a powerful electron donor, AA increased loci of growth and cell division and thus there was not only an enhancement in the nodulation but also the functional efficiency of nodules. Moreover, lower concentration of AA was more effective than higher ones and this also supports the fact that AA may function as a growth hormone in the processes of nitrogen fixation. Koch et al. (1967) reported considerable increase in molecular nitrogen fixation by AA when supplied to nodule bries and cell free bacteroid extracts.

1. EFFECT OF EXOGENOUSLY APPLIED AA ON NODULATION

1.1. Nodule initiation

In the case of Bengal gram (Cicer arietinum L.) treatments given through the rooting medium were more effective than feeding through the apex. Nodule initiation in Bengal gram treated with AA 20 ppm and 100 ppm started 75 and 74 days after sowing against 92 days in the water treated controls (Swaraj and Garg, 1970). Nodulation started 7 days earlier in berseem treated with 50 mg of AA per pot (Garg, B.K. 1972). In the case of lucerne, where the dose of AA applied was split into 4 equal instalments totalling 50, 100 and 200 mg per pot, nodule initiation took place 6 to 7 days earlier (unpublished data). These experiments were conducted in pots where a limited number of plants could be sampled each day for observing nodule initiation.

1.2. Nodule production

AA treatment greatly enhanced nodule production. Application of 20 and 100 ppm AA resulted in a 2-3 fold increase in nodule number of *Cicer arietinum* (Swaraj and Garg, 1969; 1970). Swaraj and Garg (1971) noted a spectacular increase in nodule numbers, while at 200 mg/l, the promotive effect was almost lost. Towards the last stages of growth when the controls as well as 200 mg AA treated plants showed a decline in nodule number, those treated with 20 and 100 ppm AA showed either some increase or sustained the previous gains. It can be inferred from these observations that AA treatments prolong the span of nodule formation. In the case of pea, the promotive effect of AA increased upto 200 ppm but when the treatments were given as foliar sprays (Kumar, 1973). However, the promotive effect decreased with further increase in the amount of AA applied. The promotive effect of AA on nodule number was the most prominent in case of berseem where it resulted in a 6 and 10 fold increase (Garg, B.K., 1972). Thirty to sixty percent increase in the number of nodules was observed in lucerne and groundnut when the amount of AA applied ranging between 50 and 200 mg was split into three or four equal instalments.

Tonzig and Bracci (1951) on the other hand, reported inhibition or complete suppression of nodulation on treatment with AA, but they used clearly inhibitory concentrations ranging between 0.05% and 0.1%. An inhibitory effect of high levels of exogenously supplied AA on growth of a number of plants had also been reported by Tonzig and Marre (1961), who ascribed this inhibition to increased levels of DHA. On the contrary, application of low concentrations of AA have been reported by Chinoy et al (1957) and Garg et al. (1958) to result in an increase in growth and yield characters.

2. EFFECT OF AA ON NODULE WEIGHT

The increase in nodule number with AA treatments was also accompanied by an increase in their fresh and dry weights. In Bengal gram, treatment with AA (20 and 100 ppm) resulted in a much higher nodule weight than that in the control (Swaraj and Garg, 1970). These authors (Swaraj and Garg, 1971) reported that the effects of various AA treatments did not strictly parallel with the number of nodules. The increase in weight of nodules with AA (200 ppm) was relatively much higher than in their number with this treatment. At certain stages of growth, treatment with AA 25, 50, and 100 mg per pot almost doubled the nodule weight in cowpea. Kumar (1973) observed that the promotive effect of AA on nodule weight increased even upto a concentration of 500 ppm, but 1000 ppm appeared to be supra-optimal though still promotive with reference to the control. Treatment of berseem with AA 50 mg/pot resulted in 6 to 12 fold increases in nodule weight but the promotive effect declined with further increase in the amount of AA applied (Garg, B.K., 1972). In lucerne and groundnut also, the nodule weight increased considerably on treatment with AA (Goswami and Garg, 1978).

Swaraj (1972) studied the effect of AA application on nodule size, as bigger nodules in comparison with smaller ones on a weight to weight basis are known to possess greater nitrogen fixing efficiency. It was observed that at the stage of floral initiation, the weight of the red nodules of diameters ranging from 4-6 mm, increased with all the AA treatments, i.e., AA 25, 50 and 100 mg/pot. The weight of the smaller brown nodules of 4 mm diameter did not differ by the various treatments, but the weight of the larger diameter nodules was much higher with AA treatments than the water treated controls. This could be accounted for by the early onset of nodule initiation with these treatments. With the onset of fruiting, a fall in the formation of new nodules was registered which resulted in a corresponding decline in the weight of red nodules of all diameters. At this stage also, the combined weight of red nodules with AA treatment was 3-4 times that in the control, thus showing that AA treatments also prolonged the span of fresh nodule formation. Higher weight of brown as well as green nodules in response to AA treatments showed that the overall nodule production was enhanced considerably.

3. ENDOGENOUS AA CONTENT OF THE NODULES

In all the plants studied so far (gram, pea, cowpea, berseem and lucerne), the AA content of the red functional nodules was much higher than that of the nonfunctional green nodules (Swaraj, 1968; Kumar, 1973; Goswami and Garg, 1978). The onset of browning which is the first step towards nodular senescence, resulted in a decline in their AA content leaving low to negligible AA content in the green nodules. These results have also been confirmed by histochemical studies on *in situ* localization of AA in different regions of the nodules of Cicer arietinum (Swaraj, 1972). The young developing nodule or the distal region of a well grown nodule contained a very high amount of AA, especially in the meristematic zone and the central bacteroidal region. The latter region was also rich in the red functional form of leghemoglobin and was loaded with AA which was the highest among all other regions of the nodule. The onset of browning resulted in a decrease in the amount of AA in the cortical as well as central regions of the nodule. There was hardly any AA left in the completely senescent proximal region of the nodule where leghemoglobin was completely oxidized to the nonfunctional form. The maturation of the host plant, which was accompanied by a decrease in the nitrogen fixing efficiency, resulted in a marked decrease in the AA content of even red nodules.

Exogenous application of AA increased the endogenous AA content of all classes of nodules. The promotive effect of exogenous AA on the AA content of the nodules was considerably reduced towards senescence. The AA content of the nodules of berseem and lucerne was much higher compared with that of their roots (Garg, B.K., 1972; Goswami and Garg, 1978). With the ageing of the host plant, AA content of nodules declined to a level either equal to or even lower than that of the root.

The treatment of the legume with RNA hydrolysate as well as nucleic acid bases like adenine and thymine which enhanced nitrogen fixation considerably, also increased the AA content of the nodules. Keeping the plants in the dark for 2 to 3 days markedly reduced the AA content of all its parts and during senescence, the AA content of the nodule was 2-4 times of that in the roots (Virtanen and Jorma, 1945).

It is clear from these observations that a close correlation exists between the AA content of the nodules and their nitrogen fixing ability.

4. EFFECT ON LEGHEMOGLOBIN CONTENT OF THE NODULES

The importance of leghemoglobin in its reduced functional form for nitrogen fixation has already been pointed out earlier. The pigment status of the nodules is thought to be a rough index of the functional efficiency of the nodules. AA treatments increased the leghemoglobin content of the young nodules of gram, cowpea, berseem and lucerne (Garg and his associates). The favourable effect of AA treatment on leghemoglobin was considerably reduced as the plants approached maturity by which time the AA content also decreased considerably. Nowotny-Mieczynska (1952) reported that addition of AA and cobalt chloride in suitable concentration delayed greening of leghemoglobin, thus suggesting that AA application could increase the functional span of the nodules. Similar observations on nodules of Bengal gram were made by Swaraj and Garg (1970).

5. EFFECT OF AA ON NITROGEN CONTENT OF VARIOUS PLANT ORGANS

5.1. Nodules

AA treatment resulted in prominently higher values for nodule nitrogen at all sampling stages (Garg, B.K., 1972; Swaraj and Garg, 1969; Kumar, 1973). In case of gram, a higher peak value was seen on treatment with 100 ppm compared with its lower concentration of 20 ppm but the decline occurred earlier by the higher concentration treatment. In case of berseem also, the nodule nitrogen showed 3-4 fold increase in response to AA treatments. As the plants reached maturity, the promotive effect of AA on the nodule nitrogen was considerably reduced. The effectiveness of AA treatment with the passage of time decreased as the amount applied was raised above 50 mg/pot.

In *Vigna sinensis* L., at a stage when flowering had just started, the nitrogen content of the nodules with diameters less than 4 mm increased by treatment with 100 mg AA/pot to seven times that of the control. The nitrogen content of the red nodules with diameters above 4 mm also increased considerably with AA treatments. When fruit setting started, the regularity in the effect of AA on nodule nitrogen was lost, but still the nitrogen content of all types of nodules with AA treatment was much higher than that of control. It was observed that nitrogen content of even the green nodules was very high and sometimes even higher than that of the other categories of nodules. According to Virtanen and Mietinen (1963) this appears to be due to the blocking of the vascular system of the senescent nodules and the consequent failure of translocation of already fixed nitrogen to other plant organs. Nitrogen content of both brown as well as green nodules increased with increase in their size. The differences in nodule nitrogen with various treatments diminished considerably with maturity of the plants.

5.2. Roots

The treatment with 100 mg AA/l did not show any appreciable difference from the control in root nitrogen of gram plants (Swaraj and Garg, 1970). During the period of high fixation, the root nitrogen with 20 ppm AA was consistently higher than in the control. In berseem AA treatment with 50 - 200 mg per pot resulted in a 2-4 fold increase in root nitrogen, being the highest with 50 mg per pot (Garg, B.K., 1972). A notable increase in root nitrogen with AA treatment was also observed in cowpea (Swaraj, 1972).

5.3. Stem

Starting with slightly lower values than the control, a general increase in stem nitrogen was observed with AA treatments in gram (Swaraj and Garg, 1969; 1970). A marked increase in stem nitrogen as a result of AA treatment was observed in berseem as well (Garg, B.K., 1972). On the earlier sampling stage, the nitrogen content of stem was relatively low in cowpea, as most of the nitrogen fixed by the nodules was being transported to the more actively metabolizing leaves. However, at this stage also the nitrogen content of the stem with AA treatments was much higher than that in the control. With the onset of fruiting, the stem nitrogen increased considerably under all treatments, since it coincided with rapid translocation of nitrogen from leaves to the developing fruits via the stem. At this stage AA treated plants had much higher nitrogen than controls. Kumar (1973) reported increased stem nitrogen with AA treatments as high as 50 ppm. With further increase in AA concentration, the promotive effect diminished but was still more than the control.

5.4. Leaves

AA treatments resulted in a steep increase in the nitrogen content of leaves of gram (Swaraj and Garg, 1969; 1970). The peak values for leaf nitrogen corresponded with relatively low values of stem nitrogen, thus showing rapid initial migration of the fixed nitrogen to leaves. This was followed by a notable decrease in leaf nitrogen which now coincided with a rise in stem nitrogen. This is again indicative of a rapid migration of nitrogen from leaves to reproductive parts through the stem. Thus the sinks for the fixed nitrogen shifted with the developmental stage of the plant. Similar results were obtained in Vigna sinensis L. (Swaraj, 1972). In berseem however, both leaf and stem nitrogen showed a consistent increase during the vegetative growth period, but towards the late stages of growth, the decrease in nitrogen in other organs coincided with increased accumulation in the developing fruits. The promotive effect of AA on leaf nitrogen of all these species decreased above 100 ppm. In pea also the nitrogen content of leaves on various sampling stages was also higher with AA treatments than the control (Kumar, 1973).

Ascorbic acid levels measured in leaves of pea cultivars varying in resistance to a pathogen Peronospora revealed that the decrease in AA content was much less in case of the resistant varieties in comparison to the same in pathogen resistant varieties. For example, a weak infection of the resistant variety, Pauli, the decrease in AA content was by a factor of 2 - 3.2 less than in similarly infected sensitive variety, Cherinogovskii 190. Further, Koshevskii and Basyuk (1979) attributed the decrease in the resistance of pea plants to Peronospora infection and enhancement in the pathological processes to a rapid decrease in the AA content of the leaves.

5.5. Seed nitrogen

AA treatment of 20 - 100 mg per pot almost doubled the seed nitrogen in cowpea (Swaraj, 1972). 50 mg/pot AA resulted in a 50% increase in fruit nitrogen in berseem on a per plant basis (Garg, 1972). The promotive effect decreased with higher concentrations of AA. Kumar (1973) also observed an increase in nitrogen content of flowers and fruits of pea on spraying the plants with different concentrations of AA, 500 ppm being the optimum.

5.6. Total plant nitrogen

The total plant nitrogen with AA treatments was much higher than in the control (Swaraj and Garg, 1969; 1970).

The peak value with 20 and 100 ppm was of the same order of magnitude, but was attained earlier with the latter concentration. Treatment of cowpea with AA concentrations ranging between 25 - 100 mg/pot resulted in a 30 - 100% increase in plant nitrogen (Swaraj, 1972). The nitrogen content of berseem plants treated with AA (50 mg) was twice that of the water treated ones (Garg B.K., 1972). In lucerne there was consistent increase in plant nitrogen with increase in the amount of AA applied from 50 - 200 mg per pot given through the rooting medium in four instalments (Goswami and Garg, 1978). In pea also, AA treatments increased the nitrogen content of the plants over the water treated controls (Kumar, 1973). The effect of AA in this plant increased from 50 - 500 ppm, being maximum with 500 ppm.

5.7. Growth and yield

Increase in nodulation and nitrogen fixation with AA treatment leads to a promotion of growth and yield characters as well. Treatment with AA resulted in a considerable increase in fresh and dry weight of roots, stem and leaves in gram, cowpea, pea, berseem and lucerne (Garg and his co-workers).

Pod formation started earlier in AA treated gram plants. The weight of both pods as well as seeds also increased (Swaraj and Garg, 1970). AA application between 25 and 100 mg per pot almost doubled the seed dry weight in cowpea (Swaraj, 1972). Foliar spraying of pea with AA also led to an enhancement of flower and fruit production, which was maximum with 500 ppm (Kumar, 1973). Increase in fodder yield of berseem was 5-6 fold in pot culture but under field conditions, the increase was only an average of 10% at each of the four cuttings taken, and was maximum at the second cutting.

6. ASCORBIC ACID AND NITROGEN FIXATION BY BLUE GREEN ALGAE

The blue green algae were able to grow well even in a medium lacking combined nitrogen as they were capable of utilizing molecular nitrogen. Wahal et al. (1973) observed that blue green alga Anabaena ambiqua grown with molecular nitrogen was always richer in AA than that grown with combined nitrogen. The alga grown with combined nitrogen when transferred to a medium lacking it, caused an increase in the endogenous AA. Conversely, when material cultured on the nitrogen free medium was suspended in the medium containing NH_4Cl, there was a decrease in cellular AA. Talpasayi (1967) demonstrated the presence of AA by cytochemical methods in heterocysts of Anabaena cylindrica which are considered to be seats of nitrogen fixation. Their results were later confirmed by Stewart et al. (1969). Wahal et al. (1973) showed that the ascorbic acid oxidase activity in cells grown with molecular nitrogen was 16.7% higher than that in those grown with combined nitrogen. They also reported that exogenously added AA upto 0.5 ppm increased the heterocyst frequency to 3 times that of the control. These observations clearly showed that AA metabolism was closely related with the process of nitrogen fixation in the blue green alga as well.

7. MODE OF ACTION OF AA IN NODULATION AND NITROGEN FIXATION

The precise role of this vitamin in process of nodulation and nitrogen fixation is not yet fully understood. One of the modes of action suggested for plant hormones is that it invokes enhanced gene activity leading to increased rates of nucleic acid and protein synthesis. AA has been reported to enhance protein and nucleic acid synthesis (Chinoy, 1962; Price, 1966; Garg and Kapoor, 1972) and thus increase the rate and loci of cell division (Garg, 1968; Garg, personal communication). In order to test the applicability of this hypothesis of the action of AA, treatments with RNA, RNA hydrolysate and nucleic acid bases have been tried. Considerable promotion of nodulation and nitrogen fixation on the addition of RNA and RNA hydrolysate to the rooting medium was observed in cowpea and lucerne (Goswami and Garg, 1978; Swaraj, 1972). The effect of addition of RNA was almost comparable to AA treatments. Ratner and Samoilova (1966) also reported an increase in nodulation and nitrogen fixation on treatments of soya with RNA, RNA hydrolysate and bases of RNA. Even application with single base like thymine, adenine or uracil was found to be effective in increasing the nodulation in berseem and lucerne (Garg B.K., 1972). However, the promotive effect of single base application is difficult to explain since an increase in the availability of a single base to the endogenous base pool cannot account for the enhanced rates of nucleic acid synthesis unless accompanied by a corresponding increase in the other bases.

8. ASCORBIC ACID IN RELATION TO NITROGEN METABOLISM

Virtanen (1949), Virtanen and Saubert-Von Hausen (1949) reported very favourable effect of AA on growth of plants when AA was added to nitrate solutions but upon addition of AA to ammonium solution, this type of effect was not observed. Reductones such as cysteine and glutathione also stimulated the nitrate reduction. According to these workers the redox potential within the plant was lowered by these substances and thus nitrate reduction took place. Virtanen (1950) reviewed the work done on the significance of AA and SH compounds on the nitrate reduction in green plants. AA was abundantly formed during seed germination. The SH content also rose rapidly during that time. These two substances have great significance to nitrate reduction. Addition of AA or SH compounds to the nitrate nutrient solution in which seedlings were grown could replace endosperm or cotyledon. From this result, the author concluded that the main function of the reserve nutrition of the seed was to produce reducing substances. These substances adjusted the redox potential to a suitable level so that the reduction of nitrate in the plant was made possible. Virtanen and Saubert-Von Hausen (1951) also found reduction of nitrate, mediated by uptake and metabolism of iron.

Manganese should also be considered in this connection. Tonzig and Bracci (1951; 1951 a) studied the influence of AA on the formation of root nodules of leguminous plants infected with rhizobium and nodule forming avtivity of Rhizobium leguminosarum.

Maravalhs (1951) reported that AA was responsible for the reduction of nitrate in the aerial parts but not for potato tubers. In the latter, nitrate reduction was activated in KNO_3 solution in the presence of light.

Reduction of nitrate by tubers of high reducing power during germination and by dormant discs with low reducing power, was depressed 70 - 95% in presence of 5, 10, 20, 40 mg % of AA. According to Maravalhs (1951), AA was responsible for the reduction of nitrate in the aerial parts but was not valid for potato tubers.

The role of AA, hydrogen peroxide and iron in the reduction of nitrates by Chlorella was studied by Trubachev (1968). H_2O_2 is a normal intermediate product of aerobic respiration and reduction of nitrate to ammonia is closely related to aerobic condition. In nitrite reduction, AA could serve as a hydrogen donor and nitrite as hydrogen acceptor. Gurevich and Trubachev (1964) established a chemical model for nitrite reduction in which AA participated as a hydrogen donor. H_2O_2 and ferric sulphate (as a peroxidase catalyst) also participated in this reaction which could be explained as follows: H_2O_2 activated by a peroxidase biocatalyst, monovalently oxidizes AA with the formation of its free radical which reduces the nitrite. According to the data of Einor et al (1964), AA is a natural substrate of Chlorella peroxidase and participates in oxidative processes during respiration. AA is oxidized by peroxidase within a broad range of pH. Pyrogallol is not oxidized by this enzyme in acidic medium. Kolesnikov and Einor (1964) reported the acceleration of the oxidation of $NADH_2$ by cell free extracts from Chlorella. This was accompanied by MDHA with the participation of its reductase. DHA did not accelerate the oxidation of $NADH_2$. In Chlorella, AA and AA oxidases (general) as well as H_2O_2 were present. Nitrite was present during the reduction of nitrate. Ionic iron could work as a peroxidase catalyst in induced reduction reaction. The addition of AA to nitrate mixture intensified the reduction of nitrates to ammonia in Chlorella. H_2O_2 enhanced this effect. Trubachev (1968) put forward the hypothesis that nitrates were reduced with the participation of the induced reduction reaction. H_2O_2 formed during aerobic respiration or if introduced from outside being activated by a peroxidase biocatalyst (iron ion), monovalently oxidized AA forming its free radical which reduced the hydrogen acceptor, nitrite arising in the reduction of nitrates.

Van der Ben (1971) reported that several species of desmids were able to reduce nitrates to nitrite *in vivo*. This reducing property depended upon a number of factors such as pH, temperature, light and nitrate concentration.

Stranrastrum cultures deprived of N_2 after the exponential growth phase were unable to reduce nitrites, but then nitrate reduction was enhanced. In the presence of AA the nitrate and nitrite reducing activity of the algae was often considerably modified.

Pietz (1938) suggested that variations in the redox potential of nodules towards the more oxidation side may be the cause of their greening and cessation of nitrogen fixation. A decrease in the AA content of nodules with the onset of browning and its complete disappearance from the green nodules, suggested that AA could be the substance controlling the redox potential. The presence of AA in heterocysts of the blue green alga Anabaena cylindrica (Talpasayi, 1967) and higher amount of cellular AA in Anabaena ambiqua grown with molecular N_2 than those grown with molecular NH_4Cl as the nitrogen source, indicate that higher AA content in cells was accompanied with higher nitrogen fixation.

It has been well recognized by a number of workers that ATP is necessary for nitrogen fixation to proceed. As photo-oxidation of AA has been reported by Trebst and Pistorius (1969) to be coupled with photo-synthetic phosphorylation, it might be increasing the rate of nitrogen fixation by supplying greater amounts of ATP.

Bhatt et al. (1981) showed that in shoots of all three F_1 hybrids of Sorghum bicolor (L) Moench, both chlorophyll a and b contents and AA turnover showed better parental and / or mid-parental heterosis. Since ascorbate participates in photosynthesis and nitrate reduction, it was suggested that it may be associated with the manifestation of hybrid vigour in Sorghum.

The *in vitro* studies carried out by Koch et al. (1967) suggested that AA protected the nitrogenase from denaturing effect of phenolase which led them to add AA to the extraction medium in their experiments. However, Wong et al. (1971) put forward a more direct role for AA in nitrogen fixation. Their studies on an *in vitro* system containing bacteroidal ferredoxin and active nitrogenase suggested that photo-oxidation of AA, which was linked with photosynthetic electron transport chain, resulted in reduction of ferredoxin, thus generating the necessary redox power required for nitrogen fixation. Similar suggestions were made by Smith et al (1971) for Anabaena cylindrica. A more direct link between AA and nitrogen fixation had been shown by Koch et al. (1970). They characterized and purified a nonheme iron protein from glycine nodule bacteroids that function effectively in the light dependent transfer of electrons from AA via photosystem I to purified bacteroid nitrogenase. This nonheme protein however, proved to be ineffective as an electron transport factor in the phosphoroclastic breakdown of pyruvate. The bacteroid nonheme protein also failed to function in the photoreduction of NADH in the presence of spinach chloroplasts thus differing from the ferredoxin isolated from Azotobacter. AA, therefore, might function as a reductant to nitrogenase through photosystem I.

8. SUMMARY

The significance of AA and -SH compounds in nitrate reduction within green plants was elucidated. The role of AA, hydrogen peroxide and iron in the reduction of nitrate in plants has also been outlined. Hydrogen peroxide, activated by a peroxidase biocatalyst, monovalently oxidizes AA with the formation of its free radical. This 'induced reduction reaction' participates in the reduction of nitrites formed during the reduction of nitrates. Result of several experiments suggested that AA beneficially affected nodulation in leguminous plants by promoting nucleic acid and protein synthesis and thereby increased loci of growth and cell nitrogen when AA was supplied to nodule bries and cell free bacteroid extracts. AA treatment resulted in appreciably higher values for nodule nitrogen in a number of plants at all sampling stages.

Nitrogen content of nodules, root, stem, leaf and seed of various plants was invariably higher in plants treated with AA. The total nitrogen content of the plant was also increased considerably under AA treatment.

The increase in nodulation and nitrogen with AA treatment also led to promotion of growth and yield in cowpea, gram, pea, berseem and lucerne. The blue green alga Anabaena ambigua grown with molecular nitrogen was always richer in AA than that grown with combined nitrogen. Moreover, ascorbic acid oxidase activity in cells grown in molecular nitrogen was 16.7% higher than that of cells grown with combined nitrogen. Addition of 0.5 ppm AA to the cells increased the heterocyst frequency to 3 times that of the control. Considerable promotion of nodulation and nitrogen fixation was observed in cowpea and lucerne with the addition of RNA and RNA hydrolysate. Even application of a single base like thymine and adenine or uracil had promotive effect on nodulation in berseem and lucerne. AA protected the enzyme nitrogenase from the denaturing effect of phenolase. The photo-oxidation of AA was linked with the photosynthetic electron transport chain. It results in reduction of ferredoxin thus generating the necessary redox power required for nitrogen fixation.

REFERENCES

1. Bhatt KC, Vaishnav PP, Singh YD, (Late) Chinoy JJ 1981. Ann Bot 47: 321.
2. Chinoy JJ 1962. Indian J Plant Physiol 5: 172 FG Gregory Memorial Volume.
3. Chinoy JJ, Nanda KK, Garg OP 1957. Physiol Plant 10: 869.
4. Einor LO, Tupik ND, Kolesnikov PO 1964. Dopovidi An USSR No 2, 238.
5. Garg BK 1972. M.Sc. Dissertation, Haryana Agri University, Hissar, India.
6. Garg OP 1968. Proc Interni Symp "Plant Growth Subs" Sircar SM (ed), 211, Calcutta University.
7. Garg OP, Kapoor VA 1972. J Exptl Bot 23: 699.
8. Garg OP, Chinoy JJ, Nanda KK 1958. Fyton 10: 17, III.
9. Goswami CL, Garg OP 1978. Proc. Symp Recent Advances in Plant Physiol in India to felicitate Prof. JJ Chinoy & Sem Plant Growth & Metab Jan 1-2, Ahmedabad, India, Abstr No. 23, p 42.
10. Gurevich AA, Trubachev IN, 1964. Dokl Akad Nauk SSSR, 157: 467.
11. Koch B, Evans HJ, Russell SA 1967. Plant Physiol 42: 466.
12. Koch B, Wong P, Russell SA, Howard R, Evans HJ 1970. Biochem J 118: 773.
13. Kolesnikov PA, Einor LO 1964. Biokhimiya 29: 402.
14. Koshevskii II, Basyuk GF 1979. Nauk Pr-Ukr Sil's'kogospod Akad 230: 72.
15. Kumar S 1973. Ph.D. Thesis, Haryana Agri Univ Hissar, India.
16. Maravalhs N 1951. Dusenia 2: 141.
17. Nowotny-Mieczynska A 1952. Acta Microbiol Polon 1: 43.
18. Pietz J 1938. Zentr Bakteroil Parasitank Abt II 99: 1.
19. Price CE 1966. Nature (Lond.) 212: 1481.
20. Ratner EI, Samoilova SA 1966. Soviet Plant Physiol 13: 914.
21. Smith RV, Nay RJ, Evans MCW 1971. Biochim Biophys Acta 253: 104.
22. Stewart WDP, Haystead A, Pearson HW 1969. Nature (Lond.) 224: 226.
23. Swaraj K 1968. Unpublished record.
24. Swaraj K. 1972. Ph.D. Thesis, Haryana Agri Univ, Hissar, India.
25. Swaraj K, Garg OP 1969. Indian J Plant Physiol 12: 38.
26. Swaraj K, Garg OP 1970. Physiol Plant 23: 889.
27. Swaraj K, Garg OP 1971. Proc Interni Sem Physiology of Differentiation in Plants, Simla, India, May 17-20.
28. Talpasayi ERS 1967. Current Science (India) 36: 190.

29. Tonzig S, Bracci L 1951. Nuovo Giorn Bot Ital 58: 258.
30. Tonzig S, Bracci L 1951 a. Ibid 58: 237.
31. Tonzig S, Marre E 1961. In: Plant Growth Regulation, Proc IVth释Internl Cong Plant Growth Regulation, The State Univ Press, Ames, Iowa, USA: p 725.
32. Trebst A, Pistorius E 1969. Biochim Biophys Acta 131: 580.
33. Trubachev IN 1969. Soviet Plant Physiol 15: 549.
34. Van der Ben C 1971. Hydrobiologia 38: 165.
35. Virtanen AI 1949. Experientia (Basel) V/8: 313.
36. Virtanen AI 1950. Acta Agri Scand 1: 1.
37. Virtanen AI, Jorma J 1945. Suom Kenistilehiti (B) 18: 50.
38. Virtanen AI, Mietinen JD 1963. In: Plant Physiol Steward FC (ed), Academic: New York III: 118.
39. Virtanen AI, Saubert-Von Hausen S (1949). Z Pflanzenernahr Dung V Bodenke 45: 1.
40. Virtanen AI, Saubert-Von Hausen S 1951. Acta Chem Scand (Copenh) 5: 638.
41. Wahal CK, Bhattacharya NC, Talpasayi ERS 1973. Physiol Plant 28: 424.
42. Wong HJ, Klucas R, Russell S 1971. Plant Soil, Sp Volume, p 523.

8. DROUGHT RESISTANCE AND PRESOWING TREATMENTS

Presoaking or presowing hardening (pretreatments) of seeds with various chemical compounds is now known for more than 200 years (Duhameldu-du-Monceau, 1763). May et al (1962) had surveyed the Russian literature since details on this aspect were not available to the English-speaking scientific community. This method for increasing the drought resistance was suggested by Henckel (Henkel, Genkel) and Kolotova (1934). However, there was not much interest among scientists of other countries, although it was a new technique of obvious importance in the agriculture of semiarid areas and the findings were reported from time to time (Henckel, 1946; 1952; 1961; 1964). Henckel (1964) concluded that presowing hardening was the result of extensive physiological reorganizations induced by the dehydration process. Protoplasmic colloids showed increased hydrophilic viscosity and protoplasmic elasticity. This could be explained only if the energy level of the plant was considered which was due to the compounds containing high energy bonds.

Presowing hardening of seeds may be described as a treatment preliminary to sowing during which seeds are moistened (70% water) and again dried (once or a number of times, generally three), according to Henckel, to activate certain physiological processes, viz., increased tissue hydration, redistribution of nutrient reserves, heightened respiratory activity and in consequence of this, enhanced seedling growth (Dawson, 1965; May et al., 1962; Korneev, 1963; Henckel, 1964; Salim and Todd, 1968; Chinoy et al., 1969; 1969 a; Heydecker,1973; Heydecker et al., 1973; 1974; Saxena, 1974, 1979). This would enable the resulting plant to resist overheating and dehydration, endure drought better and increase yield of crops especially under arid conditions. Similarly, the hardened plants showed induction of greater tolerance against frost. The physiology of drought resistance in wheat has been extensively studied by Chinoy (1960; 1961; 1961 a).

As germination advances, the resistance of the embryos to drying decreases and the likelihood of permanent injury from drying increases. Thus the degree of hardening induced would be greater in more advanced embryos at the time of drying. Hence an optimal stage of germination for imposing drying treatment may well be best comprise between these two conflicting tendencies (ref. May et al., 1962). Kydrev (1959) claimed that wheat plants grew most vigorously if the grain was dried when the first root was about to emerge. A slow drying resulted in greater responses than rapid drying. Slow drying was likely to continue the hardening process in the dry period. The seeds could be hydrated for eight hours at 24 - 25° C and even upto twenty hours at lower temperatures when only proteins were being synthesized and then safely dehydrate the embryo upto that stage (8 hours) though not beyond, as it was impossible for them to repair the damage once DNA replication had started. Lettuce seeds were wetted for one to twenty four hours and then again dried. After six hour imbibition, it was no longer possible to redry them without losing the advantage associated with the process of pre-wetting (Heydecker, 1973; Osborne, 1973). In many cases evidences of enhanced performance were lacking during adverse drought and extremes of temperature conditions and hence the term 'advancing' is preferred over hardening. Advancing is the same process as hardening but with the restricted aim of enabling the seed to pass through the first stage of germination, short of radicle emergence, before the seeds are actually grown.

The factors which influence the plant during the early stages of development, may also profoundly influence its subsequent life history. The size of the seeds, degree of its maturity, soaking seeds in water or in solutions of various chemicals before sowing and temperature are important factors. Similarly, it was emphasized that further work was needed on detailed changes occurring during germination and hardening treatment and on defining the developmental stages of growth and amount of moisture at which the response to the treatment was greatest (May et al., 1962). Heydecker (1973) cautioned that the initial process must be carried out under strictly controlled and hygenic conditions in order to avoid failure. Roberts (1948) emphasized that the temperature during dehydration must not exceed 20° C. The temperature of dehydration, period of treatment and stage of seedling development are critical (Crocker and Barton, 1957). Submergence of lima beans during first ten minutes of imbibition appeared to condition the axes to reduce growth and ribonuclease development at a later period and this had adverse effect even on processes leading to final growth and development (Roos and Pollock, 1971).

The loss of essential materials (metabolite, enzymes and hormones, etc.) from the seeds most probably is one of the chief causes of soaking injury. In imbibition of pea, growth hormones placed in solution under aeration during germination may partly replenish endogenous plant hormones lost in the soaking solution (Choe, 1972).

It appears that the seed pretreatment which is claimed to be a method of wide applicability is to be considered with much caution and a newer perspective. It does not influence the same character in the same direction or to the same extent even in different varieties of the same species and cannot induce hardening to the same degree as it may have depressing effect or none at all. Thus every variety and species must be investigated individually to determine the effect of such treatments in different cellular characteristics in a wide range of experimental conditions (Salim and Todd, 1968). One major problem is the selection of the right type of dosage of the correct growth regulator in every instance. Sometimes application of the bioregulants in an aqueous solution is a problem in itself. It appears that the current presowing recommendations are empirical. However, it seems clear that the technique of pretreatment uniformly improves the rate of seedling emergence (Heydecker, 1973). The practical value of presowing technique over a wide range of condition in different agroclimatic regions of a country and in different countries, needsexploration. The methods needed require information on water requirements and stage of the embryo at the time of drying so as to induce greatest degree of drought resistance with minimum injury.

Various aspects of this technique were discussed thoroughly in a panel discussion on presowing seed treatment. Hardening or repeated imbibition of seeds, short of radicle emergence, followed by redrying, causes the activation of latent physiological mechanism which enable the resulting plants to withstand certain adverse environmental conditions such as drought and extremes of temperature and finally it was concluded that seed treatments may have any or all of the following purposes:

1. Improved percentage of germination in field leading to a fuller stand; 2. Improved rate and uniformity of seedling establishment; 3. Improved seedling growth; 4. Improved performance of seeds and seedlings under inimical (or at least sub-optimal) environmental conditions; 5. Improved total or equal earlier yield; and 6. Greater uniformity at the time of and in the quality at crop maturity (ref: Heydecker, 1973).

In India presowing method was first tried by Chinoy (1942; 1947) in wheat and in rice (Parija, 1943; Parija and Pillay, 1945). The earlier work of Professor Chinoy carried out at IARI, New Delhi (Chinoy, 1942; 1947), Delhi University (Gupta, 1956; Nanda et al., 1959) and here at Gujarat University (Patel, 1967; Acharya, 1968; Chinoy et al., 1966; 1969; 1970; Dave, 1969; John, 1965; 1969; Jani, 1969), may be summarised as follows:

I. Pretreated plants have higher dry matter production as well as relative growth rates of the seedlings even under restricted water (drought and salinity combined). Pretreated seeds could extract more moisture from the atmosphere. It is due to the change in the lyophilic colloids.

II. Hydrolytic activity of enzymes especially amylase was higher in pretreated seeds indicating faster utilization of starch for enhanced seedling growth therein. Catalase was also higher in pretreated seeds.

In view of these advantages, the pretreated seeds should be able to gain an advantage over the untreated seeds, even in the matter of grain production. An increase of 25% in grain yield per acre under field conditions was obtained in plots sown with pretreated seeds (Chinoy, 1947; 1967; 1968; Chinoy et al., 1969 b; 1970; 1971; 1971 a; Chinoy and Saxena, 1978). The beneficial effect of presowing hardening treatment was most marked under adverse conditions of soil moisture. Pretreated seeds by virtue of quicker and better germination on such plots with depleted soil moisture in the first six inches were able to establish a better stand of plants, and as a consequence of that, the yield increased by more than 30% (Chinoy, 1947). Thus the greatest advantage of seed pretreatment was in the ability of the pretreated seed to extract sufficient moisture from the soil when its moisture content was too low for quick germination and finally due to accelerated growth, higher yields were obtained. This would be very useful under dry-land farming.

Ascorbic acid plays important roles in germination and growth of the plant (Chinoy, 1962; 1967; 1968; ref. Chapter 4) and hence a simple pretreatment of seeds with AA was developed. It was found to accelerate germinative processes as well as growth and yield of mature plants (Chinoy, 1967; 1968; Chinoy et al., 1969 b; 1970; 1971; 1971 a; Chinoy and Saxena, 1978).

100 gm seeds were soaked with 40 ml solution of AA (25 - 50 ppm) for 6 hours with frequent shakings. The seeds were dried again at room temperature to original dry weight. This process was repeated three times (three cycles of alternate soaking and drying) and such pretreated seeds along with distilled water pretreated seeds and untreated

(seed)-controls were sown in pots and plots using standard agronomic practices. Our earlier results have been reviewed elsewhere (Chinoy, 1967; 1968; Chinoy et al., 1969 b; 1970; 1971; 1971 a; Dave, 1969; Chinoy and Saxena, 1978).

Pretreatment with AA resulted in stimulated germination, more efficient water absorbing capacity, ability to withstand against the atmosphere drought and more vigorous growth than the untreated seeds in a number of crops, e.g., wheat, oat, barley, sesamum, ragi, and maize. Further, under wilting treatments, AA pretreatment stimulated the plant to resist the adverse conditions of drought better than the control plants (Chinoy et al., 1966; 1970; Acharya, 1968; Chinoy, 1968; Jani, 1969; John, 1969). It has also been shown that seeds pretreated with AA, when grown under saline conditions, show a comparatively higher degree of tolerance to salinity (John, 1965; 1969; John and Jani, 1969).

This method of pretreatment was also developed to a point of field applications by farmers (for details see: Chinoy, 1967, 1968; Chinoy and Saxena, 1978). Extensive pot and field experiments have shown that presowing treatments with AA brought about earliness in flowering, delayed senescence (ripening) of the grain and increased yield (Chinoy et al., 1969 b; 1970; Chinoy and Saxena, 1969 ; 1978).

The above mentioned method of presowing treatment was modified slightly to include very dilute solutions of hydrogen peroxide in the medium along with AA to facilitate production of free radicals of AA (MDHA) (Chinoy et al., 1972; 1973; 1973 a). Even DNA and RNA have been used for presowing treatment together with AA and H_2O_2. The response to pretreatment with these metabolites was appreciably better than that obtained with AA alone. A further modification was the inclusion of thermal treatments along with AA and H_2O_2. In some cases, the yields per plant were doubled (P.N. Shah, 1972; Ghesani, 1972; H.K. Shah, 1973; H.O. Shah, 1975; Kaul, 1975; J.N. Patel, 1976; D.H. Mehta, 1977; Jani, 1978; K.G. Mehta, 1978; Shah and Shah, 1979; Sharma, 1981; Varghese, 1981; Vora and Patel 1979). Vora and Patel (1975) reported beneficial effect of thermal pretreatments and GA_3 feeding on seedling growth of mung. Thus one or all of the beneficial aspects have been confirmed by various workers. On the other hand, some workers have reported negative results following presowing methods (Lang, 1965; Salim and Todd, 1968; Heydecker, 1973; Rammelt, 1973).

In view of the foregoing some more critical studies were initiated in our laboratory. Mehta (1973) worked on physiological changes during pretreatments in Sesamum indicum cv. Limbdi. Generally a three hours' period during pretreatment showed higher AA turnover (rate of synthesis, release of free AA from bound form and rate of enzymatic utilization) as compared to 6 hours (may be due to lesser hydration of protoplasm - 40 ml solution for 100 g seeds) - during subsequent periods. Catalase and peroxidase activities were appreciably higher with the duration and number of pretreatments. Second and third pretreatment did not show any beneficial effect. This might be due to the leakage of metabolites and enzymes indicating harmful effects as a result of soaking the seeds. The pretreatments enhanced the rate of germination and overall growth of the seedling. The maximum root and shoot length was obtained in the second cycle suggesting that more than 2 pretreatments were not necessary.

Thakker (1974) has examined three cycles (100 g seeds with 40 ml solution or distilled water (DW) of pretreatments with AA (25 ppm); H_2O_2 (0.01 vol) and AA + H_2O_2 in Sesamum indicum cv. Kundla. The data showed an appreciable reduction in vegetative and reproductive characters (especially 1000-seed weight) with increase in the pretreatment cycles. DW exhibited an inhibitory effect after second cycle. Single pretreatment with AA + H_2O_2 was beneficial in obtaining increased yields. Saxena (1973 - 1976; 1974; 1979; 1981; 1982; 1983) critically investigated the effect of sprays and pretreatments with different hormones in Sesamum, soyabean, berseem, mung, okra (lady's finger) and wheat.

Plants were hardened against moisture stress by giving presowing soaking treatment to the seeds with free radical scavengers, growth hormones and salt solutions with the basic idea of protecting the membrane against moisture stress damage as the primary step followed by triggering of basal metabolism to support growth (Ahluwalia and Srivastava, 1977; Srivastava, 1978; Srivastava and Singh, 1979). Presowing chilling treatment has been shown to enhance growth and yield in pea and Vicia faba (Banerji and Kausal, 1980).

Basu (1976) has been advocating that presowing treatments of the seeds with different organic solvents, phenols, inorganic salts was quite beneficial for better storage of seeds and maintenance of viability. On the other hand, Harrington (1973) had shown that pretreating the seeds with salts curtailed their life span.

In our studies on the residual effect of presowing treatments in next generation it was concluded that growth and yield performance of wheat plants raised from pretreatments with sucrose - 1%, AA - 25 ppm and distilled water were not significantly different as compared with the untreated controls and thus the beneficial effect of pretreatment was not carried over in the next generation (Chinoy et al., 1969 b; Gurumurti et al., 1969). Similar findings on carry over effects of salinity in wheat were reported by Singh and Vijaykumar (1974).

These aspects were also emphatically highlighted in recent publications (Heydecker and Coolbeor, 1977; Srivastava, 1978; Scott, 1978; Thomson, 1979; Malik, 1981; Parmar et al., 1981). There is lot of scope for future work which might enable the seeds of doubtful quality to perform better, besides improving the performance of good quality seeds.

Hausen (1936) reported an increase of 35 - 75% in dry matter of plants due to AA treatment. The most significant enhancement in dry matter was observed during the flowering period and AA content of the treated plants also increased significantly. Increase in the dry matter was specifically due to AA because a corresponding treatment with glucose did not increase growth of the plants. Ovcharov (1958) has also reported the beneficial effect of treatment with vitamins. Sechet (1953) found that soaking seeds of Lupinus in lactoflavin, AA, neurin, nicotinic acid and nicotinamide enhanced vernalization effect. The presowing treatment with AA noticeably enhanced the growth of axillary branches (Ovcharov, 1963). Even soaking in AA solution alone, significantly enhanced growth of the axillary branches but no stimulation of root growth occurred. However, application of heteroauxin as well as tri-iodobenzoic acid (TIBA) did not inhibit the growth, and the latter in combination with AA, stimulated growth of axillary branches to a considerable extent. Gutmanis (1967) showed that pretreatment of seeds with 0.1% vitamin C solution increased pea yield by 14%. The best result in the improvement of the chemical composition of the pea seed was also achieved by the same dose of AA.

Pretreatment with AA was found to accelerate germinative processes as well as growth and yield of crop plants (Chinoy et al., 1975). It also increased the yield of wheat by 7 to 33%. Abraham et al. (1969) have studied the effect of AA pretreatment on the grain yield of hybrid bajri (Pennisetum typhoides), maize (Zea mays cv. Godhra), ragi (Eleusine coracana Gaertn cv. CO.2 and CO.7). Presowing treatment with AA (50 mg/l) resulted in a higher total spike weight and 25% increment in grain yield over the control (untreated). AA treated plants also gave significantly higher yield compared to that of plants pretreated with distilled water. AA pretreated maize plants gave 32.6% more yield of grain compared to that of the untreated ones. In the case of the two cultivars of ragi cv. CO.2 and CO.7, pretreatment with AA increased the grain weight per spike over the control by 32.3 and 32.0% respectively (Fig. 8.1; Tables 8.1, 8.2).

Presowing treatment of wheat seeds with AA and sucrose had beneficially affected growth and yield characters

Table 8.1. Drought intensity during wilting period (Chinoy et al., 1969 b)

Variety	Growth period-days	Pre-treatment	Drought at tiller initiation			Drought at shooting stage		
			Average soil moisture	Mean temperature	Drought intensity	Average soil moisture	Mean temperature	Drought intensity
		C	13.3		12.9	8.9		21.1
N.P. 21	96-106	DW	12.0	21.3	14.2	8.4	23.6	22.5
		AA	13.1		13.0	10.3		18.4
		C	11.7		14.6	8.8		21.4
Betzes 6398	108-119	DW	14.4	21.3	11.9	9.0	23.6	20.9
		AA	12.4		13.8	8.1		23.3
		C	14.0		12.2	11.1		18.8
Spartan-5027	115-123	DW	13.3	21.3	12.9	11.7	26.0	17.7
		AA	13.3		12.8	8.2		25.3

C = Control DW = Distilled Water AA = Ascorbic Acid

Table 8.2. % survival values in plants after wilting at different growth stages (Chinoy et al., 1969 b)

Variety	Growth period-days	Pre-treatment	Drought at tiller initiation: Total No. of Plants			Drought at shooting stage: Total No. of Plants		
			Before wilting	After wilting	% Survival	Before wilting	After wilting	% Survival
N.P. 21	96-106	C	134	128	95.5	141	24	17.0
		DW	146	142	97.3	121	14	11.6
		AA	137	135	98.5	93	19	20.4
Betzes-6398	108-119	C	114	110	96.5	106	15	14.2
		DW	98	91	92.8	92	19	20.6
		AA	109	105	96.3	98	29	29.6
Spartan-5027	115-123	C	113	106	93.8	129	37	28.7
		DW	124	118	95.1	120	37	30.8
		AA	107	105	98.1	116	51	44.0

C = Control DW = Distilled Water AA = Ascorbic Acid

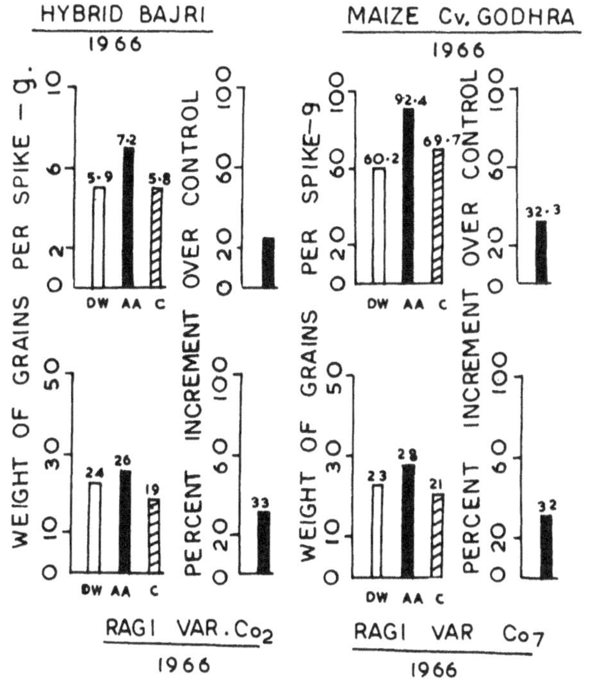

FIGURE 8.1. Effect of AA pretreatment on grain yield per spike of hybrid bajri (Pennisitum typhoides), maize (Zea mays) and two cultivars of ragi viz., ragi CO.2 and ragi CO.7 (Abraham et al., 1969).

(Abraham, 1969), brought about early flowering and a delay in senescence (Fig. 8.2) which in turn favourably affected the grain filling processes. The plants pretreated with AA and sucrose showed higher values in total number of grains, total grain weight as well as 1000 kernel weight compared with those of plants pretreated with distilled water and control (Fig. 8.3). Plants raised from seeds pretreated with AA and sucrose showed higher total AA content especially during the period of reproductive differentiation. This had a bearing on the consequent effect upon growth and development as well as increased crop production (Fig. 8.4). Pretreatment of the seeds with AA was found

1

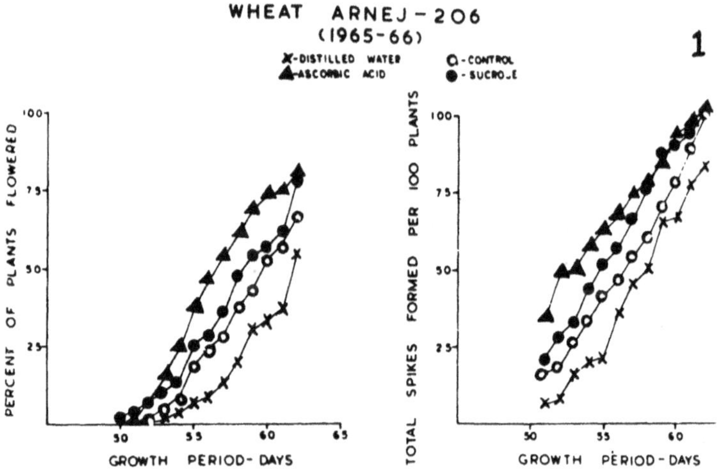

2

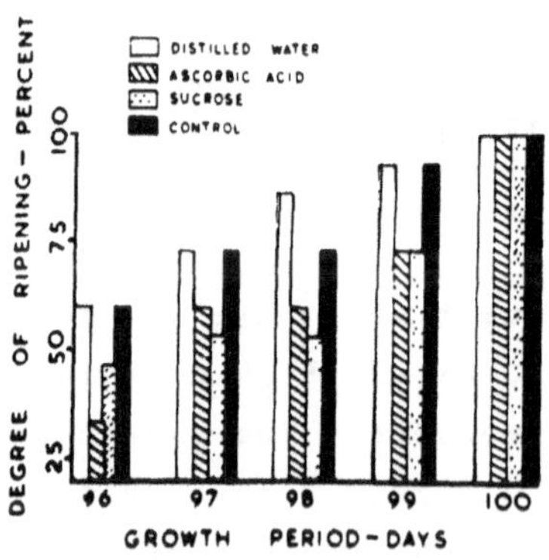

FIGURE 8.2. Effect of pretreatment of wheat seed with DW, AA and sucrose on flowering, spike formation and ripening of grain (Abraham, 1969).

to be helpful in accelerating the hydrolytic activity of enzymes, in extracting moisture from the atmosphere more efficiently, and in promoting better growth and yield

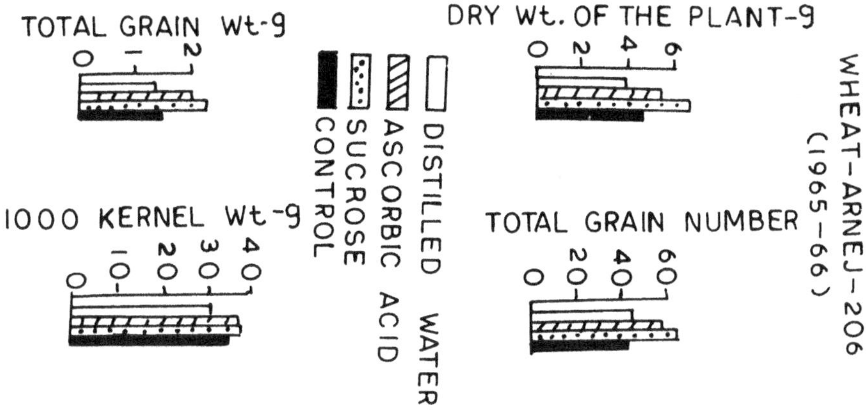

FIGURE 8.3. Effect of pretreatment of wheat seed with DW, AA and sucrose (1%) on total grain number, total grain weight and 1000 kernel weight (Abraham, 1969).

Abraham et al., 1969; Chinoy et al., 1975). The data presented here showed (Fig. 8.4) an inductive effect of AA on its own biosynthesis and consequent stimulation of growth and metabolism. Similar results have been reported earlier (Patel, 1967; Abraham et al., 1968 a, b, c; Korneev, 1963).

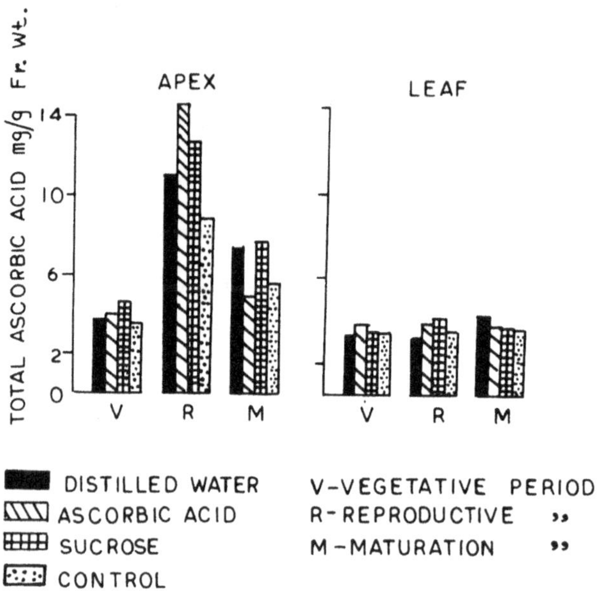

FIGURE 8.4. Effect of pretreatment of wheat seed with DW, AA and sucrose (1%) on total AA contents of apex and leaf during the vegetative (V), reproductive (R) and maturation (M) phases (Abraham, 1969).

1. EFFECT OF PRETREATMENT ON ASG CONTENT IN BARLEY

ASG content was higher in the apex and leaf of control plants compared with that of treated ones except in the case of vegetative apex. In leaves, the ASG content of AA pretreated plants remained at a lower level throughout the growth period as compared with that in untreated ones. Thus pretreatment with AA helped in releasing the AA from bound state during important phases of plant growth and development (Vyas, 1971) (Figs. 8.5 to 8.7).

A survey of literature revealed a paucity of work on AA turnover and its role in the metabolism of seeds under thermal pretreatment. Therefore, work was undertaken on thermal pretreatment of dry seeds (Shah, 1975), which revealed increased rate of germination and productivity of seeds. The seeds of wheat and soyabean are given thermal

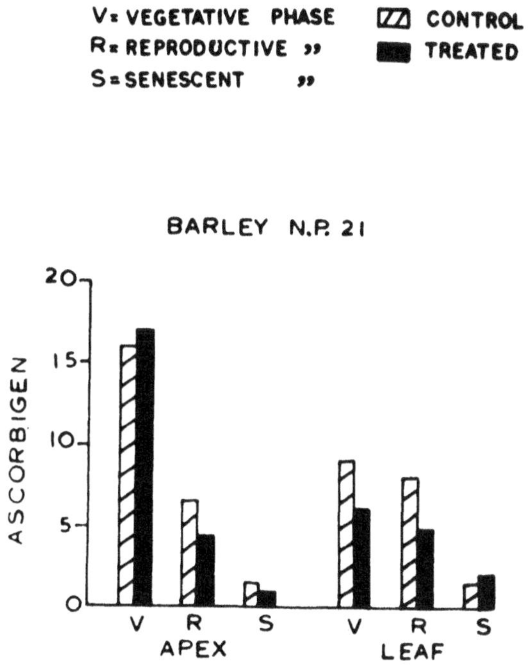

FIGURE 8.5. Effect of pretreatment with AA on ASG content of barley (Vyas, 1971).

pretreatment at 50°, 70° and 100° and AA turnover as well as some enzymic activities were studied. The temperature was either raised gradually over a period of time to the requisite level (50° G, 70° G, 100° G) (G = gradual) or the seeds were directly placed in the oven which was previously heated to the requisite temperature (50° D, 70° D, 100° D) (D = direct). In wheat N.P. 718, there was an increase in AA content as well as AAU and -SH groups of the apical organ and the leaf of some of the temperature pretreated plants (Fig. 8.8). Levels of ASG and AA - MM complexing were also raised by thermal pretreatment (Fig. 8.9). With the enhancement of AA turnover, activities of enzymes like amylase, peroxidase, AA - FR peroxidase, catalase, invertase, RNAse and protease were enhanced.

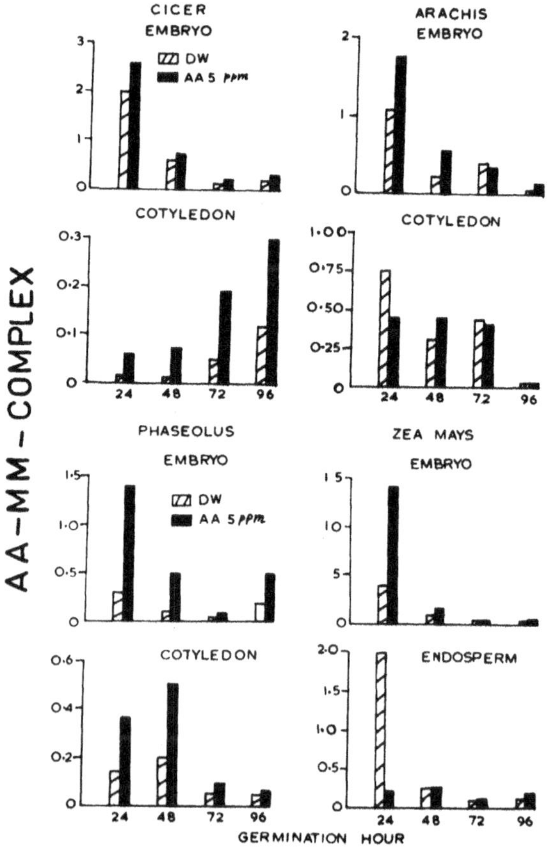

FIGURE 8.6. Effect of pretreatment with AA on AA - MM complexing in <u>Cicer</u>, <u>Arachis</u>, <u>Phaseolus</u> and <u>Zea</u> (Sutaria, 1972; Mehta, 1972; Patel, 1972, Shah, 1972).

The content of metabolites like DNA, RNA, proteins, histones and reducing sugars were accelerated as a result of thermal pretreatment (Figs. 8.11 to 8.13). The enzymic activity and contents of metabolites were generally at a higher level during the period of reproductive differentiation (rd) in both the plant organ compared with that during vegetative differentiation.

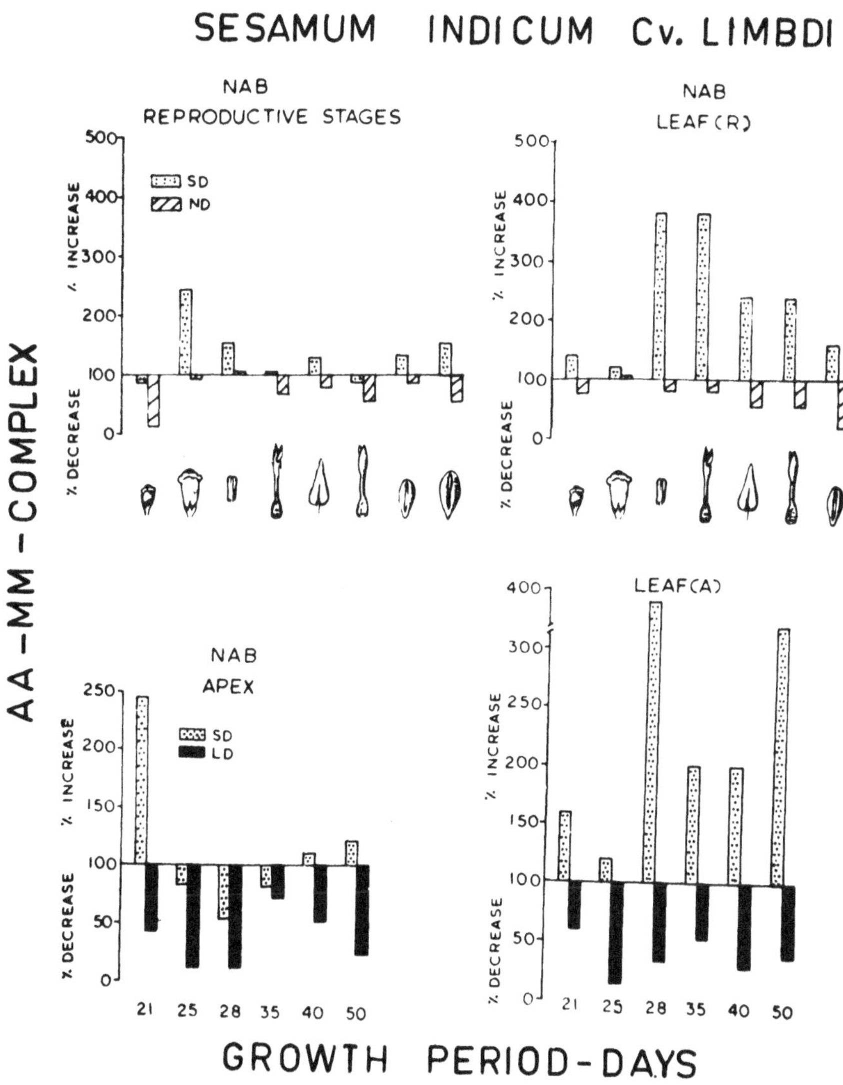

FIGURE 8.7. Effect of pretreatment with AA on AA - MM complexing in Sesamum indicum cv. Limbdi (Chinoy et al., 1970).

2. EFFECT OF THERMAL PRETREATMENT ON ASG CONTENT OF WHEAT

The seeds of wheat cv. N.P. 718 were used for this experiment and thermal treatments with different temperatures

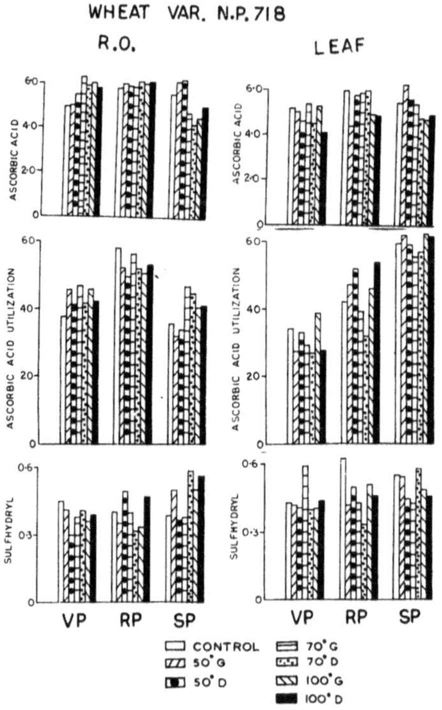

FIGURE 8.8. Effect of different thermal pretreatments on AA, AAU and SH content of the shoot apex or the reproductive organ (R.O.) and of the corresponding leaf of wheat cultivar N.P. 718 during its vegetative (VP), reproductive (rd) and maturation (SP) phases (Shah, 1975).

were given to them. The control and thermally treated seeds were sown in pots and ASG content was determined from apex and leaf during different phases. The ASG content was low in the vegetative phase of N.P. 718 wheat plants raised from preheated seeds and then it increased during reproductive differentiation. Reverse trend was found in the embryo axis of control plants (Shah, 1975) (Fig. 8.14). In the senescence phase, ASG content increased in control 50° G (gradual) and 50° D (direct), 100° G and 100° D; while it decreased in 70° G and 70° D. In the case of leaf, ASG content was low in the control, 50° G, 50° D and 70° G at

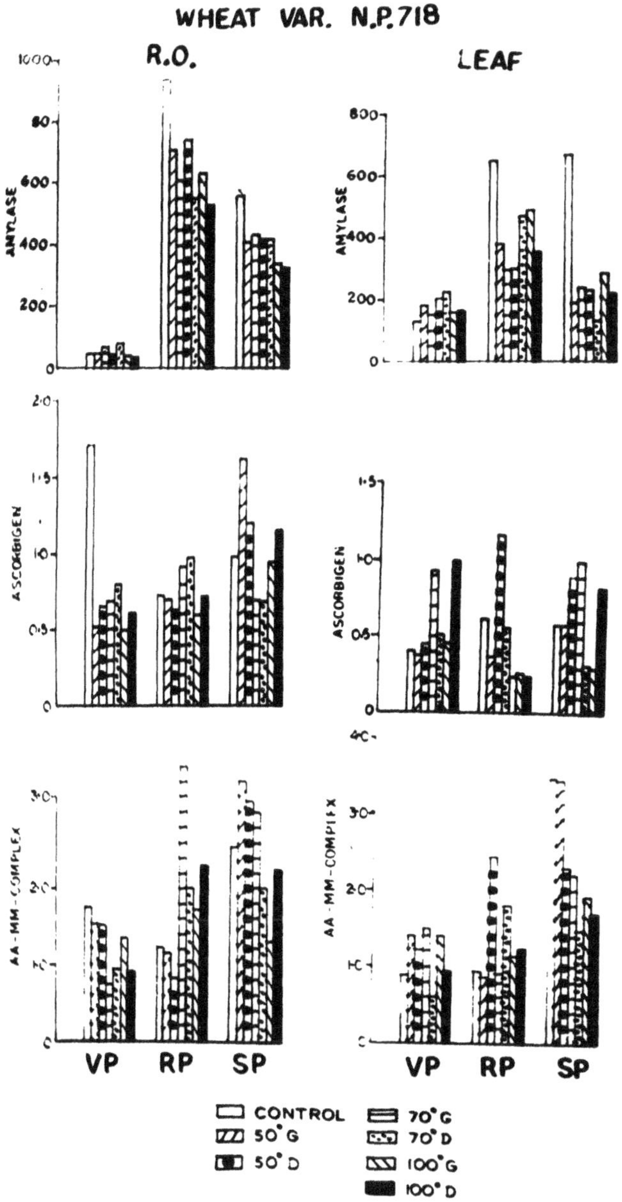

FIGURE 8.9. Effect of different thermal pretreatments on amylase activity, ASG content and AA - MM complexing of the shoot apex or the reproductive organ (R.O.) and of the corresponding leaf of wheat cultivar N.P. 718 during its vegetative (VP), reproductive (RP) and maturation phases (SP) (Shah, 1975).

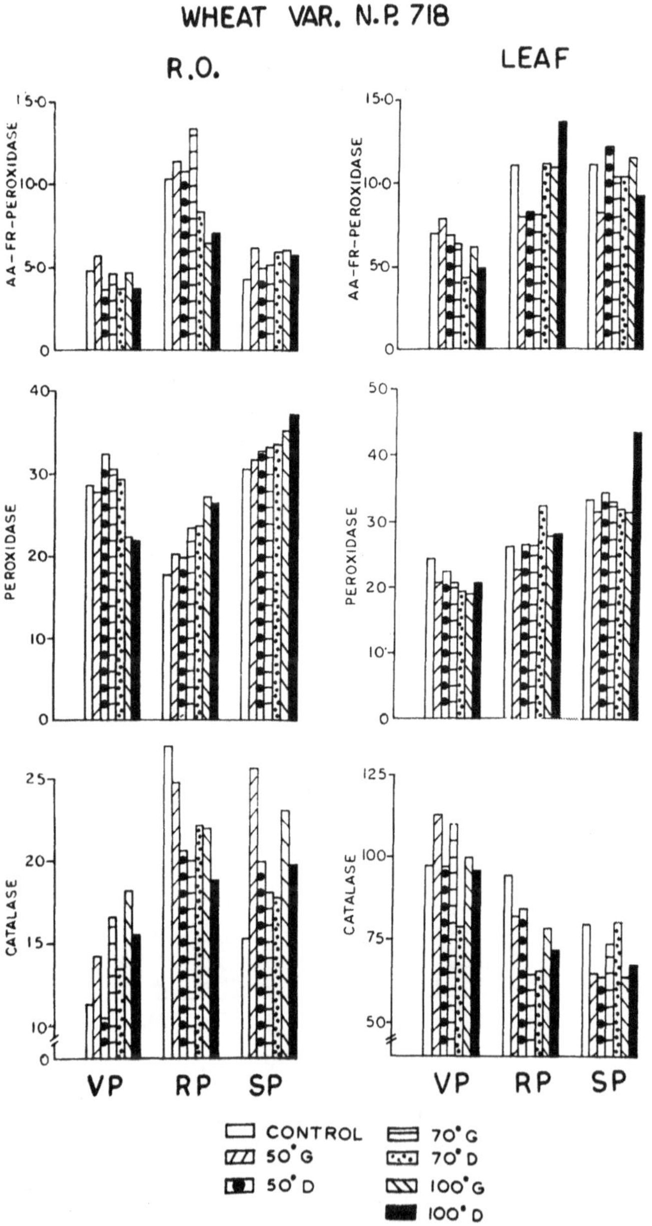

FIGURE 8.10. Effect of different thermal pretreatments on AA - FR peroxidase, peroxidase, and catalase activities of the shoot apex or the reproductive organ (R.O.) and of the corresponding leaf of wheat cultivar N.P. 718 during its

vegetative (VP), reproductive (RP) and maturation (SP) phases (Shah, 1975).

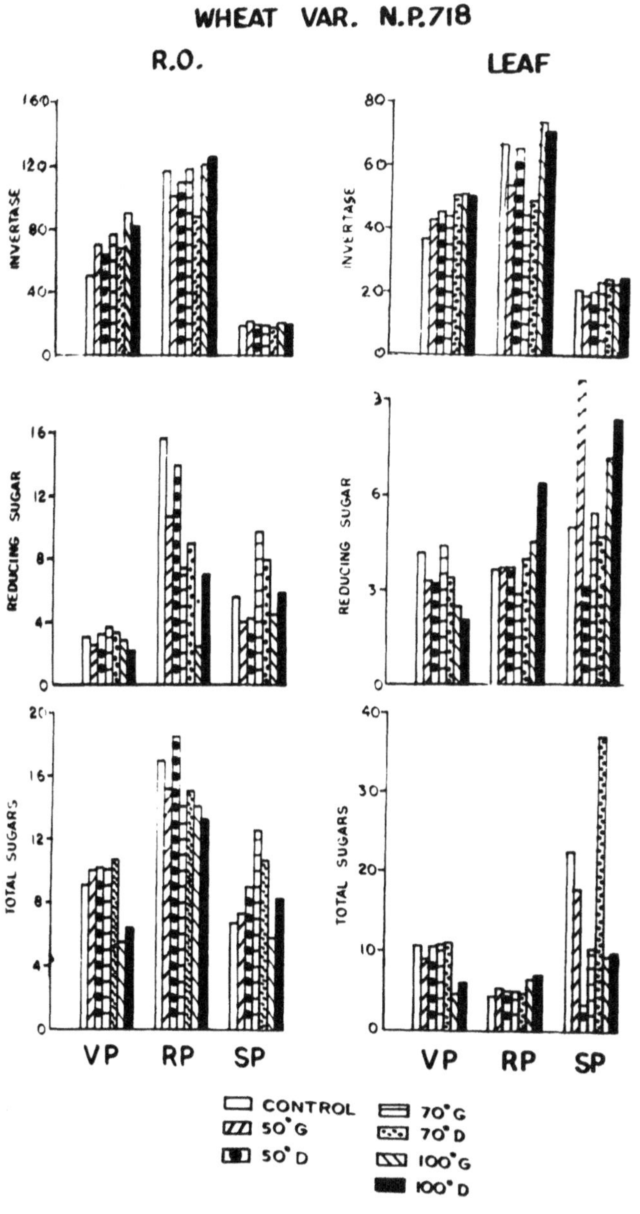

FIGURE 8.11. Effect of different thermal pretreatments on invertase activity, reducing sugar and total sugars of the shoot apex or the reproductive organ (R.O.) and of the

corresponding leaf of wheat cultivar N.P. 718 during its vegetative (VP), reproductive (RP) and maturation (SP) phases (Shah, 1975).

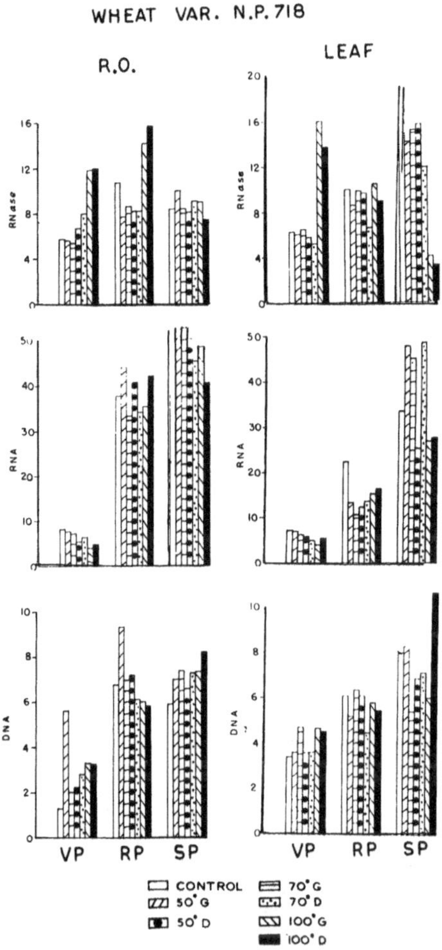

FIGURE 8.12. Effect of different thermal pretreatments on RNAse activity, RNA and DNA of the shoot apex or the reproductive organ (R.O.) and of the corresponding leaf of wheat cultivar N.P. 718 during its vegetative (VP), reproductive (RP) and maturation (SP) phases (Shah, 1975).

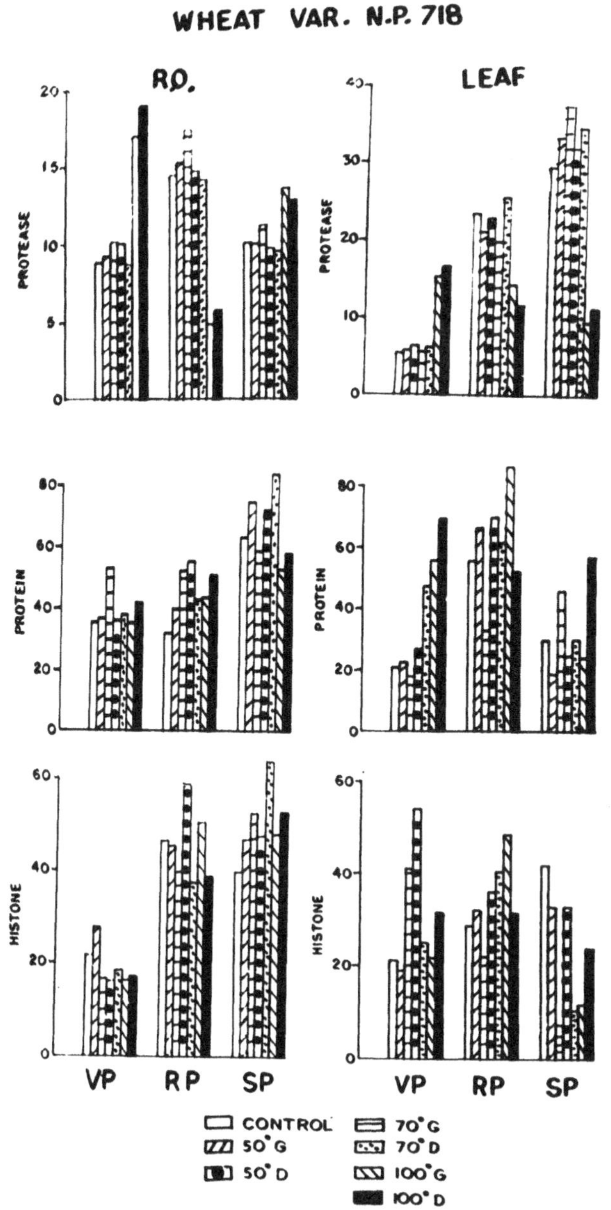

FIGURE 8.13. Effect of different thermal pretreatments on protease activity, protein and histone contents of the shoot apex or the reproductive organs (R.O.) and of the corresponding leaf of wheat cultivar N.P. 718 during its vegetative

(VP), reproductive (RP) and maturation (SP) phases (Shah, 1975).

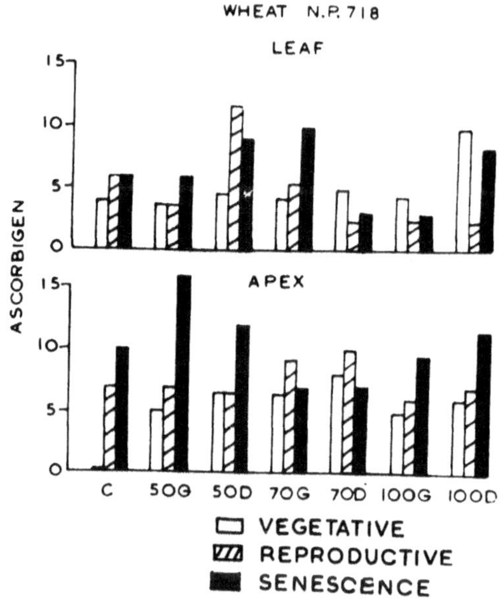

FIGURE 8.14. Effect of thermal pretreatment on the ASG content of apical meristem and leaf of wheat during vegetative, reproductive and senescence phases (Shah, 1975).

the time of vegetative phase, but it increased during the reproductive one. Reverse results were reported for 70° D, 100° G and 100° D. In the control and temperature pretreated plants, ASG was found at a higher level during the senescence.

Hydrogen peroxide is one of the products of aerobic respiration which interacts with AA to produce its free radical, monodehydroascorbic acid (MDHA). The above mentioned facts have been made use of in developing a method of presowing treatment of seed with AA plus hydrogen peroxide (H_2O_2), with a view to determining the effect of free radical (FR) of AA on germination, juvenile and adult growth as well as yield (Chinoy et al., 1973). Seeds of Brassica juncea cv. Varuna after a thermal treatment (60° C for 24 hours) were given a pretreatment with a solution containing 100 ppm AA and 0.01 vol H_2O_2 according to Chinoy et al. (1975). The results obtained on percent germination, fresh weight, percent moisture content and AAU and also AA - FR peroxidase activity during germination showed that concomitant with the superiority of the seedlings raised from the pretreated seeds in growth parameters, AAU and AA - FR peroxidase activity were also at significantly higher level. Increased AAU, an important metabolic index for an accelerated energy tranduction was stimulated by pretreatment with high temperature + H_2O_2, the latter providing free radicals of AA during germination which starts with imbibition of water as in Brassica. It is implied that pretreatment with AA + H_2O_2 provides free radicals of AA during germination. This process activates AA - FR peroxidase at a higher level indicating a greater amount of free radicals of AA being generated during the germination processes (Patel et al., 1980).

3. ASCORBIC ACID AND WATER STRESS

The ascorbic acid content increased slightly under moderate water stress (50% field capacity) in young oat plants (Eisele, 1944). Under severe water stress, however, there was a sharp increase in AA content of stressed plants. The respiratory rate and the concentration of hexoses were increased under water stress, as a result of hydrolysis of polysaccharides resulting in the synthesis of increased amount of AA. Hunter et al. (1950) were also able to demonstrate an appreciable enhancement of AA content in the upper leaves of Brassica rapa plants grown under low soil-water potential. Other workers have also reported increased AA concentration in plants during a period of severe water stress (Solomon, 1955; Novitskaya, 1958). Similarly, Farkas and Rajhathy (1955) found an enhanced concentration of AA in tomato plants grown under dry conditions. An increasing water stress in consecutive leaves from below upwards was found to be correlated with increasing concentration of AA. Tombesi (1951; 1952) on the other hand, found that plants of Pelargonium zonale grown under dry conditions registered a decrease in AA as well as glutathione contents. Concomitant with that there was an enhancement in oxidase actvity, thus reducing the ratio: AA/enzymatic activity. This investigator worked with a number of species like Beta vulgaris, Medicago sativa, Solanum lycopersicum, Ricinus communis, and Nicotiana tabacum. The author has ascribed the high oxidative capacity of these plant tissues to high rate of respiration. Grzesink et al (1979) did not find any correlation between levels of peroxidases, proteases, glutathione and AA and drought resistance in winter rye.

Stocker (1960; 1961) has reviewed work on the role of AA in metabolism of plants undergoing stress. Contradictory results on AA content of plants grown under water stress have been reported. Stocker (1956, 1960, 1961) has reported an enhancement in AA content of plants which were subjected to severe water stress. He showed that the loosening of the protoplasmic structure which occurred during the reaction phase was generally followed by an increase of the hydrolytic and oxidative activity. On the other hand, the strengthening of the protoplasmic structure during the restitution phase brought about an enhancement of the synthetic and reductive activity. The author has determined the oxidation-reduction states (quotient) by studying the ratio of oxidized AA/ reduced AA in plants undergoing water stress. One could utilize it as a relative unit.

AA concentrations were determined during the reaction phase on four different conditions with increasing intensification of drought conditions, which differed in degree, on the last two days, according to whether the plants had been grown under moist or dry conditions. In the first condition with no drought effects, the metabolism proceeded in anabolic direction. Respiration and the oxidation-reduction quotient were low. This ratio was even lower at mid-day. The total amount of AA was low as there was no supply of sugars. On the second occasion also, as intensive drought conditions were absent, the redox ratio remained low although there was some increase in AA concentration due to the abundance of sugars formed by photosynthesis. The third determination fell within a drought period and therefore the metabolism proceeded in a hydrolytic direction. Due to the presence of large quantities of sugars, considerable quantity of AA was synthesized and the redox ratio reached a high level. The ratio was appreciably higher at mid-day in plants of "dry" culture (30% field capacity) compared with that in the "moist" cultures (85% field capacity). On the fourth occasion, when very severe drought conditions existed, there was a pronounced oxidative-catabolic phase manifested by increased respiration and a corresponding decrease in the photosynthesis. The total amount of AA was lower as compared with that in the previous occasion on account of the shortage of the substrate. The redox ratio (quotient) was, however, very high and even exceeded 1. During the restitution phase, i.e., in plants grown with a permanently low water content of the soil, the redox ratio diminished with the decrease of soil moisture concomitantly with respiration. In extreme dry conditions, however, the ratio increased due to pathological causes. Under moderately dry conditions, a slight increase in the total AA concentration occurred. There was an extraordinary upsurge in the total AA content along with abnormally high oxidation-reduction ratio and very high level of respiration under extreme drought conditions suggesting a complete breakdown of the metabolism (Chinoy, 1968; Chinoy et al., 1969 b). The other workers (Sheherbakov, 1963; Ivanov and Karakash, 1964; Lukicheva, 1968) also found increased AA concentrations in plants during periods of extreme water stress. The content of DHA increased in the bast tissue of apple trees under the influence of insufficient water supply (Ivanov and Karakash, 1964). In the case of wheat, Lukicheva (1968) observed increased intensity of respiration and catalase activity along with increased AA content. Startseva (1963) was able to show that presowing treatment of seeds with microelements enhanced endogenous AA content which enabled the plants to overcome the adverse effects of drought.

Work has been carried out not only on AA concentration in relation to water stress but also on mobilization of the bound ascorbigen, AAU and AA - MM complexing in relation to water stress (Chinoy et al., 1968; 1969 b; 1971; 1971 a; 1973; 1973 a; Acharya, 1968; Jani, 1969). Chinoy et al. (1968) subjected seedlings of Hordeum vulgare L. cv. N.P. 13 and cv. Munsing to desiccation in chambers maintained at low humidity at 25 ± 3° C for five and ten days respectively. At the end of each desiccation period, seedlings were revived by moistening the sterilized sand of the perti dish. AA and AAU were enhanced during desiccation. Embryo axis receiving 10- day desiccation showed consistently higher concentration of AA. The AAU of the embryo axis was also higher in both the cultivars at all stages of growth compared with that of undesiccated or revived embryo axis (Figs. 8.15, 8.16). The ASG content of seedlings revived after 10 days desiccation was the highest in both the cultivars followed by those seedlings which were revived after undergoing a 5-day desiccation. ASG content of 5- and 10- days desiccated embryo axis was appreciably lower compared with that in revived seedlings (Fig. 8.16).

An enhancement in AA content and AAU with a concomitant reduction in ASG during the period of desiccation, and a reduction in free AA, but an increase in ASG content during the subsequent period of revival, are suggestive of important protective and electron generating roles of AA during water stress.

A study of the effect of desiccation on AA turnover as well as on some enzyme systems of seedlings of Triticum, Eleusine and Sesamum revealed that AA content increased with advance in seedling growth and during desiccation period as well as after revival in all the three crop plants (Acharya,

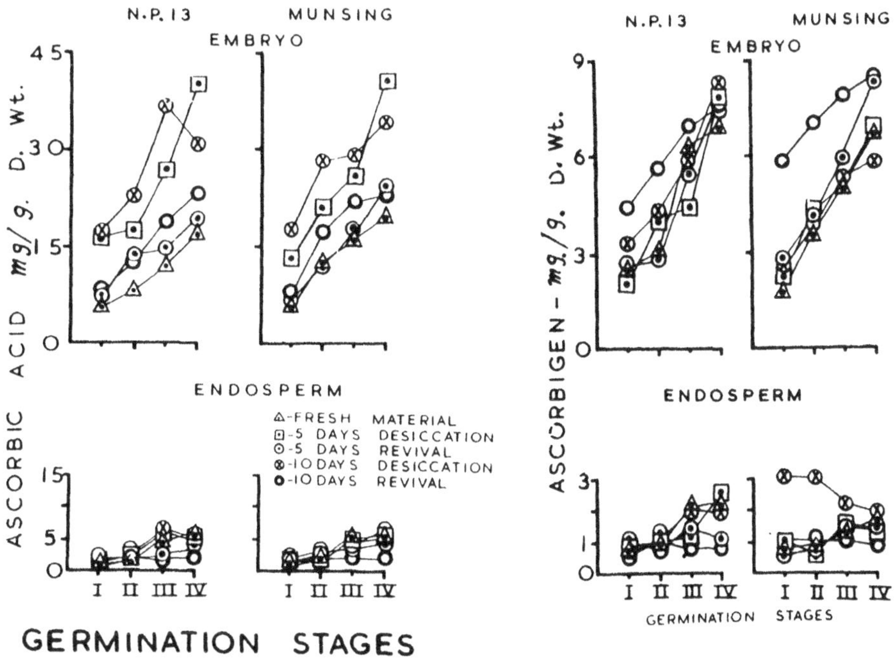

FIGURE 8.15. AA and ASG contents of desiccated and revived embryo axis as well as endosperm of two cultivars of barley, N.P. 13 and Munsing (Chinoy et al., 1968, 1969 b).

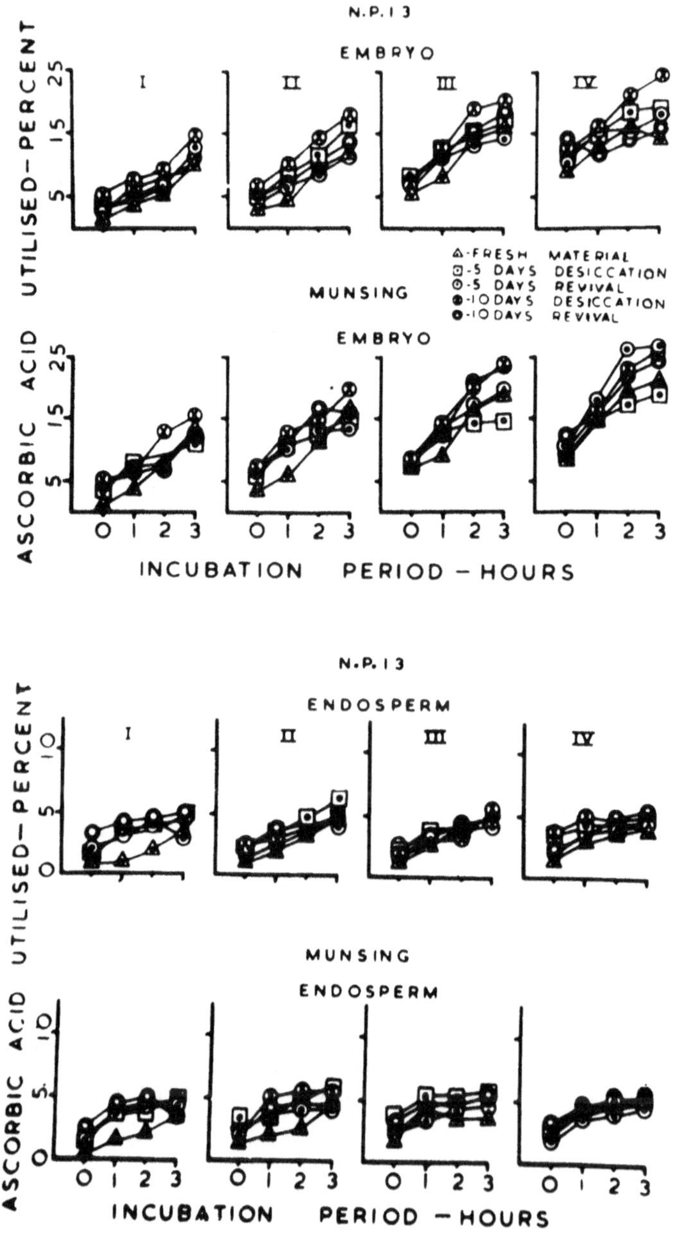

FIGURE 8.16. AAU in desiccated and revived embryo axis and endosperm of two cultivars of barley N.P. 13 and Munsing (Chinoy et al., 1968; 1969 b).

1968). During desiccation, a slight reduction in the utilization of AA occurred which became quite significant after revival. ASG content was higher in the embryo axis of *Triticum* and *Eleusine* (Finger millet) and also in the whole seedling of *Sesamum* after revival. Pretreatment with AA enhanced amylase activity of all the three plants during both the desiccation treatments. In desiccated embryo axis of *Hordeum vulgare*, cv. N.P. 13 and cv. Munsing, the concentration of AA and its utilization were higher at all stages of growth compared to those of undesiccated or revived ones. The ascorbigen concentration of embryo axis (revived after desiccation treatment) increased appreciably over that of the control one (Jani, 1969).

4. EFFECT OF CONTINUOUS WATER STRESS ON AA TURNOVER

Wilting treatment decreased ASG content in barley, while rewatering increased it under 15 - 20% soil moisture in barley cultivars, N.P. 20 and Peatland. Continuously water stressed plants contained higher AA than adequately watered ones. Generally there was no effect of wilting treatment when plants were raised under inadequate soil moisture regime (Vyas, 1971) (Fig. 8.17).

Chinoy et al (1969 b) have shown that during wilting treatment ASG content decreased in both the cultivars of barley. On revival, it surpassed that of control in N.P. 21 (early), while in Spartan (late), there was no difference in ASG content of control and revived plants. Wilting treatment at the anthesis also decreased ASG. On rewatering, it reached the level of control plants (Chinoy et al., 1969 b). The data indicate that during wilting treatment AA may be released from its bound form which may maintain proper redox state of plants under the unfavourable conditions. The increased ASG content on revival suggests that it may take part in regulating the normal life of plants after passing through adverse conditions.

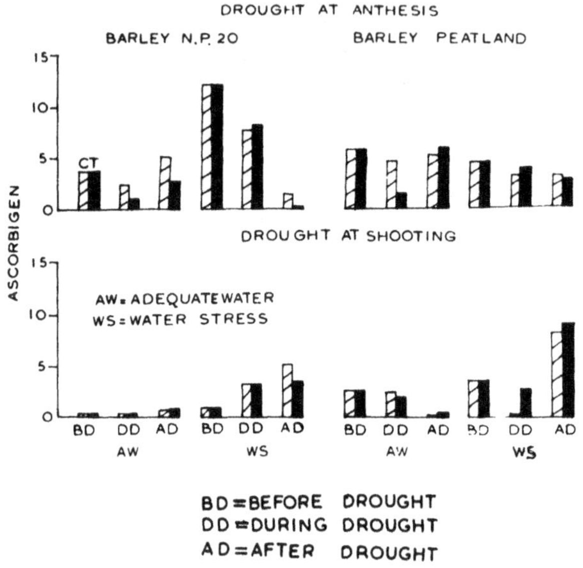

FIGURE 8.17. Effect of continuous water stress on ASG contents in barley (Vyas, 1971).

In both the cultivars of Hordeum vulgare, N.P. 18 and Munsing, wilting at shooting stage under long day (LD) photoperiod decreased ASG content while reverse trend was found for normal day (ND) in case of control (Fig. 8.18). Rewatering increased ASG content over the control under both the photoperiodic treatments. This LD treatment may release AA from the bound state during wilting treatment and which may provide protection against drought condition (Chinoy et al., 1969 b). Increase in ASG during revival indicate its significant role in the revival mechanism of barley plants. During adverse condition such as water stress, AA may be released from the bound state, thus creating a reductive atmosphere enabling the plants to resist the adverse conditions.

Plants of N.P. 21 and Spartan were raised under two photocycles, i.e., long and short day (LD and SD). Drought treatments were given after 20 and 40 days of growth by

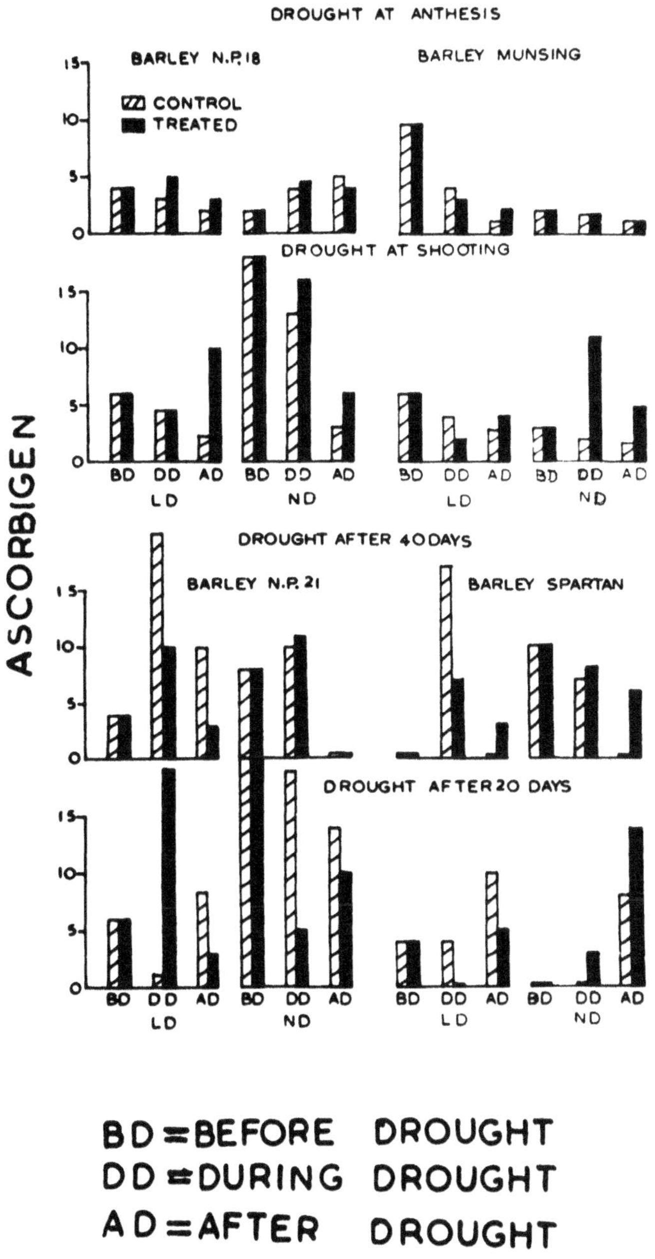

FIGURE 8.18. Effect of photoperiods (LD and ND) and wilting treatments on the ASG content of barley cultivars before, during and after drought (Raj, 1971).

witholding the water for six days. ASG content was determined before wilting treatment, on sixth day of wilting and on rewatering. In both the cultivars of barley, ASG content increased when plants were wilted for 6 days after 20 days of growth under LD and ND. A reverse trend was found on revival. After 40 days of growth, wilting treatment decreased ASG while rewatering increased it under both photocycles (Raj, 1971) (Fig. 8.18). Generally, wilting treatment given at earlier stages of plant growth was found to be beneficial, resulting in better growth and yield. In this experiment, plants were wilted after 20 days of growth, moreover, wilting was given only for six days and therefore, ASG content did not decrease as there was no necessity for providing protection. On the other hand, when drought was given after 40 days of growth, AA was released from ASG and this provided the necessary protection against the unfavourable condition.

A study of the effect of different periods of desiccation on AA turnover, enzymatic activities and paramagnetic behaviour during the juvenile phase of Cicer arietinum L. cv. Chafa (Chinoy et al., 1971; 1971 a; 1974) showed that desiccation for 3 or 6 days caused a reduction in AA content, whereas it sharply increased during the revival period. In the undesiccated cotyledon AA content was more or less constant, while the desiccated ones had enhanced AA concentration (Fig. 8.19). The ASG showed a fluctuating trend in the fresh embryo axis. It was maintained at a very low level during desiccation, whereas it increased considerably during the period of revival both in the embryo axis as well as in the cotyledon (Fig. 8.19). In both the embryo axis and the cotyledon, the AAU was maintained at an enhanced level during desiccation throughout the course of germination. On revival, the high level of AAU was maintained. The cotyledons showed a similar trend. There was however, an increase in AAU in the cotyledon with advance in germination. On the whole, the activity was substantially higher in the embryo axis, compared to that

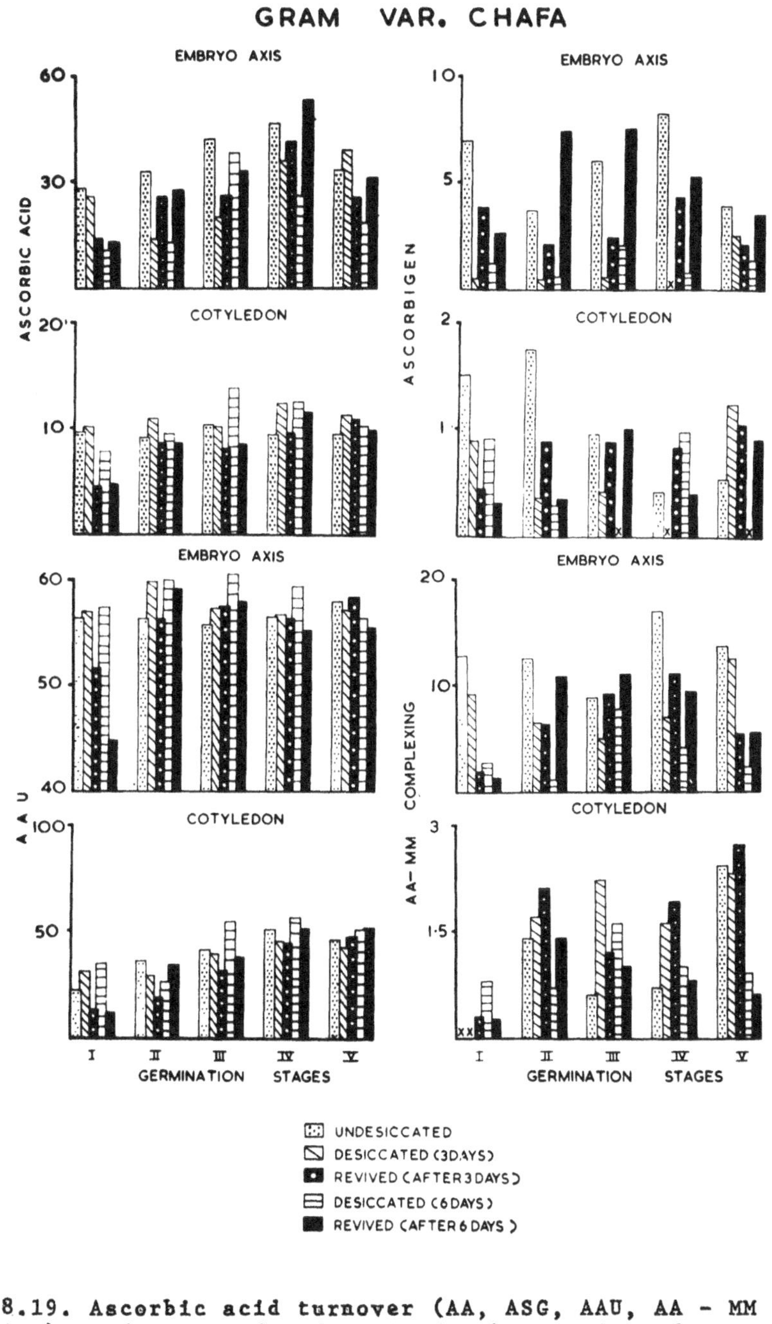

FIGURE 8.19. Ascorbic acid turnover (AA, ASG, AAU, AA - MM complexing) in desiccated and revived embryo axis and cotyledon of *Cicer arietinum* cv. Chafa at different stages of germination (Chinoy et al., 1971; 1974).

in the cotyledon (Fig. 8.19). AA - MM complexing showed a fluctuating trend with the highest complexing ability at 4th stage of germination in the fresh embryo axes. In the 3 and 6 day desiccated embryo axes and cotyledon, a decline and increase respectively in AA - MM complexing was noted during periods of desiccation as well as revival at I to IV stages of germination. On revival, the seedlings showed higher complexing ability than the desiccated ones except in stage I. The same trend was observed in the cotyledon. Seedlings revived after 3 day desiccation gave higher complexing ability compared to those revived after 6 days of desiccation. It may be pointed out here that the revival time to reach the next stage of germination was only 24 to 40 hours after the removal of water stress (Fig. 8.19). The highest AA - FR peroxidase activity was found in fresh embryo axis, which declined to a low level during 3 and 6 day desiccation periods at I and II seedling stages. In the later stages, reduction in this enzymic activity was not so pronounced. The activity was however, restored to a considerable extent in reviving embryo axis. The recovery was more pronounced in embryo axis which was subjected to 4 day desiccation as compared to the one subjected to 6 day desiccation. The overall trend in cotyledons was the same except that it showed a much lower level of the enzymic activity as compared to that of the embryo axes. In later stages of germination, the enzymic activity was significantly higher in both the desiccated seedlings as well as in those which had revived compared to that in undesiccated ones (Fig. 8.20).

Joshi and Hegde (1978) have reviewed their work on AA and stress physiology. They have shown that AA stimulated even chlorophyll synthesis and rate of carbon utilization in rice and sugarcane. AA was also found to induce resistance to salt stress.

The role of AA in nitrogen fixation, nodulation and leghemoglobin of cowpea, lucerne and groundnut (Bharti et al., 1978; Garg, 1978; Goswami and Garg, 1978; Lal et al.,

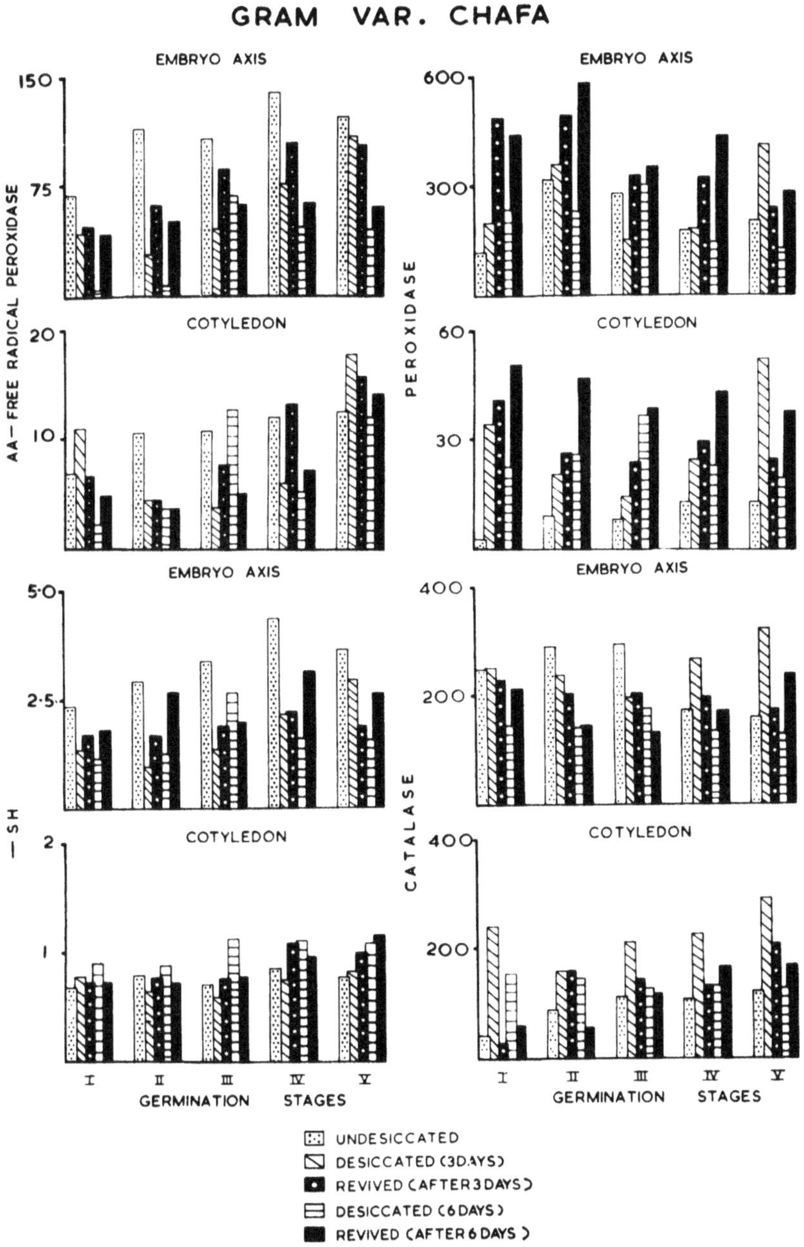

FIGURE 8.20. AA - FR peroxidase, general peroxidase and catalase activities as well as -SH content of desiccated and revived embryo axis and cotyledons of Cicer arietinum cv.

Chafa at different stages of germination (Chinoy et al, 1971; 1974).

1978) have shown that AA largely counteracted the effect of stress.

The revival after water stress, enhanced the AA content in barley cv. N.P. 20 and Peatland. Rapid AA turnover helps in maintaining normal metabolic activity in the growing plants under water stress (Chinoy et al., 1973; 1973 a). During the adult phase, AA and AAU increased during wilting period and also after revival in wheat (Acharya, 1968). AA content was considerably enhanced in wheat by wilting treatment at the shooting stage. In ragi it showed good increase in plants wilted at the tillering stage. In both wheat and ragi, AA content decreased in plants revived after wilting at the anthesis stage. AA was utilized more in plants wilted at tillering and shooting stages than those undergoing wilting treatment at anthesis. In Sesamum AA content increased considerably in plants subjected to wilting treatment at flower-bud initiation stage. Here AAU was the highest in plants at fertilization stage. Wheat and ragi plants raised from seeds pretreated with AA had more ASG content compared to both the controls i.e., pretreatment with distilled water and without pre-treatment. ASG content was higher in plants subjected to wilting treatment at the tillering and shooting stages compared with that of anthesis. It was observed that during wilting period and after revival, ASG content increased considerably in wilted plants as compared with that in fully watered ones.

From the EPR spectra of excised embryo axis as well as of cotyledons, the free radical content of fresh and desiccated material and of seedlings revived after desiccation was determined on dry weight basis (Fig. 8.21). Two kinds of signals were recorded: Class I signal which was most probably due to charge transfer complexes involving a series of actively metabolizing biological molecules; and

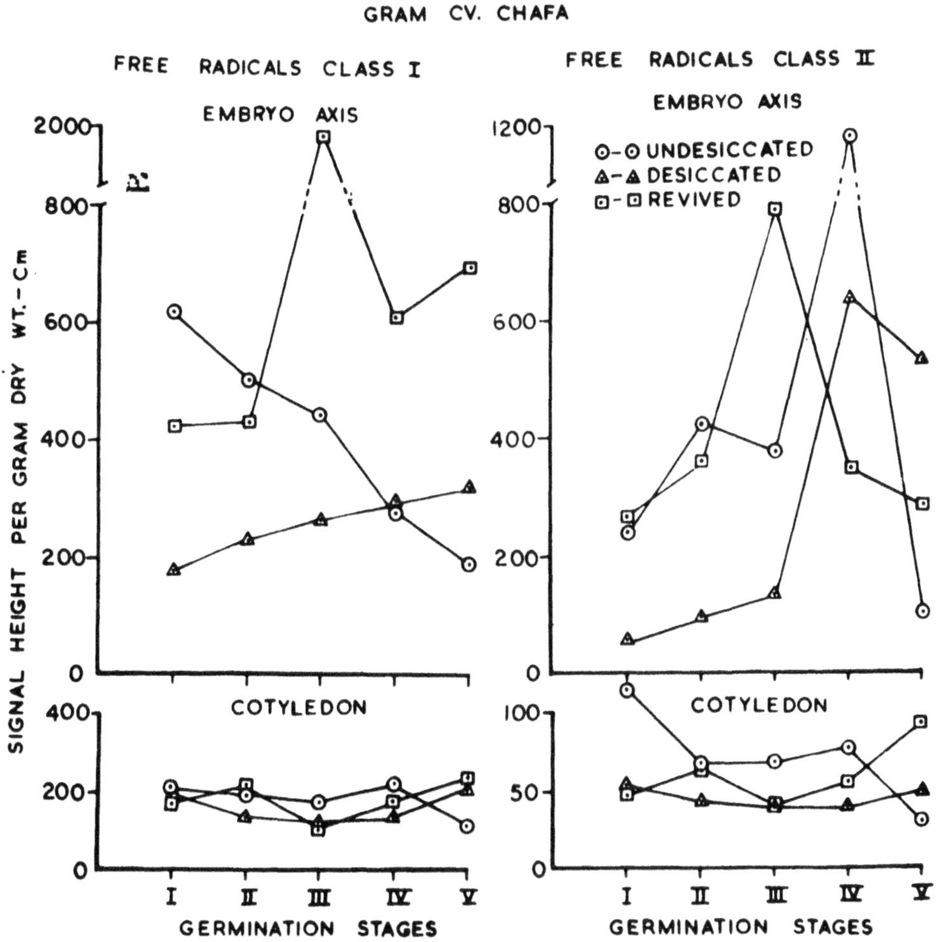

FIGURE 8.21. Free radical content (signal height) of desiccated and revived embryo axis and of cotyledon of Cicer arietinum cv. Chafa at different stages of germination.
Class I signal arising most probably from the breakup of charge transfer complexes of actively metabolizing biological molecules.
Class II signal arises due to a single species of organic free radical (Chinoy et al, 1971; 1974).

Class II signal due to a single species of organic free radical. In both the cases, the free radical content of the embryo axis decreased during desiccation. On revival, however, there was a significant upsurge in the free radical content. In the cotyledon there was a slight increase in the free radical content after revival (Chinoy et al., 1971; 1971 a; 1974).

Seeds of four cultivars of barley, viz., N.P. 13, N.P. 18, Presto and Munsing were pretreated with distilled water and 25 mg/l AA solution, sown in pots and subjected to four drought treatments: (i) Full water (15 - 18% soil moisture); (ii) drought at tillering stage (8 days); (iii) drought at shooting state (8 days) and (iv) continuous water stress 4 to 5% soil moisture (Chinoy et al., 1966). The results clearly indicated that height, dry weight, and number of tillers were all favourably affected by pretreatment with AA. Another point worthy of note was that the levels of all these growth characters after recovery from drought at tiller initiation stage, were the highest in the case of plants raised from seeds pretreated with AA (Fig. 8.22). AA - pretreatment also increased the grain yield and the 1000 kernel weight of all the four cultivars of barley especially in early ones like N.P. 13 and N.P. 18. In the case of yield and 1000 kernel weight also, the AA - pretreated plants which were revived after wilting at tiller initiation stage gave the highest values (Fig. 8.23).

In a review (Chinoy, 1977) on the importance of presowing treatment of the seeds for inducing drought as well as salinity resistance, it was concluded that: (1) AA - pretreated plants gained an appreciable advantage in growth over the control during the seedling stage. This advantage was brought about by acceleration in the dry matter production as well as in the enhanced relative growth rates of seedlings not only after full water conditions, but also under restricted watering and under combined drought and salinity; (2) The results clearly indicated the superiority of pretreated seeds over the untreated ones in their capacity to extract moisture from the atmosphere more efficiently. It appeared that presowing hardening treatment

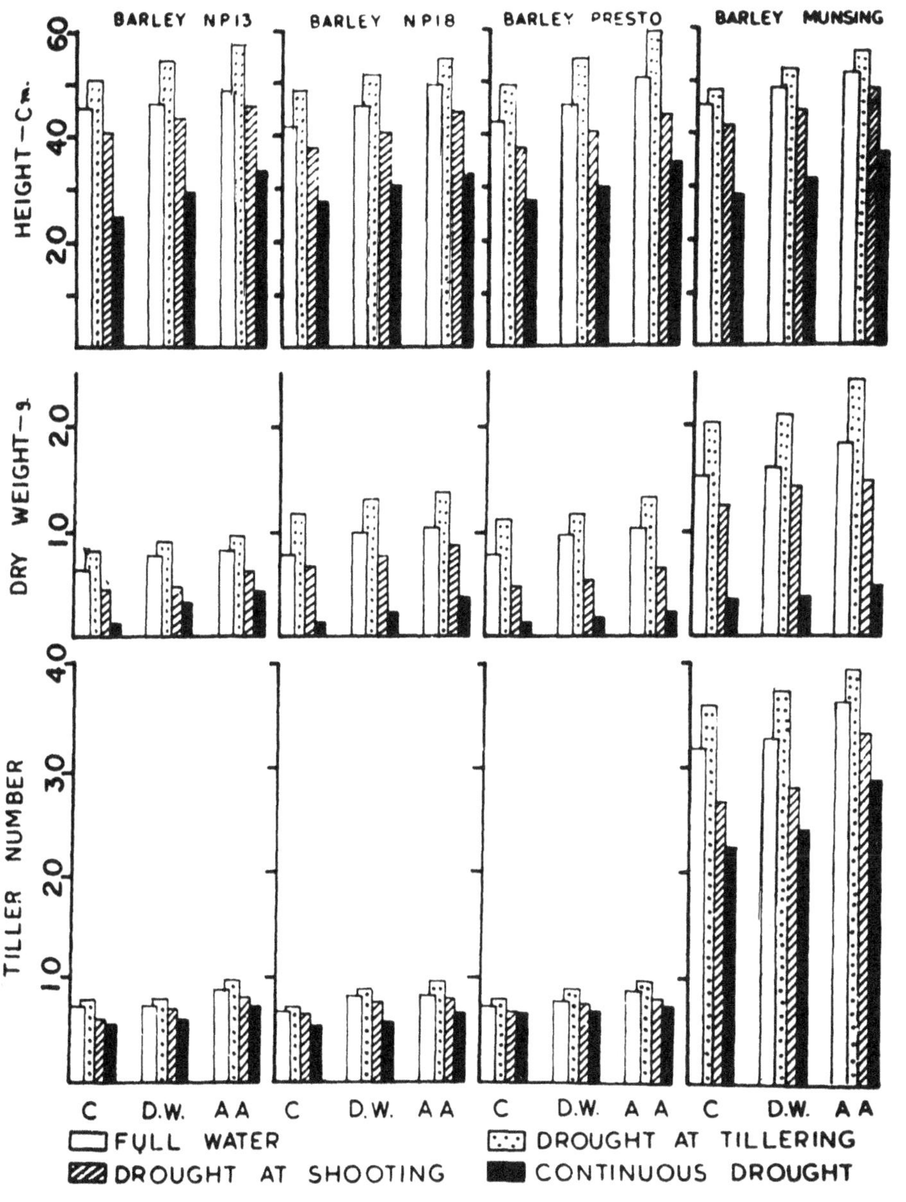

FIGURE 8.22. Effect of water stress at different stages of growth as well as that of pretreatment of seed with distilled water (DW) and AA on height, dry weight and tiller number in

four cultivars of barley, N.P. 13, N.P. 18, Presto and Munsing (Chinoy et al., 1966).

with AA brought about some change in the lyophilic colloids of the seeds, enabling them to imbibe water even when it was at a higher tension; (3) The enhanced hydrolytic activity of amylase in the AA - pretreated seeds throughout the course of the experiment revealed undoubtedly that hydrolysis of starch occurred at a much faster rate in the pretreated seeds as compared to the control ones; (4) The sharp fall in the starch contents of the pretreated series compared to the control lends further support to the above conclusions. This was also confirmed by an accelerated increase in the quantity of soluble carbohydrates like sucrose in the germinating seeds of the pretreated series, elucidating a faster mobilization of reserve carbohydrates for growth of the seedlings.

John and Jani (1969) investigated the effect of presowing treatment of seeds of Hordeum vulgare, cv. N.P. 21, Betzes and Spartan, the early, medium and late flowering cultivars respectively, with 25 mg/1 solution of AA and with distilled water (DW). The drought intensity and survival values (per cent plants fully surviving after revival from drought) were determined.

AA pretreatment increased the survival values of plants especially under water stress given at the shooting stage both under saline and nonsaline conditions. This enhancement in survival values of AA - pretreated plants over those of the control and DW - pretreated ones was especially noticeable in the case of Betzes and Spartan, the medium and the late flowering cultivars respectively, both under saline and nonsaline conditions.

Increase of phosphorus level increased sugar and AA concentration and decreased the nitrogen content in plant tissues, especially under water deficit (Popovskaya, 1957). AA content was found to be inversely related to the nitrogen concentration and the amount of water in the plant tissues.

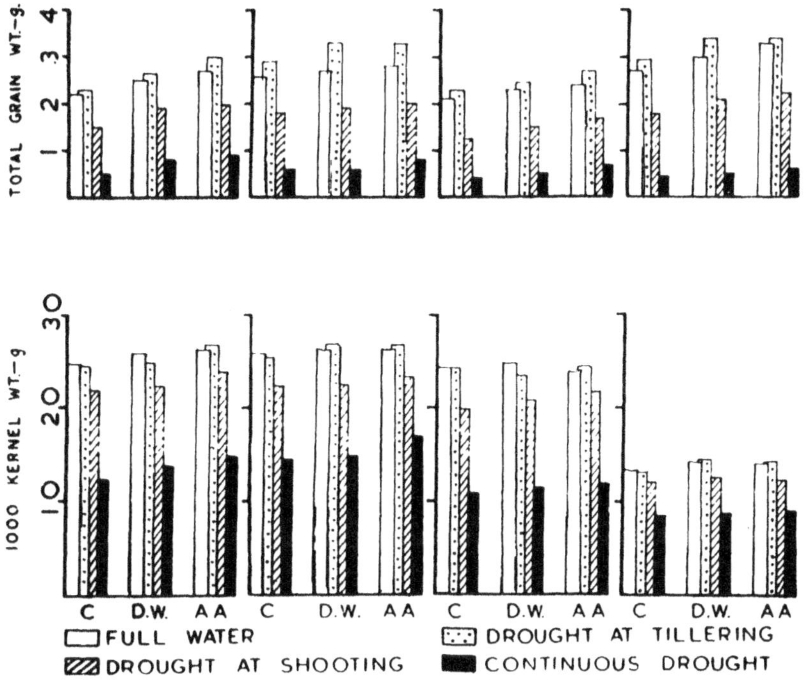

FIGURE 8.23. Effect of water stress at different stages of growth and pretreatment of seed with DW and AA on grain yield per plant and 1000 kernel weight in four cultivars of barley, N.P. 13, N.P. 18, Presto and Munsing (Chinoy et al., 1966).

Pea seeds were pretreated with aqueous solution of Mn, Cu, Co, Mo and Vn salts. Control seeds were treated with water. In the case of all the trace elements Vitamin C increased at the time of blossoming. Best results were obtained with Mo and Vn. Maximum increase in foliar mass, pea pod and kernel weight were found in Mo treated plants. Maximum chlorophyll and Vitamin contents were found at the blossoming phase. Chlorophyll content was highest in Mo and Mn treated plants. Mo, Mn and Co increased protein content in pea kernels (Kamynina , 1965). Kobakhidze et al. (1971) reported that Mn intensified the plant respiration, enzymic activity and ascorbic acid in maize. Proteins and lipids were also accumulated in kernels under the influence of Mn. Yield was also increased by Mn application. When boron and Mo were added to the soil, the intensity of transpiration increased. Water deficit was lowered and ascorbic acid content of plants was also improved by both the trace elements. They also enhanced growth and productivity of the plant (Guseinov et al., 1965).

Cicer arietinum (chickpea) raised in soil with 0, 2, 4 and 6 mmhos/cm of salinity levels were spray-treated with GA_3, kinetin and L-AA, thrice, 42, 52 and 62 days after sowing. Salinity adversely affected the growth parameters of leaves, stem and root which could be ameliorated with the help of kinetin and L-AA treatment (Singh and Singh, 1980).

It would be well to note that when seeds of diverse crop plants like wheat, barley, ragi and Sesamum were germinated under optimum moisture conditions upto a given seedling growth stage and then subjected to desiccation at almost zero humidity for a given period of time, there invariably occurred an enhancement in AA content during desiccation. On revival, there was a fall in AA concentration with a concomitant rise in ASG content (Acharya, 1968; Chinoy et al., 1969 b). On the other hand, when seeds of barley were germinated and grown in low moisture levels from the beginning it was found that continuous water stress thus created from the inception of imbibition significantly lowered the AA content of water stressed embryo axis compared to that of fully watered ones. Metabolism was, therefore, at a low ebb in continuous water stressed seedlings.

In the adult phase also, when plants were subjected to a continuous low moisture regime (7 - 10%) their AA turnover, growth, metabolism and yield were reduced considerably as compared with that in plants receiving adequate water (15 to 20%). Although plants grown under continuous water stress (7 - 10%) did not suffer actual wilting even once during the growth period, they showed adverse effects of low moisture level. On the other hand, plants undergoing permanent wilting for 8 to 10 days especially during the juvenile phase, not only recovered from the adverse effect of wilting on restoration of normal water regime, but also surpassed in growth, yield and AA turnover compared to those grown under normal watering throughout their growth period.

On the basis of the extensive data on AA-turnover, AA - FR - peroxidase activity, AA - MM complexing and free radical content as well as taking into consideration enhanced growth, earlier development and increased yield of crop after restoration of the normal water regime, it was postulated (Chinoy et al., 1971 a, b), that a period of wilting of a short duration during the early stage of plant growth stimulated the redox environment of the cell towards the reductive side thus protecting enzymic activity of the cell. Further, the decrease in the free water content of the cell due to wilting brought about the molecules of AA and macromolecules like DNA and RNA closer to each other, thus facilitating the formation of a charge transfer complex (Chinoy, 1969, 1969 a). Enhancement in the activity of AA - FR peroxidase during the period of revival produced the free radical of AA (Chinoy, 1969, 1969 a). On restoring the normal water regime, both these processes enhanced electron energy which was utilized for faster synthesis of cell constituents leading to rapid growth, earlier development and better yield of plants which were wilted during the juvenile phase for a short duration and subsequently returned to normal water regime. Plants under continuous water stress as well as those which were wilted for a short period at the shooting and the flowering stages not only did not follow the above pattern but showed reduction in the growth and yield.

5. SUMMARY

Contradictory results have been reported on AA content of plants undergoing water stress. Some workers have reported an enhancement in AA concentration under moderate water stress, while others showed that even under severe water stress there was a sharp increase in AA content of leaves and other plant parts. On the other hand, plants of several species grown under dry conditions registered a decrease in AA and reduced glutathione contents.

An enhancement in AA content and AAU with concomitant reduction in ASG during the period of desiccation as well as a lowering of free AA content, but an increase in ASG levels during the subsequent period of revival, are suggestive of important protective and electron generating roles of AA during water stress.

The free radical content of the embryo axis was reduced by desiccation, but on revival, a significant upsurge in its content occurred. However, in the cotyledon, a slight increase was observed after revival.

On the basis of the available evidence it is suggested that when seeds of diverse crop plants like wheat, barley, ragi and Sesamum were germinated under optimum moisture conditions upto a given growth stage of the seedling and then subsequently subjected to desiccation at almost zero humidity for a given period of time, there was invariably an enhancement in AA content during desiccation. On revival, a fall in AA content occurred with a concomitant rise in ASG level. Under continuous water stress (7 - 10%), the plants did not show any wilting during the entire period of growth, but they revealed adverse effect of low moisture level. However, plants undergoing 8 to 10 days 'permanent' wilting especially during the juvenile phase not only recovered from the adverse effect of wilting on restoration of normal watering (15 - 20%), but also surpassed the AA turnover, growth and yield of plants grown under normal watering throughout. Plants under continuous water stress as well as those surviving after wilting for a short period of time at the shooting and flowering stages showed reduction in growth and yield.

Numerous workers from various countries have reported beneficial effect of exogenous application of AA to seeds of different crop plants in their growth and yield and also induced resistance to drought as well as salinity. There is a paucity of work on AA turnover and its role in the metabolism of seed under thermal pretreatment. High temperature treatment (50° C, 70° C, 100° C) given to air-dry seeds of wheat and soyabean accelerated AA turnover and consequently activities of various enzymes and levels of metabolites which in turn increased metabolic processes and thus resulted in enhanced growth and yield.

It has been shown that increased free radical content in many plants brought about an enhancement in growth and productivity. The above mentioned fact has been made the basis of a method of pretreatment of seeds with AA + H_2O_2. The addition of H_2O_2 in traces to AA helped to generate MDHA, the free radical of AA, whose effect was to enhance germination, juvenile and adult growth as well as yield.

REFERENCES

1. Abraham PG 1969. Ph.D. Thesis, Botany Department, University School of Science, Gujarat University, Ahmedabad - 380009, India.
2. Abraham PG, Dave IC, Saxena OP, Pandya RB 1968 a. Abst 37 th Annual Session Nat Acad Sci India p 45.
3. Abraham PG, Dave IC, Pandya RB, Saxena OP 1968 b. Abstr Ibid p 48.
4. Abraham PG, Gurumurti K, Dave IC, Pandya RB, Saxena OP 1968 c. Abstr Ibid p. 49.
5. Abraham PG, Gurumurti K, Dave IC, Pandya RB, Saxena OP 1969. Vidya, Gujarat University Journal 12: 194.

6. Acharya UH 1968. Ph.D. Thesis, Botany Department, University School of Sciences, Gujarat University, Ahmedabad - 380009, India.
7. Ahluwalia A, Srivastava AK 1977. Indian J Ecol 4: 157.
8. Banerji D, Kausal SK 1980. J Indian Bot Soc (in press).
9. Basu RN 1976. Seed Research 4: 15.
10. Bharti S, Garg OP, Prabha C 1978. Indian Soc Plant Physiol Symp, Ahmedabad Jan 1-2, Abstr p 40.
11. Chinoy JJ 1942. Curr Sci 11: 400.
12. Chinoy JJ 1947. Indian Farming 8: 72.
13. Chinoy JJ 1960. Fyton 14: 147, VI.
14. Chinoy JJ 1961. Fyton 16: 131, VI.
15. Chinoy JJ 1961 a. Fyton 17: 1, IX.
16. Chinoy JJ 1962. Indian J Plant Physiol 5: 172 FG Gregory Memorial Volume.
17. Chinoy JJ 1967. Poona Agr Coll Mag 57: 1.
18. Chinoy JJ 1968. Vidya 11: 138.
19. Chinoy JJ 1969. Proc Symp "Planning for Drought Areas" New Delhi, 4-6th May, Abstr Section IV p 34.
20. Chinoy JJ 1969 a. Proc. IIIrd Int Cong of the Int Union for Pure and Applied Biophysics. Cambridge, Massachusetts USA, Aug 29-Sept 3, Abstr No. II K 3 p II K.
21. Chinoy JJ 1977. Indian J Plant Physiol 20: 1.
22. Chinoy JJ, Saxena OP 1969. Sci & Cult 35: 606.
23. Chinoy JJ, Saxena OP 1978. Botanical Progress 1: 6.
24. Chinoy JJ, Jani BM, John D 1966. Proc Sem "Growth and Development of Plants" Bot Dept Panjab Univ Chandigarh Dec 28-30, 1965, p 169.
25. Chinoy JJ, Jani BM, Raj KPS 1968. Symp "Natural Resources of Rajasthan" Jodhpur Univ & UGC, Jodhpur Oct 23-26 Abstr No. A 19, p A 17.
26. Chinoy JJ, Saxena OP, Abraham PG, Pandya RB, Dave IC, Gurumurti K, Sastri K 1969 a. Report PL 480 Research Project FG-In-182 (A7-CR-87), Botany Department, University School of Sciences, Gujarat University, Ahmedabad - 380009, India, p 1-350.
27. Chinoy JJ, Patel HD, Jani BM, John D, Vyas AV, Raj KPS 1969 b. Report PL 480 Research Project FG-In-208 (A7-CR-108), Botany Department, Gujarat University, Ahmedabad - 380009, India, p 1-450.
28. Chinoy JJ, Abraham PG, Pandya RB, Saxena OP, Dave IC 1970. Indian J Plant Physiol 13: 40.
29. Chinoy JJ, Singh YD, Gurumurti K 1971. J Indian Bot Soc 50: 462.
30. Chinoy JJ, Singh YD, Gurumurti K 1971 a. Indian Agriculturist 15: 33.
31. Chinoy JJ, Singh YD, Laloraya MM, Saxena OP 1972. Sem "Current Trends in Plant Sciences" Delhi Oct 3-10.
32. Chinoy JJ, Ghesani PI, Saxena OP 1973. Acta Bot Indica 1: 11.
33. Chinoy JJ, Vyas AV, Raj KPS, Jani BM 1973 a. Ind J Plant Physiol 16: 35.
34. Chinoy JJ, Singh YD, Gurumurti K 1974. Biologia Plantarum (Praha) 16: 301.
35. Chinoy JJ, Patel HD, Patel JN 1975. Vidya 58: 69.

36. Choe HT 1972. Hort Science 7: 476.
37. Crocker W, Barton LV 1957. Physiology of Seeds Chronica Botanic Waltham Mass.
38. Dave IC 1969. Ph.D. Thesis, Botany Department, University School of Sciences, Gujarat University. Ahmedabad - 380009, India.
39. Dawson MJ 1965. Indian J Plant Physiol 8: 52.
40. Duhamel-du-Monceau 1763. Èlements d' agriculture Paris 1.
41. Eisele G 1944. Diss Darmstadt as cited by Stocker O 1948. Planta (Berlin) 35: 445.
42. Farkas GL, Rajhathy T 1955. Planta (Berlin) 45: 535.
43. Garg OP 1978. Indian Soc Plant Physiol Symp Ahmedabad Jan 1-2, Abstr p 21.
44. Ghesani PI 1972. Ph.D. Thesis, Botany Department, University School of Sciences, Gujarat University, Ahmedabad - 380009, India.
45. Goswami CL, Garg OP 1978. Indian Soc Plant Physiol Symp Ahmedabad, Jan 1-2, Abstr p 42.
46. Grzesink S, Letki J, Sojka E 1979. Zesz Nauk Akad Roln Tech Olsztynie Roln 26: 15.
47. Gupta SM 1956. Ph.D. Thesis, Botany Department, Delhi University, Delhi.
48. Gurumurti K, Sastri K, Pandya RB, Saxena OP. Abraham PG, Shah PN 1969. National Sem "Plant Physiology and Crop Production" Solan May 26-31, Abstr VI.
49. Guseinov BZ, Mamedova ZYu, Mamedova AM 1965. Izv. Akad. Nauk Azerb SSR, Ser Biol Nauk 4: 7.
50. Gutmanis K 1967. Izv Akad Nauk Lat SSR 5: 105.
51. Harrington JF 1973. Panel Discussion p 525, In: Seed Ecology Heydecker W (ed) London : Butterworths.
52. Hausen Saubert-Von VS 1936. Ann Acad Sci Fenn Ser A. 46: 1.
53. Henckel PA 1946. Truol Inst Fiziol Rast Timiryazeva 5: 237.
54. Henckel PA 1952. Field Crop Abstr 5: 588.
55. Henckel PA 1961. Proc Madrid Symp Paris: UNESCO p 167.
56. Henckel PA 1964. Ann Rev Pl Physiol 15: 363.
57. Henckel PA, Kolotova SS 1934. Imp. Bur Plant Genet. Abverstwyth & Cambridge.
58. Heydecker W 1973 (ED) Seed Ecology London: Butterworths.
59. Heydecker W, Coolbeor P 1977. Seed Sci Technol 5: 354.
60. Heydecker W, Higgins J, Gulliver RL 1973. Nature (Lond.) 246: 42.
61. Heydecker W, Higgins J, Turner YJ 1974. Proc. 17th Intern Seed Testing Assoc Congress Walsa June 17-22, 1974 38: 1.
62. Hunter AS, Kelly WC, Somers GF 1950. Agron J 42: 96.
63. Ivanov SM, Karakash LA 1964. Kartya Moldovenyaske: Kishnev 21;30, Zh Biol 9 G 59.
64. Jani BM 1969. Ph.D. Thesis, Botany Department, University School of Sciences, Gujarat University, Ahmedabad - 380009, India.
65. Jani DJ 1978. Ph.D. Thesis, Botany Department, University School of Sciences, Gujarat University, Ahmedabad - 380009, India.

66. John D 1965. Proc Sem "Sea, Salt and Plants" Bhavnagar, Gujarat, India, p. 79.
67. John D. 1969, Ph.D. Thesis, Botany Department, University School of Sciences, Gujarat University, Ahmedabad - 380009, India.
68. John D, Jani BM 1969. Proc Natl Acad Sci India 39 B: III: 217.
69. Joshi GV, Hegde BA 1978. Indian Soc Plant Physiol Symp Ahmedabad, India p 21.
70. Kamynina LM 1965. Agrokhimiya 10: 123.
71. Kaul M 1975. Ph.D. Thesis, Botany Department, University School of Sciences, Gujarat University, Ahmedabad - 380009, India.
72. Kobakhidze KV, Chkoniya TA, Dzhavshnanashvili IZ 1971. Soobshch Akad Nauk Gruz SSR 62: 153.
73. Korneev PK 1963. Sov Plant Physiol 9: 493.
74. Kydrev TG 1959. Soviet Pl Physiol 6: 179.
75. Lal H, Swaraj K, Garg OP 1978. Indian Soc Plant Physiol Symp Ahmedabad, India, p 85.
76. Lang A 1965. Effects of some internal and external conditions on seed germination In: Encycl Pl Physiol Ruhland W (Ed) Berlin: Springer-Verlag, 15: 842.
77. Luckicheva EL 1968. Trudy Inst Bot Alma-Ata 25: 23.
78. Malik CP 1981. Proc 8th Ann Conf Soc Advan Bot Chandigarh, India, March 24-25.
79. May LH, Milthorpe EJ, Milthorpe FL 1962. Field Crop Abstr 15: 93.
80. Mehta BH 1973. M.Sc. Dissertation, Botany Department, University School of Sciences, Gujarat University, Ahmedabad - 380009, India.
81. Mehta DH 1972. M.Sc. Dissertation, Botany Department, University School of Sciences, Gujarat University, Ahmedabad - 380009, India.
82. Mehta DH 1977. Ph.D. Thesis, Botany Department, University School of Sciences, Gujarat University, Ahmedabad - 380009, India.
83. Mehta KG 1978. Ph.D. Thesis, Botany Department, University School of Sciences, Gujarat University, Ahmedabad - 380009, India.
84. Nanda KK, Chinoy JJ, Gupta SM 1959. Fyton 12: 153.
85. Novitskaya YUE 1958. Trudy Bot Inst Akad Nauk SSSR Sec 4: 74.
86. Osborne DJ 1973. Panel Discussion Presowing seed treatments In: Seed Ecology Heydecker W (ed). London: Butterworths, p 521.
87. Ovacharov KE 1958. Akad Nauk SSSR Press The role of Vitamins in Plants (in Russian).
88. Ovcharov Kyu 1963. Soviet Plant Physiol 10: 65.
89. Parija P 1943. Curr Sci 12: 88.
90. Parija P, Pillay KP 1945. Proc Nat Acad Sci India 15 B: 6.
91. Parmar U, Singh P, Malik CP 1981. Proc. 8th Ann Conf Soc Advan Bot Chandigarh, India, March 27-28, p 101.
92. Patel HD 1967. Ph.D. Thesis, Botany Department, University School of Sciences, Gujarat University, Ahmedabad - 380009, India.

93. Patel JN 1972. M.Sc. Dissertation, Botany Department, University School of Sciences, Gujarat University, Ahmedabad - 380009, India.
94. Patel JN 1976. Ph.D. Thesis, Botany Department, University School of Sciences, Gujarat University, Ahmedabad - 380009, India.
95. Patel JN, Jadeja PD, Patel KM, Joshi AK, Chinoy JJ 1980. Indian J Plant Physiol 23: 33.
96. Popovskaya EM 1957. Fiziol Rastenii 4: 326.
97. Raj KPS 1971. Ph.D. Thesis, Botany Department, University School of Sciences, Gujarat University, Ahmedabad - 380009, India.
98. Rammelt R 1973. In: Proc Colloq The Winter Hardiness of Cereals Rajki S (ed) Hungarian Acad Sci Martonvasal p 71.
99. Roberts WO 1948. Jour Agr Sci 38: 458.
100. Roos EE, Pollock BM 1971. Crop Sci 11: 78.
101. Salim MH, Todd GW 1968. Agron Jour 60: 179.
102. Saxena OP 1974. Planto 4: 1.
103. Saxena OP 1973 - 1976. Gujarat Univ Final Research Progress Report.
104. Saxena OP 1979. In: Current Trends in Plant Reproductive Biology Malik CP (ed) New Delhi & Ludhiana: Kalyani Publishers p 323.
105. Saxena OP 1981. Proc 8th Ann Conf Soc Advance Bot Chandigarh March 27-28 B - 37: 57.
106. Saxena OP 1982. GAAS Sem GAU Ahmedabad Feb 28, p 243.
107. Saxena OP 1983. Adv. in Pl. Reprod. Biol. Malik CP (ed). New Delhi: Kalyani Publishers, p 1.
108. Scott TK (ed) 1979 Plant Regulation and World Agriculture. New York: Plenum.
109. Sechet I 1953. C R Acad Sci 237: 434.
110. Shah HK 1973. Ph.D. Thesis, Botany Department, Gujarat University, Ahmedabad - 380009, India.
111. Shah HO 1975. Ph.D. Thesis, Botany Department, Gujarat University, Ahmedabad - 380009, India.
112. Shah PN 1972. Ph.D. Thesis, Botany Department, Gujarat University, Ahmedabad - 380009, India.
113. Shah SK 1972. M.Sc. Thesis, Botany Department, Gujarat University, Ahmedabad - 380009, India.
114. Shah CK, Shah HO 1979. Proc Symp Form Structure and function in plants.
. Shah JJ (ed) New Delhi: Today & Tomorrow Publishers p 47.
115. Sharma NS 1981. Ph.D. Thesis, Botany Department, Gujarat University, Ahmedabad - 380009, India.
116. Shcherbakov BI 1963. Trans Doklady Akad Nauk SSSR 152: 231.
117. Singh G, Singh H 1980. Indian J Agric Sci 50: 22.
118. Singh OS, Vijaykumar KR 1974. Seed Research 2: 13.
119. Startseva AV 1963. Kazanskogo Filiola Akad Nauk SSSR, Ser Biol Nauk 9: 68.
120. Solomon LL 1955. Biol Abstr 30: 9553, 1956.
121. Stocker O 1956. Handbuch der Planzenphysiologie Berlin: Springer-Verlag 3: 696.

122. Stocker O 1960. In: Plant water relationships in arid and semi-arid conditions: Arid zone research, Paris: UNESCO 15: 63.
123. Stocker O 1961. Indian J Plant Physiol IV: 87.
124. Srivastava AK 1978. Seed physiology in relation to Crop Production, Panjab Agri Univ Ludhiana March 20-22, p 32.
125. Srivastava AK, Singh G 1979. Indian J Exp Biol 17: 1280.
126. Sutaria MA 1972. M.Sc. Dissertation, Botany Department, Gujarat University, Ahmedabad - 380009, India.
127. Thakker PS 1974. M.Sc. Dissertation, Botany Department, Gujarat University, Ahmedabad - 380009, India.
128. Tombesi L 1951. Durante il Ciclo Vegetative e la Cura Tobacco No 618.
129. Tombesi L 1952. Ann Sper Agr Roma N S 6.
130. Thomson JR 1979. An Introduction to Seed Technology, Leonard-Hill Bishopbriggs.
131. Varghese S 1981. Ph.D. Thesis, Botany Department, Gujarat University, Ahmedabad - 380009, India.
132. Vora AB, Patel DV 1979. Botanique 10: 119.
133. Vora AB, Patel MK 1975. Curr Sci 44: 749.
134. Vyas AV 1971. Ph.D. Thesis, Botany Department, Gujarat University, Ahmedabad - 380009, India.

9. RADIATION

Nuclear energy has been utilized in this field of biology and agriculture for the development of high yielding and disease resistant crop mutants, physiological stimulation using low doses of radiation for higher yield of crops, forages and horticultural plants, lengthening the storage period or post harvest half-life and understanding the basic mechanisms in life sciences e.g., through the radiolytic products of important biomolecules under both *in vivo* and *in vitro* conditions.

The effect of radiation on AA metabolism in plants has been dealt with in this chapter.

1. EFFECTS OF RADIATION ON AA TURNOVER

Low ^{60}Co radiation doses stimulated growth, development and synthesis of essential metabolites in plants, such as vitamins B and C, phosphatides, RNA and DNA. After treatment with small doses of radiation, the stimulating effect on these vitamins was reduced due to storage (Chou Liu et al., 1965). Seven day old winter wheat, barley and bean seedlings grown at 12 - 15° C, 25° or 40° C were irradiated with 100 - 1000 r doses of ^{60}Co γ-rays. Low doses of γ-irradiation stimulated respiration in barley while high doses of 1000 r inhibited it. Wheat seedlings reacted differently than the barley seedlings. The intensity of respiration became normal 30 minutes after irradiation. Sugar content increased and AA content decreased within 20 minutes of the post-irradiation period (Batygin and Skalinova, 1969). The AA content of beans, corn and cotton increased sharply due to 500, 1000, 5000 r doses. Radiation of moderate doses improved the quality of some vegetable products and increased the vital tone of the plant tissues. Peroxidase and polyphenol oxidase also rose sharply under optimal dose (Tagizáde and Guseinov, 1964). Irradiation of grape seeds in a gamma cell (^{60}Co) in a range of 200 - 3000 r increased the AA content of leaves. This increase was accompanied by an increase in enzymic activities of catalase, peroxidase and polyphenol oxidase (Kikacheishvili et al., 1979).

Chinoy et al. (1973) reported the inductive effect of low doses of gamma irradiation on the biosynthesis of AA and -SH compounds. The seeds of Triticum aestivum cv. N.P. 718 were irradiated with 20, 40, 50, 100, 150 and 200 r from ^{60}Co. The control and irradiated seeds were germinated at 28 $\pm$ 2° C and the embryo and endosperm were analyzed for AA, ASG and AAU. Irradiation treatment enhanced AA content in the embryo axis. In the endosperm, irradiation treatment showed a fluctuating trend. The mean AA content of the endosperm was reduced by irradiation except in the case of 40 and 50 r treatments. Irradiation treatment reduced the ASG content in the embryo axis. Reduction was more pronounced in 20, 50 and 200 r treatments. In the endosperm, the ASG content was reduced by irradiation of 20, 40, 50 and 150 r, while with 100 and 200 r, ASG showed a higher value than the control. Dry matter production of the embryo axis was higher in irradiated seeds, while endosperm of the seedlings showed faster depletion of dry matter compared with that in the control ones. The low doses of ionizing radiations stimulated the release of AA from its bound form to scavenge the harmful radicals produced in the system. Thus there was a net increase in the free AA which was correlated with higher dry weight of the growing axis by accelerating metabolic processes.

Triticum vulgare, Hordeum vulgare and Zea mays seedlings were exposed to β-irradiation emitted from small doses of $H_3{}^{32}PO_4$ or ^{132}I and their AA contents were determined. The results showed that the radiation may either inhibit the biosynthesis of AA or accelerate its destruction during seed germination (Baraket et al., 1966).

Samples of seeds of Vicia faba, Vigna sinensis, Phaseolus vulgaris, Trigonella - Foenum graecum, Lupinus termis were mixed with aqueous solutions containing varying quantities of

$H_3{}^{32}PO_4$, or of $Na^{131}I$ for 24 hours. Other samples were irradiated with 5,000 to 10,000 r (^{137}Cs) and varying doses of γ-irradiation from ^{60}Co. AA content was measured for initial 30 days post-irradiation germination. In general, the concentration of AA was slightly less in irradiated seeds. The difference increased proportionally to the amount of irradiation and varied according to the seed types. It was concluded that both types of irradiation either decreased AA or accelerated its destruction in plants (Baraket et al., 1964). Mutations induced in tomato and radish by fast neutrons were observed only in the second generation. The amount of fruit produced was increased by 100% with 20 to 25% increase in sugar and AA contents. It was concluded that electromagnetic waves, as well as X-, α- and β-irradiations and fast and thermal neutrons produced similar biological, biochemical, biophysical and genetic changes (Giese, 1968). The growth of wheat radicle and grapevine cuttings was retarded by the action of X-rays (1000 - 3000 r), the formation of AA being suppressed as well. Additional irradiation with UV rays (15×10^5 erg/sq. cm to 31×10^5 erg/sq. cm) immediately after X-irradiation, restored the growth of plants and AA accumulation in them (Kezeli et al., 1973). In most, but not all the cases, the presence of ^{32}P and ^{56}Fe increased the AA content of plant leaves. The effect was especially favourable with 1 and 25 Ci. The effectiveness of ^{56}Fe decreased with the increasing dose (Vezirova, 1963).

Pisum sativum and *Vigna unquiculata* (red seeded variety) were exposed to 10 - 30 Kr γ-rays from ^{60}Co γ-irradiation which increased the AA in the seedlings obtained in subsequent germination. The increase was more pronounced when seeds were grown in light. This induced increase was partially reversed when irradiated seeds were treated with 1000 ppm GA_3 before germination. The action of GA_3 was more pronounced for *Vigna* than for *Pisum* (Pathak and Simlot, 1970). The γ-irradiation of seeds stimulated the growth, development and crop of tomato fruit. At doses 500, 1000 and 2000 r, AA content increased by 16.2, 24.3 and 29.7% and sugars by 8.0, 20.0 and 36.0% respectively (Abdullaev and Berezina, 1968).

2. RADIOPROTECTIVE AND RADIOSENSITIZING PROPERTIES OF AA

Inactivation of lambda phage by X-irradiation was greatly reduced in the presence of protective substances like cysteine, AA and others. An increase in the radioresistance of the phage was recordge in anoxia only in presence of cysteine and AA (Hradecna and Toufarova, 1966). X-ray induced physiological damage in crop plants was prevented with cysteine, $FeSO_4$, vitamin C or Na-ascorbate, (ATP was also effective). The induced chromosome abberations were reduced by K compounds, cysteine and reducing agents in both pre- and post-treatments. Among them, reducing agents and cysteine decreased the number of chromosome bridges proportionally to that of fragments in the stages of anaphase to telophase in the first mitotic division (Biswas et al., 1965). There was a 22% reduction in the rate of aberrant mitosis caused by 100 r X-rays with the application of 10^{-2} M AA amongst other substances (Bartheiness et al., 1968). The soaking of rice and barley seeds in solution of AA, Na-ascorbate and others prior to X-irradiation afforded protection against physiological damage. Soaking after irradiation had no effect (Biswas and Matsuo, 1966).

Hazama (1961) found that growth of irradiated embryos with doses from 10 - 40 Kr (at 40 Kr/hour) in agar medium was slightly suppressed. In AA medium, growth was apparently suppressed as concentration of AA ranged from 10^{-2} M to 10^{-4} M. AA decreased day by day. The investigator believed that growth suppression was due to the remaining AA after oxidation. AA at 10^{-6} to 10^{-3} µg/ml, exhibited antiradiation activity on both germinating and dry <u>Allicon fistulosum</u> seeds. Its protective effect against ionizing radiation and the decrease in chromosome mutation was greatest at 10^{-6} and 10^{-3} µg/ml respectively (Selimbekova, 1969). Albumin synthesis induced by X- or UV-radiations was strongly inhibited by the presence of cysteine and AA. The protective substances were added to the solution, immediately before irradiation (Flemming, 1964).

AA has been used in model experiments to investigate the possible role of reducing agents in radioprotection and radiosensitization (Rennert and Antoszsewska, 1970). The repair of organic radicals by their reaction with AA was demonstrated by using pulse radiolysis (Redpath and Willson, 1973). Radiometric measurements showed that the penetration and absorption of radioactive ions like ^{137}Cs, ^{87}Sr and ^{144}Ce were considerably decreased after binding with AA. Impregnation of packing foil with this compound could be a protective measure against radio contamination of foods (Stankovic and Veljkovic, 1968). The protective action of AA against the UV irradiation damage in algae, Chlorella and Chlorococcus was ascribed to its radical scavenging properties (Vijaya and Vaidya, 1980).

Ascorbic acid is known to protect seed germination and plant growth against radiation damage (Conger, 1975; Naslund et al., 1976; Pitillo and Lucas, 1967). The treatments of seeds with AA alone or as post-irradiation treatment are known to be injurious to the germination of barley seeds (Conger, 1973; 1975). A comparative protective role was observed only when the seeds were treated with AA prior to irradiation (Conger, 1975). Pre-irradiation treatment of seeds of wheat and maize with AA showed relatively increased germination, seedling height and chlorophyll content as compared to post-irradiation treatment with AA. The protection rendered by pretreatment with AA may be due to AA acting as a free radical scavenger and thus protecting seeds from radiation induced damage (Tewary and Mookerjee, 1982).

3. EFFECT OF RADIATION ON AA

A 0.5 mg/l aqueous solution of AA was stored for 5 days with 0.5% addition of albumin or gelatin in citrate buffer at pH 2, 4 and 6. In γ-irradiated solutions, the yield of radiation-induced decomposition increased with increasing concentrations of H^+ ions. Albumin and gelatin inhibited this reaction at ≤ 4 pH and were without effect at pH 6. In all cases, the protective properties of albumin were more pronounced than in case of gelatin (Kulesza and Wilska - Jeszka, 1970).

There is evidence to show that a back reaction: BH + HOH = 2B + H_2O leads to a reconstitution of the base where B represents a base. AA is effective at low concentration in blocking this reaction by virtue of H atom transfer reaction: BH + R'H = H_2B + R', where BH_2 represents the dihydrogen derivative of the pyrimidine nucleus. The yield of radiolytic reduction of the 5, 6-double bond of cytosine is essentially quantitative in the presence of AA at low concentration (Holian and Garrison, 1969). AA in the solid state underwent no or little breakdown upon irradiation with γ-rays. Even in solutions with active component, content over 2%, γ-irradiation can be used safely. However, a great change is observed in aqueous solution (0.05 - 0.5%), the loss being 1 - 2%. Parameters used as criteria were colour change, melting point, UV absorption spectra, pH, and polarographic behaviour. Irradiation was carried out with 2.5 megarad at 100,000 r/hr intensity. The best study of degradation products is possible using concentration of 0.1% (Farkas and Nagykaldi, 1970).

Effect of ^{60}Co γ-radiation of AA in deuterated aqueous solution was studied without addition of an anti-oxidant. The reactivity of H atom with AA was also detected by a scavenger method under the experimental conditions (1 mM solution, pH 3.6) G (- AA) = 3.6, while yields of DHA and H_2 were 1.9 and 0.49 respectively, decomposition of AA and formation of H_2 were constant within the concentration range 0.5 - 10 mM for dose upto 6×10^{18} eV ml^{-1} (Ogura et al., 1970). The protective effect of various organic acids and other organic compounds against the radiation decomposition of ascorbic acid in aqueous solution was examined. The decomposition value of AA in aqueous solution per M rad under the atmosphere was increased proportionally with the concentration of AA. On the other hand, the decomposition value for M rad under N_2 was almost constant regardless of the concentration of AA (Umeda et al., 1970).

4. SOME ASPECTS OF FREE RADICAL REACTIONS OF CARBOHYDRATES, ASCORBIC ACID AND DNA STRAND BREAKS

The free radicals are highly reactive species and may be long or short-lived in nature. In a living system they are generated enzymically in a controlled orderly manner as and when required. It is rather difficult to get a verv correct picture of their generation, reactivity and their ultimate fate using *in vivo* systems exclusively. The radiation techniques employed to study them are superior to other radical-generating reactions as except for a little amount of H_2O_2, no interfering product is generated, neither is any additive required. In aqueous system the γ-radiolysis products include -OH and H radicals, solvated electrons (e^- aq) and H atoms arising from H_2O.

$$H_2O \xrightarrow[N_2O]{\gamma\ rays} {}^{\cdot}OH,\ H^{\cdot},\ H_2,\ H_2O_2,\ H^+,\ OH^- \text{ and } e^- \text{ aq.}$$

$$e^- \text{ aq.} + N_2O \longrightarrow {}^{\cdot}OH + N_2 + OH^-$$

H' and 'OH radicals have the capacity to abstract hydrogen atoms attached to carbon atoms of the carbohydrates (Asmus et al., 1973).

Schuchmann and Sonntag (1978) have reviewed the free radical reactions of carbohydrates in general and L-AA in particular. The important events taking place during radical generation include principally the radical reactions. These are characterized by elimination and rearrangement reactions, radical-radical reactions, radical-scavenger reactions and peroxyl-radical reactions using EPR spectroscopic studies. The 'OH' radical was much more acidic (PK_a = -0.45) than L-AA per se (pK_a = 4.1) (Fessenden and Verma, 1978). This property is due presumably to a powerful charge delocalization than in L-AA itself. Fessenden and Verma (1978) studied the oxidation of AA by -OH radical in aqueous solution, employing time resolved EPR spectroscopy. They observed an absolute mono-anion radical within < 2 μS of radiolysis pulse. In acidic solution, (pH 3.45 N_2O): saturated) three radicals - an ascorbate monoanion radical, an OH adduct (also seen in steady state EPR studies) decaying with $t_{1/2}$ of ~ 100 μS and another OH adduct at C - 2 with the main spin density around C-3 decaying with a $t_{1/2}$ ~ 10 μS were observed. The authors concluded that the addict at C-2 loses water rapidly to form the ascorbate radical.

Perhaps the only study on the radiation induced free radicals in crystalline L-AA (McDearmon and Moulton, 1982) suggest models for four species of radicals. Single crystals of L-AA and partially deuterated L-AA were irradiated with either X-rays or laser radiation at 77° and 330° K. At 77° K two oxidation products, radical I is MDHA and radical IV is generated with a loss of H attached to C-4 of the furan ring. Radical III is a reduction product. At ~ 140° K, MDHA decays into DHA with the loss of a second H, and the adduct radical is produced in the process i.e., radical II. On warming upto 300° K, no other changes were observed. Radicals II, III and IV are also absent in crystals irradiated at 300° K, but in different proportions than in crystals irradiated at 77° K and warmed upto 300° K. The unpaired electron in all the cases are in II molecular orbital on the furan ring as evidenced from the eigenvector of the minimum g. eigenvalue.

AA possesses the property of causing DNA breaks in all probability through its free radical(s) (ref: Chapter 1). The free radical induced reactions at the deoxyribose moiety of DNA could result into DNA strand breaks. These reactions involve dephosphorylation of the sugar phosphate esters. One of the mechanisms is the creations of a sub-structure in the DNA chain resulting in the alkali labile sites (Brown and Todd, 1952), ultimately leading to elimination of even the bases.

The free radical of AA is formed by splitting of H atom or electron from -OH group at C-3 of AA. The experimental results were supported by a calculation of the II electron density of AA, its monoanion and its radical (Kluge et al., 1967). The absolute specificity of AA radical formed by a pulse radiolysis of aqueous solution of AA, saturated with N_2O was recorded and rate of decay seen. The transient absorption was found at wavelengths greater than those of AA and its permanent reaction product. It was due to a radical intermediate in the oxidation of AA to DHA (Bielski and Allen, 1970; Bielski et al., 1971). Kalus and Filby (1977) have shown that presence of additives like heavy metal ions and sulphite considerably affect the autooxidation via its free radical formation. The nature of the effects is either catalytic or protective in relation to the radical decay. The decay rate was remarkably dependent on the presence of sulphite (10^{-4} M) which protected the radical against decay by a factor of 14, while 10^{-6} M Cu^{2+} / Na_2SO_3 accelerated the decay by $\sim$ 4. These reactions exhibit strong pH dependence, and generally follow overall second order kinetics, being however of first order when H_2O_2 is present. The authors suggest that either the prevention of anion radical A^- formation completely or its stabilization sufficiently long so as to allow its reduction to AA is of importance in fruit preservation where the metabolic activity is at its lowest ebb.

Stable species are formed during the oxidation of L-AA with different reagents and unstable m-species are also formed. Thus when AA is treated with Ti^{3+} -2 H_2O_2, 1- and m- species are observed. The m-disappears when Fe^{3+} is added. The disappearance of m-species are observed when Fe^{3+} is replaced by Ce^{4+}. The 1- and m- species are different in defined radicals (Kirino and Kwan, 1971; 1972). The absorption spectra of L-AA free radicals generated in aqueous solution (pH, 0.3 to 11) by pulse radiolysis were studied. Variation in Extinction Coefficient with pH suggested that the radicals can exist in a protonated form (AH_2^+), as a neutral species (AH) and as anion radicals (A^-).

The radicals decayed by the second order kinetics throughout the entire pH range. The radicals produced in the radiolysis of aqueous solution of AA, arabo-AA, reductic and hydroxytetronic acids were examined by the *in situ* hydrolysis EPR method.

5. SUMMARY

Several workers have reported that high doses of α, β, γ- and X-ray irradiations and electromagnetic waves caused a reduction in AA content as well as retarded growth and metabolism. Low doses γ-irradiation generally stimulated AA synthesis, moderate doses improved the quality of some vegetable products. UV-irradiation 1 to 4 hours prior to soaking brought about an enhancement in AA content after germination for 2 - 9 days. The activity of AA oxidase, polyphenol oxidase and peroxidase in the seedlings were related to changes in the vitamin C concentration.

Radiation induced physiological damage to crop plants could be prevented by applying reducing substances, especially AA. The repair of organic radicals by their reaction with AA was demonstrated by using pulse radiolysis. It was suggested that chronic irradiation shortened the metabolic life cycle. AA was effective at low concentrations in blocking the reaction, BH + BOH = 2 B + H_2O (where B represents a N base). The yield of radiolytic reduction of 5, 6-double bond of cytosine was essentially quantitative in the presence of AA at low concentration. Stable and unstable species of radicals were formed during the oxidation of AA with different reagents. The variation in extinction coefficient with pH suggests that radicals could exist in a protonated form (AH_2^+), as a neutral species (AH) and as an anion radical (A^-). The radicals decayed by second order kinetics throughout the entire pH range.

Several investigators have suggested the possible role of reducing agents in radioprotection and radiosensitization.

REFERENCES

1. Abdullaev MA, Berezina NM 1968. Dokl Akad Nauk Azerb SSSR 24: 38.
2. Asmus KD, Mockel H, Henglein A 1973. J Phys Chem 77: 1218.
3. Baraket MZ, Abdel-Wahab MF, Megahed YM 1964. Monatsh Chem 95: 1613.
4. Baraket MZ, Abdel-Wahab MF, Megahed YM 1966. J Sci Food Agri 17: 205.
5. Barthelness A, Huber H, Fuerst H 1968. Strahlentherapie 136: 116.
6. Batygin NF, Skalinova NP 1969. Sb Tr Agron Fiziol No 17, 106 (In Russian).
7. Bielski BHJ, Allen AO 1970. J Amer Chem Soc 92: 3793.
8. Bielski BHJ, Comstock DA, Bowen RA 1971. J Amer Chem Soc 93: 5624.
9. Biswas S, Yamamoto K, Sato Y, Ukai Y, Onozawa Y, Matsuo T 1965. Radioisotopes (Tokyo) 14: 411.
10. Biswas S, Matsuo T 1966. Radiation Bot 6: 575.
11. Brown DM, Todd AR 1952. J Chem Soc: 52.
12. Chinoy JJ, Gurümurti K, Singh YD 1973. Ind J Plant Physiol 16: 42.
13. Chou Liu, Mei Chi Hsu, Chin Tao Liu, Kuang Chao 1965. Yuan Tzu Neug No 3 280.
14. Conger BV 1973. Radiation Bot 13: 375.
15. Conger BV 1975. Ibid 15: 39.
16. Farkas PE, Nagykaldi A 1970. Pharmazie 25: 254.
17. Fessenden RW, Verma NC 1978. Biophys J 24: 93.
18. Flemming K 1961. Naturwissenschaften 48: 555 (In German)
19. Giese AC 1968. Cell Physiology, 3rd Ed. Tokyo, Japan, Toppan Co. Ltd., p. 201.
20. Hazama K 1961. Effects of ionizing radiation on seeds. Vienna: Internl Atom Energy Agency Pub p 411.
21. Holian J, Garrison WM 1969. Nature (Lond.) 221: 57.
22. Hradecna Z, Toufarova D 1966. Folio Biol (Prague) 12: 124.
23. Kalus WH, Filby WG 1977. Internl J Vit Nutr 47: 258.
24. Kezeli TA, Gvamichava NE, Tarasashvili KM, Piranshvili NS 1973. Soobshch Akad Nauk Gruz SSR 69: 665.
25. Kikacheishvili RM, Shavishvili KN, Dzhincharadze TG, Murachashvili IA. Nauchno Issled Inst Sadovad Vinograd Vinodol Tiflis USSR.
26. Kirino Y, Kwan T 1971. Chem Pharm Bull 19: 381.
27. Kirino Y, Kwan T 1972. Chem Pharm Bull 20: 2651.
28. Kluge H, Rasch R, Brux B, Frunder H 1967. Biochim Biophys Acta 141: 260.
29. Kulesza J, Wilska-Jeszka J 1970. Zesz Nauk Politech Lodz Chem Spozyw No 17 33 (In Polish).
30. McDearmon GF, Moulton GC 1982. Radiation Res 89: 468.
31. Naslund M, Ehrenberg L, Djalali-Behjadi GC 1976. Int J Rad Biol 30: 95.
32. Ogura H, Murata M, Kondo M 1970. Radioisotopes 19: 89.

33. Pathak GL, Simlot MM 1970. Radiation Radioisotope Soil Stud Plant Nutr Proc. Kamath VA (Ed) p. 255.
34. Pitillo RF, Lucas MB 1967. Rad Res 31: 63.
35. Redpath JL, Willson RL 1973. Internl J Radiation Biol 23: 51.
36. Rennert A, Antoszsewska L 1970. Radiobiologiya 10: 580.
37. Schuchmann MN, Sonntag Cvon 1978. J Chem Soc Perkin Trans 2, 1958-1963.
38. Selimbekova DA 1969. Dokl Akad Nauk Azerb SSR 25: 78 (In Russian)
39. Stankovic S, Veljkovic S 1968. Acta Vet (Belgrade) 18: 203.
40. Tagizade AKh, Guseinov SG 1964. Izv Akad Nauk Azerbaidzh SSSR Biol Med Nauk 6: 91.
41. Tewary BK, Mookerjee AC 1982. Indian J Plant Physiol 25: 258.
42. Umeda K, Takano H, Sato T 1970. Nippon Shokuhin Kogyo Gakkai-Shi Japan 17: 55.
43. Vezirova NB 1963. Dokl Akad Nauk Azerb SSR 19: 53.
44. Vijaya CM, Vaidya BS 1980. Indian J Exptl Biol 18: 304.

10. POLYPHENOLS

Polyphenols are substances having more than one hydroxyl group in the nucleus. They include a large variety of substances from catechol, having two hydroxyl groups, various tannins, hydroxycinnamic acids, flavonoid pigments and betalains, all of which are widely distributed in plants.

1. REGULATORY ACTION OF POLYPHENOLS WITH AA

Ascorbic acid (AA) shows structural and chemical affinity with the polyphenols in the possession of a dienol group which is responsible for many of the chemical properties of the two groups of substances, such as the reducing property controlling redox reactions, donor in peroxidase reactions, chelation of metals, etc. In plant tissues and cells, these are known to coexist and control important redox reactions. Polyphenols have been shown to regulate the oxidation of AA and have both a protective as well as a stimulatory action on oxidative systems (Clemetsen and Andersen, 1966; Stenlid and Samorodova-Bianki, 1969). Thus they play a key role in controlling the hormonal levels of tissues and lead to growth and differentiation of cells and organs (Laloraya et al., 1972).

2. POLYPHENOLS AND THE OXIDATION OF ASCORBIC ACID

Clemetsen and Andersen (1966) have tabulated in some detail, the anti-oxidant property of various plant polyphenols for AA. Different polyphenols exert differential action as anti-oxidants for AA. In general, the flavonols, flavonolglycosides and dihydroflavonols showed a high degree of anti-oxidant activity, and among the flavone and flavanone aglycones, only those with two adjacent acidic hydroxyl groups in the β-ring showed any significant anti-oxidant activity. The best anti-oxidant activity was shown by rutin and its aglycone quercetin. It was, however, pointed out by Clemetsen and Andersen (1966) that flavonols with a catechol configuration in the β-ring and caffeic acid solutions show an optimal concentration above which less protection is obtained. Addition of cupric ions to the system required increased concentration of the flavonol to give maximum protection. This elucidated that anti-oxidant activity was related mainly to the chelation of heavy metals with little direct anti-oxidant action. The anti-oxidant activity of polyphenols (flavonoids) for AA is thus due to the presence of chelatogenic pairs of groups viz., 3' 4'- catechol couplet of the β-ring and the 3-hydroxyl, 4-carbonyl couplet of the gamma pyrone ring.

3. CO-EXISTENCE OF POLYPHENOLS AND ASCORBIC ACID IN CELLS AND TISSUES

Since polyphenols possess anti-oxidant property for AA and are widely distributed in plants, the level of reduced AA in cells may be maintained due to the presence of various polyphenols. Thus the polyphenols and AA may co-exist in cells having high concentration of (reduced) AA which have been recorded in flowering apex, floret primordia, tapetum, middle layers of anther wall and fully mature pollen grains (Chinoy, 1962; Chinoy et al., 1969; 1976). Since the major pigments of flowers are the flavonoid pigments, it would be expected that the transforming floral primordia would actively synthesize them. All the different parts of the flower are known to contain high levels of flavonoid pigments (Sheen, 1969). Even during microsporogenesis, during and immediately after the separation of the microspores from the tetrads, a marked increase occurred in the production of flavonols. Both the pollen and the tapetum have been reported to show high concentration of flavonols (Wierman, 1969).

Stomatal and trichome initials known to accumulate polyphenolic pigments, show high reducing activity towards silver nitrate, a larger part of which is obtained even after $CuSO_4$ treatment. It has been suggested that a considerable part of the reducing activity may indeed be due to the flavonoid pigments present in these cells. A survey of large number of plants have shown the presence of high reducing activity in stomata and trichomes (Laloraya et al., 1972). Cells and tissues having high AA content are also characterized by the accumulation of polyphenolic substances.

3.1. Co-occurrence of phenols, ascorbic acid and peroxidase in cells

Chinoy et al. (1976) reported that during the transformation of the vegetative apex to the reproductive apex, concomitant with the increase in AA, there also occurred an increase in the peroxidase activity. The stomatal and trichome initials of a large number of plants which showed high reducing activity, also revealed intense localization of peroxidase (Laloraya et al., 1972).

Peroxidase catalyses the rapid oxidation of AA (Yamazaki, 1962) and both AA as well as phenols serve as donors in peroxidase mediated reactions. The observation that AA accumulates in a cell having high peroxidase activity indicates that the enzymic activity is being regulated by some other factors like flavonoid pigments present in the cell. It was reported that the flavones of the leaves and roots of wheat sprout and anthocyanins of the coleoptiles accelerated the peroxidase oxidation of AA (Kolesnikov and Zore, 1964). Moreover, morin also accelerated the peroxidase oxidation of AA (Kursanov et al., 1950). A detailed study on the effect of flavonoids and oxidation of AA in plants has been conducted by Stenlid and Samorodova-Bianci (1969). These authors have compared the effect of flavonoid on different enzymic and non-enzymic systems, viz., I by copper ions; II by AAO; III Cytochrome oxidase-cytochrome system; IV tyrosinase, laccase or peroxidase system with phenols as donors. Their results have shown that while the Cu (II) and AAO mediated oxidation of AA were almost similarly affected by flavonoid pigments, the oxidation of AA was being inhibited by flavonoids to varying degrees, but those coupled to tyrosinase, laccase and peroxidase were generally stimulated. The same flavonoid may bring about opposite effect depending upon the oxidizing system. Thus fisetin, which depresses both the Cu (II) AA-oxidase mediated oxidation of AA, is a powerful stimulator in tyrosinase as well as in peroxidase systems. Quercetin also, which is a powerful inhibitor in systems I and II, promotes oxidation in systems III and IV. Hence, depending on the nature of flavonoid present and the system operating, the oxidation of AA may vary in different cells and tissues and polyphenols seem to regulate the oxidation of AA in plant cells.

The work of Chinoy (1962), Chinoy et al. (1969) and Laloraya et al. (1972) has shown that almost all types of plant differentiations are linked with the presence of peroxidase and reducing activity in the differentiated cells. Laloraya et al. (1972) have suggested that peroxidase and polyphenol- AA play a basic role in regulating the redox state and thereby the energy levels of dividing cells and also hormonal imbalances resulting in differentiation. These authors have proposed a scheme suggesting redox state and hormonal levels in plants (Fig. 10.1).

That semiquinone produced in the peroxidase reaction of phenols can readily oxidize substances like NADH, $NADPH^+$, GSH, AA and IAA has been known for a long time. However, the fact that some polyphenols exert differential action on the oxidation of AA and IAA, the two important growth regulating substances, their (polyphenols) presence may result in causing an imbalance of the two substances at a given time in a tissue. Since the free radical of AA is highly reactive and can complex with macromolecules like DNA, it may trigger reactions essential for differentiation.

4. SUMMARY

Polyphenols have been shown to regulate the oxidation of AA as well as exercise protective action and stimulate oxidative systems which play a key role in controlling the hormonal levels in tissues leading to growth and differentiation of cells and organs. Polyphenols and AA may coexist in the cell. A survey of a large number of plants has shown

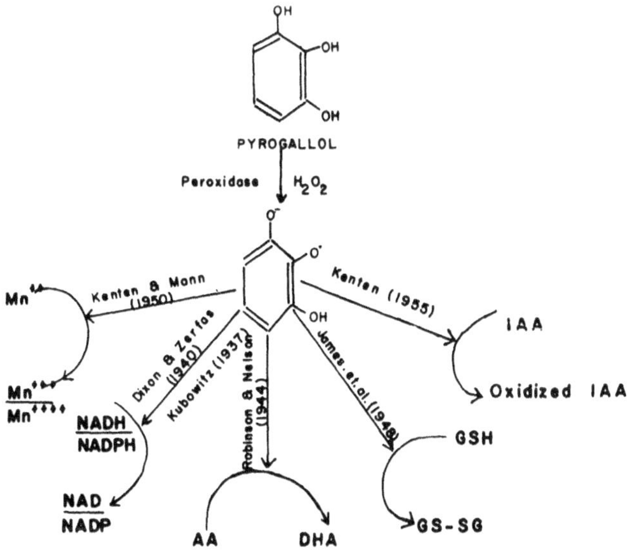

Proposed Role of Peroxidase and Phenolic substances in differentiation.

FIGURE 10.1. Proposed role of peroxidase and phenolic substances in differentiation.

the presence of high reducing activity in developing stomata and trichomes, part of which is due to polyphenolic substances. The tissues and cells containing high concentrations of AA and polyphenols are also characterized by high peroxidase activity. Peroxidase catalyzes the rapid oxidation of AA and both AA and phenols serve as donors in peroxidase mediated reactions. Accumulation of AA in cells of high peroxidase activity indicates that the enzymic activity is being regulated by flavonoid pigments present in the cell.

REFERENCES

1. Chinoy JJ 1962. Indian J Plant Physiol 5: 172 (FG Gregory Memorial Volume).
2. Chinoy JJ 1970. Indian J Plant Physiol 13 (2): 115.
3. Chinoy JJ, Saxena OP, Abraham PG, Pandya RB, Dave IC, Gurumurti K, Sastry K 1969. P.L. 480 Research Project Report, Project FG-In-182 (A7-CR-87) p. 1-350, Department of Botany, University School of Sciences, Gujarat University, Ahmedabad - 380009, India.
4. Chinoy JJ, Singh YD, Laloraya MM 1974. Vth Int Symp Mag Resonance. Jan 14th - 18th, Bombay, India, Abstr 17A 64.
5. Chinoy JJ, Singh YD, Gurumurti K 1976. Indian J plant Physiol 19: 122.
6. Clemetsen CAB, Andersen L. 1966. Ann N Y Acad Sci 136: 339.
7. Kolesnikov PA, Zore SV 1964. Fiziol Rast 11(3): 522.
8. Kursanov AL, Dukin VN, Povolotskaya KL, Zaprometov MN 1950. Biokhimiya 15: 337.
9. Laloraya MM, Srivastava HN, Bakshi N, Shah N, Pandya KJ, Singh YD 1972. All India Symp Biol Land Plants Meerut, India.
10. Sheen SJ 1969. Phytochemistry 8 (10): 1839.
11. Stenlid G, Samorodova-Bianki GC 1969. Lantbrukshogskolans Annaler 35: 837.
12. Wierman R 1969. Planta 88 (4): 311.
13. Yamazaki I 1962. J Biol Chem 237 (1): 224.

ACKNOWLEDGEMENTS

The help rendered by Prof. M.M. Laloraya, Head, Department of Life Sciences, Indore University, Indore, in the writing of this Chapter is gratefully acknowledged.

11. ENZYME ACTION

Considering the role of AA in enzyme systems, it appears that there are very few investigations in which AA is assigned the function as a coenzyme like a number of derivatives of B vitamins do in metabolic reactions. Szent-Györgyi (1931) had suggested that AA may serve as a respiratory carrier in a system coupled with glutathione in plants. Further studies showed that hydrogen may be transferred from reduced glutathione to DHA by the enzyme DHA-reductase in plants (Ettlinger et al., 1961) as follows:

$$1.\ \text{DHA} + 2\ \text{GSH} \xrightarrow{\text{DHA-reductase}} \text{AA} + \text{GSSG}$$

and from reduced NAD or NADP to glutathione (GSH) in animals, plants, yeast and bacteria by glutathione reductase (Mapson and Goddard, 1951; Mapson, 1958; Van Heyninger and Pirie, 1953; Racker, 1955; Foyer and Halliwell, 1976).

$$2.\ \text{GSSG} + \text{NADPH} + H^{+} \xrightarrow{\text{GSH-reductase}} 2\ \text{GSH} + \text{NADP}$$

An alternate route of hydrogen transfer from reduced NAD directly to oxidized AA has been suggested by a number of workers (Kern and Racker, 1954; Mathews, 1951; Nason et al., 1954; Kersten et al., 1958). In this reaction MDHA is considered to be the actual acceptor of electrons from reduced NAD (Burns and Ashwell, 1960). AA is a determining factor in establishing the equilibrium between SH and SS compounds. It protects SH against oxidation in the living cells (Aberg, 1958) and AA as well as GSH behave as catalysts of oxidation-reduction processes. The activity of certain intracellular enzymes is dependent on the presence of definite oxidation - reduction potentials. AA has not been observed to act as a prosthetic group but its potential connection with enzyme action is well established. In view of the reversible oxidation - reduction of AA, it has been suggested that it may function as a regulator of cellular oxidation - reduction potential. By altering rates of reactions, the oxidation - reduction state of the cell could be altered. It is assumed that oxidizable - reducible groups of the cells are in some sort of an equilibrium with one another in a similar manner as the ionizable groups are in equilibrium with free hydrogen ions; and poising of the redox state is referred to by analogy with buffering of the pH. It is not clear that this assumption of equilibrium is justified. Barron (1951) suggested that large amounts of AA and GSH found in the reduced state could conceivably act as poising agents.

The influence of AA, DHA, and D-arabo- AA either alone or together with various metal ions, and other agents on various enzyme activities in plant, microbial and animal systems are listed in Table 11.1. Most of the enzyme activities are stimulated, whereas others are inhibited and some are not affected.

1. MECHANISM

The action of AA on -SH groups has been mentioned above. Avigliano et al. (1978) have shown that in the absence of air, AAO is fully reduced by four mol of AA as recorded by optical and EPR spectra. The inhibition of some enzymes is caused by reducing the valency of the metal ion i.e., in xanthine oxidase, inactivation is achieved by changing cupric to cuprous state. However, the copper inactivation was reversible.

A number of enzymatic processes associated with hydroxylations, such as $NADH_2$ / NAD redox interactions, showed significant catalysis when AA was added. These activities were attributed to primarily oxidation - reduction involving the components of the ascorbic system. But recent work has revealed that in a number of reactions the effect may be interpreted not in terms of direct redox activities, but rather in terms of the effect of AA on enzymatic aggregation and / disaggregation as well as in contribution to biosynthesis of enzyme components.

1.1. Enzymatic catalysis of redox reactions involving the ascorbate system

AA participates in hydroxylation reactions, distinct from its oxidation reduction activity, e.g., hydroxylation of proline and lysine components of collagen (ref: reviews: Barnes and Kodicek, 1972; Udenfriend et al., 1954). In hydroxylation reactions electron transfer and energy transfer occur which often involve the AA free radical. One electron transfer is involved in $Fe^{++} \rightleftharpoons Fe^{+++}$ and $Cu^{+} \rightleftharpoons Cu^{++}$ whereas, AA $\rightleftharpoons$ DHA involves a 2 electron transfer.

1.2. Enzyme activation by oxidation product of AA

Refer Kern and Racker (1954), Kersten et al. (1958), Nason et al. (1954), Schneider and Staudinger (1965) and Iyanagi and Yamazaki (1969).

1.3. Inhibitory effect of AA on some enzyme activities

AA has inhibitory effect on some enzyme activities (Table 11.1).

2. BIOSYNTHETIC ACTIVITIES

AA is implicated via its free radical in various enzymatically controlled activities in the formation of gamma globulins, adrenalin and nonadrenalin, collagen and steroids (Lewin, 1976; Agrawal and Laloraya, 1977).

Numerous studies on seeds, seedlings, embryo axis and cotyledon of various crop plants viz., *Arachis hypogea* cv. Punjabi, *Sesamum indicum* cv. Limbdi, *Cicer arietinum*, *Zea mays*, *Phaseolus mungo* L. cv. local variety, barley and millet in this laboratory have revealed the stimulatory effect of exogenously applied AA on different enzymes, i.e., lipase, catalase, peroxidase, AA - FR peroxidase, protease, amylase, RNAse invertase, etc. In general, the lower concentrations of AA were more effective in accelerating enzyme activity than the higher ones (Table 11.1) (Mehta, 1972; Shah, 1972; Mehta, B.H., 1973; Chinoy et al., 1970; 1970 a, b; 1973; 1969; 1969 a; Shah S.K., 1972; Jani, 1973; Patel, 1972; Saxena, 1969). Methew et al. (1978) also reported enhancement of RNAse by AA in the shoot and root of germinating barley seedlings. The mechanism of action of AA in enhancing enzyme activities is most probably via the accelerated endogenous biosynthesis of AA (Michniewicz, 1960; Chinoy, 1967; Chinoy et al., 1965; 1967; Brugovitzky and Popovici, 1966; Saxena et al., 1969), which in turn stimulated the enzymic machinery controlling the mobilization of reserve nutrients (Chinoy et al., 1965; 1969; Chinoy and Saxena, 1972). In fact it has been suggested from this laboratory that AA serves as an enzyme mobilizing hormone (EMH) (Saxena et al., 1969).

The regression of lipase, peroxidase, AA - FR peroxidase on AA x AAU product (Fig. 11.1) clearly elucidates that the linear relationship between the given enzyme activity and AA x AAU product is highly significant. This suggests that intensification of enzymic activity is brought about by the increase in AA turnover which actively participates in germinative processes.

3. STIMULATION OF RNAse ACTIVITY BY THE FREE RADICAL OF AA

An *in vitro* experiment for demonstrating the inductive effect of MDHA on the enzyme RNAse was carried out using pure AA, RNA and RNAse (Chinoy et al., 1973 a). In the reaction medium containing AA, RNA and RNAse, the enzymic degradation of RNA was slow as seen in Fig. 11.2. In another medium which contained only AA and RNA together with a few drops of H_2O_2 (but not RNAse), the velocity of RNA breakdown was considerably faster than that of the first medium. Further, in a third reaction medium having AA, RNA, RNAse and H_2O_2 at the same concentration as in the previous two media, the hydrolysis of RNA was more than seven times faster than that in the first medium (Fig. 11.2).

From the foregoing evidence it appears that AA stimulates RNAse activity only in the presence of H_2O_2. This is brought about by the formation of the free radical of AA, viz., MDHA. Further studies supported the marked stimulation of AA - FR special peroxidase activity by exogenous application of AA + H_2O_2 (Chinoy et al., 1973). Similarly, Chinoy, N.J. (1978) and Chinoy, N.J. et al. (1982) have also reported stimulation of numerous enzymes in different animal tissues by exogenous feeding of AA which involved the AA free radical.

4. ASCORBOZYME

On the basis of the work carried out in this laboratory as well as reviewing the literature, it appears that in general, AA has a stimulating effect upon all the enzyme systems studied and it also accelerates the biosynthesis of macromolecules and many other molecules presumably due to its oxidation - reduction property.

Taking the above mentioned facts into consideration and the fact that AA is universally present, it can be postulated that AA complexes with a special isozyme of peroxidase which catalyzes the formation of its free radical. The presence of such an isozyme of peroxidase in actively growing and differentiating organs of plants and animals was first discovered by Gurevich (1963). Later on, its presence in animal tissues was also demonstrated histochemically and biochemically (Chinoy, N.J., 1970; 1973; 1978). The reaction can be written as:

$$P + A \longrightarrow PA \longrightarrow P + A^{0}$$

where, P represents the AA - FR peroxidase, A represents AA, and A^{0}, the free radical of AA.

The complexing of an enzyme (E) and its substrate (S) is energized by the free radical of AA (A^{0}) thus

$$E + S \xrightarrow{A^{0} \longrightarrow D} ES \longrightarrow E + \text{product}$$

After imparting its electron energy to the formation of the complex, A^{0} is converted into DHA (D). That this concept is workable is clear from the work of Diliberto and Allen (1981) and Wong et al. (1981). The role of AA as conceived in relation to mechanism of β-hydroxylation has been modified (Diliberto and Allen, 1981) by introducing semidehydro-ascorbate as the enzyme - oxidation product of ascorbate.

In both the plant and animal systems there occur mechanisms which enable the DHA (D) to be reconverted to L - AA (A). Thus in plants there are reductase enzymes and in

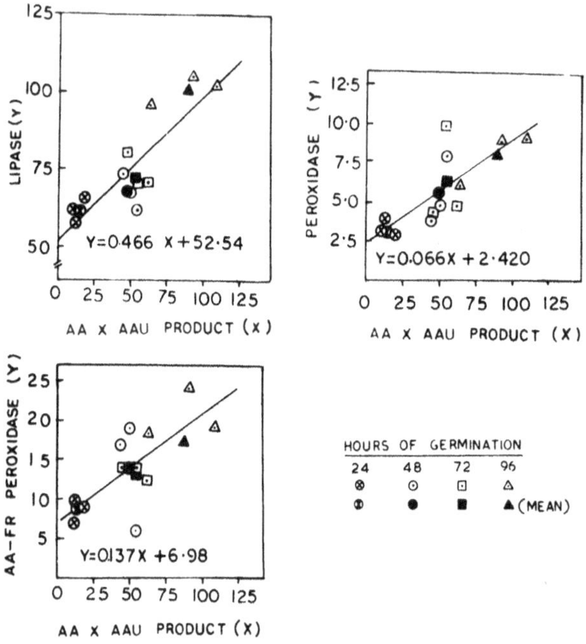

FIGURE 11.1. Linear regression of lipase, peroxidase and AA - FR peroxidase activities on AA - AAU product (Shah, 1972).

animals, glutathione reduces DHA to AA. The AA so formed can again participate in the production of its free radical (A^0) with the help of AA - FR - peroxidase (P) (Fig. 11.3).

DHA and free radicals of AA could be reconverted to AA again by two cognate enzymes -(DHA reductase (GSH. DHA - oxidoreductase) and AA - FR reductase (NADH: AA FR - oxidoreductase = NADH: mono DHA - oxidoreductase, E. C. 1. 6. 5. 4) respectively. The data collected on these enzymes from plants and animals show that while AA - FR reductase is ubiquitous and is present in cytosol, mitochondria, microsomes and chloroplasts (with maximum activity in cytosol), DHA reductase seems to be confined to the cytosol (Arrigoni et al., 1981). Foyer and Halliwell (1977) have purified the enzyme DHA reductase from spinach leaves. A high concentration of DHA inhibited the enzyme activity.

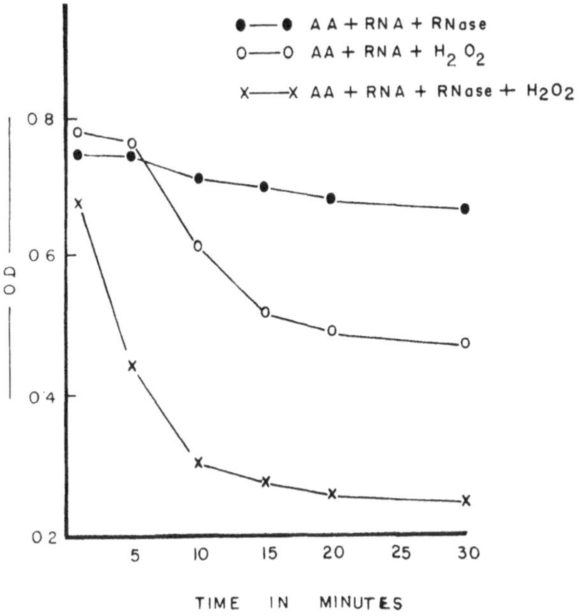

FIGURE 11.2. Stimulation of RNAse activity by the free radical of AA (Chinoy et al., 1973 a).

5. SUMMARY

The activity of a number of enzymes was activated by AA, either by exogenous application to plants or by feeding it to animals. The activating effect of AA on enzymes is due to the protection of its -SH groups from oxidation. There are a few cases of inhibition of enzymic activity which is generally ascribed to be through the oxidation - reduction of associated metal ions. The direct role of AA in enzyme inhibition is not yet established. External feeding of AA to germinating seeds increased the endogenous synthesis of AA, which in turn stimulated the enzymic machinery controlling the mobilization of reserve nutrients.

On the basis of the data presented, it is evident that AA has a stimulating effect upon almost all the enzyme systems studied. The rates of biosynthesis of macromolecules and other molecules were also accelerated presumably due to its oxidation - reduction property. AA is also known to possess

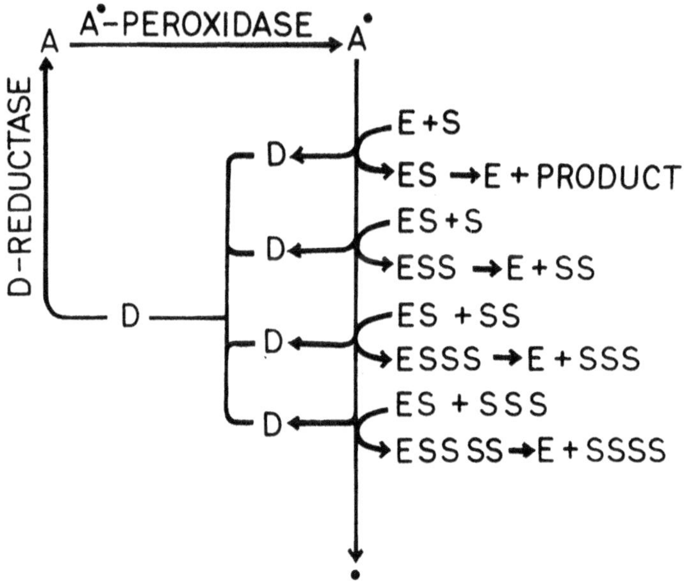

FIGURE 11.3. The mechanism of action of AA via the formation of its free radical.

a unique property of complexing with macromolecules as well as smaller molecules and is universally present in cells, especially at sites of intense metabolic activity. The above evidence leads to the postulation that AA complexes with a special isozyme of peroxidase which catalyses the production of the free radical of AA. The latter imparts its electron energy to the complex and is subsequently converted to DHA. In both plant and animals there are systems which enable the DHA to be reconverted to AA. The AA so formed is again capable of participating in the production of its free radical with the help of the AA - FR peroxidase.

REFERENCES

1. Aberg B 1958. In: Handbuch der Pflanzenphysiologie. Ruhland W (ed) Berlin: Springer Verlag 6: 479.
2. Abraham PG, Dave IC, Pandya RB, Saxena OP 1970. J Indian Bot Soc 49: 41.
3. Agrawal P, Laloraya MM 1977. Biochem J 166: 205.

4. Arrigoni O, Dipierro S, Borracino G 1981. FEBS Lett 125: 242.
5. Avigliano L, Rotilio G, Urbanelli S, Mondovi B, Finazzi A 1978. Arch Biochem Biophys 185: 419.
6. Avtandilov GG, Vyazmina TN, Stibneva OV 1979. Deposited DOC VINITI 3650.
7. Axelrod B, Saltman P, Bandurski RS, Baker RS 1952. J Biol Chem 197: 89.
8. Badinand A 1937. Compt Rend Soc Biol 125: 283.
9. Barnes MJ, Kodicek E 1972. In: Vitamins and hormones. Harris RA, Munson PL, Glover J, Dikzfalusy E (eds). 30.
10. Barron ESG 1951. Adv Enzymol, New York: Interscience, 11: p 201.
11. Bergel F, Bray RC 1959. Biochem J 73: 182.
12. Blumberg WE, Goldstein M, Lauber E, Peisach J 1965. Biochim Biophys Acta 99: 187.
13. Brugovitzky E, Popovici G 1966. Naturwissenschaften 53: 312.
14. Burns JJ, Ashwell G 1960. In: The enzymes, Boyer PD, Lardy H, Myrback K (eds) New York: Academic, 3: p 387.
15. Chauhan JS, Srivastava HS 1980. Plant Biochem J 7: 25.
16. Chinoy JJ 1967. Indian J Plant Physiol 11: 202.
17. Chinoy JJ, Saxena OP 1972. Proc 16th Int Seed Test Assoc (ISTA) Congr Washington DC, June 7-12, 1971, 37: 903.
18. Chinoy JJ, Shah CK, Patel HT, Suthar HK 1965. Proc Internl Sem "Tissue Culture" Baroda, India, Jan 21-28, Ramakrishan CV (ed), p 377.
19. Chinoy JJ, Shah CK, Patel HT, Suthar HK 1967. Biol Plantarum 9: 182.
20. Chinoy JJ, Patel HD, Jani BM, John D, Vyas AV, Raj KPS 1969. Final Report PL 480 Research Project, Botany Department, University School of Sciences, Gujarat University, Ahmedabad - 380009, India.
21. Chinoy JJ, Saxena OP, Abraham PG, Pandya RB, Dave IC, Gurumurti K, Sastri K 1969 a. Final Report PL 480 Research Project FG-In-182 (A7-CR-87) p. 1-350.
22. Chinoy JJ, Saxena OP, Gurumurti K 1970. ISPP Symp Plant Growth Regulators, Kharagpur, India, Abstr p. 13.
23. Chinoy JJ, Saxena OP, Gurumurti K 1970 a. Ibid, p 14.
24. Chinoy JJ, Shah PN, Ghesani PI 1970 b. ICAR Final Research Progress Report, Botany Department, Gujarat University, Ahmedabad - 380009, India.
25. Chinoy JJ, Ghesani PI, Saxena OP 1973. Acta Bot Indica 1: 11.
26. Chinoy JJ, Gurumurti K, Singh YD 1973 a. Indian J Plant Physiol 16: 42.
27. Chinoy NJ 1970. Stain Technol 45: 99.
28. Chinoy NJ 1973. Ind Natl Sci Acad 39 B: 119.
29. Chinoy NJ 1978. J Anim Morphol Physiol Silver Jubilee Volume: 68.
30. Chinoy NJ, Sheth KM, Seethalakshmi L, Parmar PY, Sanjeevan AG, Rao MV et al. 1982. Comp Physiol Ecol 7: 325.
31. Cooper 1961. as cited by Lewin 1976. see ref.68, p.57.

32. Crandall DI 1955. In: Symp "Amino Acid Metabolism", McElroy WD, Glass B (eds), Baltimore: Johns Hopkins Press p 867.
33. Crandall DI 1955 a. J Biol Chem 212: 565.
34. Dickman SR, Cloutier AA 1950. Arch Biochem Biophys 25: 229.
35. Dickman SR, Cloutier AA 1951. J Biol Chem 188: 379.
36. Dickman SR, Speyer JF 1954. J Biol Chem 206: 67.
37. Diliberto EJ Jr. Allen PL 1981. J Biol Chem 256: 3385.
38. Doisy RJ, Richart DA, Westerfeld WW 1955. J Biol Chem 217: 307.
39. Eldbacher SK, Leuthardt F 1933. Z Physiol Chem 217: 89.
40. Eldbacher S, Leuthardt F 1933 a. Klin Wochschr 12: 1843.
41. Engelhardt VA 1946. In: Enzymology Nord FF (ed). New York: Interscience 6: p 147.
42. Ettlinger MG, Dateo G, Harrison B, Maloy T, Thompson C 1961. Proc Natl Acad Sci US 47: 1875.
43. Feigelson P 1952. J Biol Chem 197: 843.
44. Foyer CH, Halliwell B 1976. Planta 133: 21.
45. Foyer CH, Halliwell B 1977. Phytochemistry 16: 1347.
46. Fruton JS, Bergmann M 1939. J Biol Chem 130: 19.
47. Fruton JS, Irving GW, Bergmann M 1941 a. J Biol Chem 141: 763.
48. Giri KV, Seshagiri Rao P 1944. Nature (Lond.) 153: 253.
49. Godeh M, Udvardy J, Farkas GL 1981. Planta 152: 408.
50. Grob K, Matile P 1980. Z Pflanzenphysiol 98: 235.
51. Gurevich AA 1963. Dokl Biol Sciences Sect 145: 674.
52. Hanes CS 1935. Biochem J 29: 2588.
53. Hellerman L, Parkins ME 1934. J Biol Chem 107: 241.
54. Herr EB, Summer JB, Yesair DW 1956. Arch Biochem Biophys 62: 29.
55. Hirose Z, Kidani Y 1980. Chem Pharma Bull 28: 3189.
56. Isao T, Tadao H 1968. Agri Biol Chem (Tokyo) 32: 1420.
57. Isao T, Tadao H 1968 a. Ibid 32: 1425.
58. Iyanagi T, Yamazaki I 1969. Biochim Biophys Acta 172: 370.
59. Jani RB 1973. M.Sc. Dissertation, Botany Department, University School of Sciences, Gujarat University, Ahmedabad - 380009, Indoa.
60. Kendal LP 1949. Biochem J 44: 442.
61. Kern M, Racker E 1954. Arch Biochem Biophys 48: 235.
62. Kersten H, Kersten W, Staudinger H 1958. Biochem Biophys Acta 27: 598.
63. Lardy H 1962. In: The Enzymes 6: 67.
64. Leuthardt F, Koller F 1934. Helv Chim Acta 17: 1030.
65. Levine RL, Oliver CN, Fulks RM, Stadman ER 1981. Proc Natl Acad Sci USA 78: 2120.
66. Lewin S 1976. Vitamin C: Its Molecular Biology and Medical Potential, New York: Academic.
67. Lilyana B, Yanko E 1968. Nauch Tr Vissh Selskostop Inst Sofiya Agron Fak Ser-Rast 19: 283.
68. Malik MA, Afzal Z, Islam M 1976. Pak J Biochem 9: 15.
69. Mapson LW 1946. Biochem J 40: 240.

70. Mapson LW 1958. Ann Rev Pl Physiol 9: 119.
71. Mapson LW, Goddard DR 1951. Biochem J 49: 592.
72. Maschmann E 1934. Z Physiol Chem 228: 141.
73. Maschmann E, Helmert E 1934. Hoppe-Seyler's Z Physiol Chem 223: 127.
74. Mason HS 1957. Advan Enzymol 19: 79.
75. Mathews MB 1951. J Biol Chem 189: 695.
76. Mathew T, Dave IC, Gaur BK 1978. Z Pflanzenphysiol 86: 23.
77. Mazarean HH, Kovacs E 1965. Experientia 21: 515.
78. Mehta BH 1973. M.Sc. Dissertation, Botany Department, Gujarat University, Ahmedabad - 380009, India, p 41.
79. Mehta DH 1972. M.Sc. Dissertation, Botany Department, Gujarat University, Ahmedabad - 380009, India, p 50.
80. Melnikova MP, Neifakh YeA 1954. Biokhimiya 19: 425.
81. Michniewicz M 1960. Zesz Nauk Univ Mikolaja Kopernika Torun Biol 4: 103.
82. Moffat AC, Patterson DA, Currey AS, Owen P 1972. European J Toxicol 5: 160.
83. Muszynska G, Reifer I 1970. Acta Biochim Pol 17: 247.
84. Nagashima Z, Uchiyama M 1959. Bull Agr Chem Soc Japan 23: 555.
85. Nason A, Wossillait WD, Terrell AJ 1954. Arch Biochem Biophys 48: 232.
86. Niven CF, Smiley KL, Sherman JM 1942. J Bact 43: 651.
87. Ohtsuru M, Kawatani H 1979. Agric Biol Chem 43: 2249.
88. Orr CWM 1966. Biochem Biophys Res Commun 23: 854.
89. Orr CWM 1967. Biochemistry 6: 2995.
90. Orr CWM 1970. Methods Enzymol 18: Pt A 59.
91. Pasquier F, Polonovski 1957. Bull Soc Chim Biol 39: 1465.
92. Patel JN 1972. M.Sc. Dissertation, Botany Department, Gujarat University, Ahmedabad - 380009, India, p 44.
93. Purr A 1934. Biochem J 28: 1141.
94. Purr A 1935. Ibid 29: 5.
95. Purr A, Weil L 1934. Biochem J 28: 740.
96. Quastel JH 1943. Nature (Lond.) 152: 215.
97. Racker E 1955. J Biol Chem 217: 855.
98. Rowe AW, Wezel CE 1958. Cereal Chem 35: 289.
99. Sabri MI, Mohan Rao VK 1967. Indian J Biochem 4: 40.
100. Saxena OP 1969. Ph.D. Thesis, Botany Depaarment, Gujarat University, Ahmedabad - 380009, India.
101. Saxena OP, Ghesani PI, Shukla SP, Chinoy JJ 1969. Indian J Plant Physiol 12: 1.
102. Schepartz B 1953. J Biol Chem 205: 185.
103. Schepartz B 1953 a. Federation Proc 12: 265.
104. Schneider W, Staudinger HJ 1965. Biochim Biophys Acta 96: 157.
105. Seshagiri Rao P, Giri KV 1942. Proc Indian Acad Sci 16 B: 190.
106. Seshagiri Rao P, Giri KV 1952. Proc Indian Acad Sci 35: 132.
107. Shah PN 1972. Ph.D. Thesis, Botany Department, University School of Sciences, Gujarat University, Ahmedabad - 380009, India.

108. Shah SK 1972. M.Sc. Dissertation, Botany Department, University School of Sciences, Gujarat Universiry, Ahmedabad - 380009, India.
109. Srivastava GC, Sirohi GS 1969. Indian J Exptl Biol 7: 129.
110. Suda M, Takeda Y 1950. J Biochem (Japan) 37: 381.
111. Suda M, Takeda Y 1950 a. Med J Osaka Univ 2: 37.
112. Suda M, Takeda Y, Sujishi K, Tanaka T 1951. J Biochem (Japan) 38: 297.
113. Sutaria M 1972. M.Sc. Dissertation, Botany Department, University School of Sciences, Gujarat University, Ahmedabad - 380009, India.
114. Suzuki S, Uematsu I, Uchida M 1952. J Biochem (Japan) 39: 41.
115. Szent Györgyi A 1931. J Biol Chem 90: 385.
116. Tauber H 1938. Ergenn Enzymforsch 7: 301.
117. Takenga H, Ito S 1962. Tokyo Nogyo Daigaku Nogaku Shuhu 7: 149.
118. Udenfriend S, Clark CT, Axelrod J, Brodie BB 1954. J Biol Chem 208: 731.
119. Van Heyningen R, Pirie A 1953. Biochem J 53: 436.
120. Von Euler H, Karrer P, Zehender F 1934. Helv Chim Acta 17: 157.
121. Vyas DN, Patel KC 1969. J Amer Oil Chem Soc 46: 176.
122. Waldschmidt-Leitz E, Purr A 1931. Z Physiol Chem 198: 260.
123. Williams JN Jr 1951. J Biol Chem 191: 123.
124. Williams JN Jr 1951 a. Ibid 192: 81.
125. Wong DL, Masover SJ, Ciaranello RD 1981. J Biol Chem 256: 695.
126. Yamauchi N, Hamaguchi S, Ogata K 1980. Ergei Gakkai Zasshi 49: 414.
127. Yasunori, Michiko K 1967. Chem Abstr 68: 11709 m, 1968.
128. Yuldashev PKh, Rakhimov MM, Aizikov EJ 1971. Khim Prir Soedin (Tashk) 7: 803.
129. Ziff M 1944. J Biol Chem 153: 25.

Table 11.1. Enzyme activities influenced by AA.

Sr. No.	Enzyme	Stimulation (S) / Inhibition (I)	References
1.	Papain	S (in presence of Fe^{++})	Maschmann, 1934; Maschmann & Helmert, 1934; Purr, 1934, 1935; Hellerman and Parkins, 1934.
2.	Arginase	S (AA + Fe Complex)	Eldbacher & Leuthardt, 1933; 1933 a; Tauber, 1938; Purr, 1934; 1935; Purr and Weil, 1934; Waldschmidt-Leitz and Purr, 1931; Niven et al., 1942.
		I (in presence of N derivatives of Chlorogenic acids). Reactivated by AA, GSH, mercaptoethanol, L-cysteine).	Muszynska and Reifer, 1970.
3.	Cathepsin (animal)	S (AA + Fe Salts)	Same as under 2(S). Von Euler et al., 1934; Fruton & Bergmann, 1939; Fruton et al., 1941 a.
4.	Catalase (micro-organisms, plants, liver)	S	Mazarean & Kovacs, 1965; Maschmann, 1934; Leuthardt and Koller, 1934; Badinand, 1937; Lilyana and Yanko, 1968.
		S (by AA)	Chinoy et al., 1970 b; Mehta, 1973; Jani, 1973; Shah, 1972.
		I (presence of Cu^{2+})	Orr, 1967.

Table 11.1. contd.

Sr. No.	Enzyme	Stimulation (S) / Inhibition (I)	References
5.	Urease	S (by Mg salts)	Orr, 1967; 1966; 1970; Mapson, 1946; Purr, 1934; Seshagiri Rao & Giri, 1952; Quastel, 1943; Giri & Seshagiri Rao, 1944.
		I (by AA, AA + Cu^{2+}) eliminated by addition of cysteine.	
		I (by DHA reacting with -SH groups)	
		I (eliminated by cysteine).	
6.	β-amylase (plants)	S	Chinoy et al., 1970 b; Shah, 1972;
		S (by AA + papain)	Malik et al., 1976, Saxena, 1969. Takenga & Ito, 1962.
		I	Giri and Seshagiri Rao, 1944; Rowe & Wezel, 1958.
		I (by AA + Cu)	Purr, 1934; Hanes, 1935.
	β-amylase (animals)	S	Purr, 1934; Seshagiri Rao & Giri, 1942.
7.	Protease	S (in presence of O_2 and by D-arabo-AA)	Yasunori & Michiko, 1967.
		S (AA)	Chinoy et al., 1965; 1970; 1970 a; b; 1973; 1969; Chinoy & Saxena, 1972; Patel, 1972; Saxena, 1969; Sutaria, 1972.
		I (by DHA)	

Table 11.1. Contd.

Sr. No.	Enzyme	Stimulation (S) / Inhibition (I)	References
8.	Phenolase	S (non-specific)	Kendal, 1949.
9.	Choline dehydrogenase	S (in presence of folic acid inhibitor e.g., Aminopterin)	Williams, 1951; 1951 a.
10.	α-amylase	Not affected	Purr, 1934; Saxena, 1969.
11.	Myrosinase	S	Isao and Tadao, 1968 a; Ohtsura & Kawatani, 1977; Grob & Matile, 1980.
		I	Isao & Tadao, 1968.
12.	General peroxidase	S	Lilyana & Yanko, 1968; Chinoy et al., 1969; 1969 a; Chinoy et al., 1970 b; Shah, 1973, Abraham et al., 1970.
		I/S	Yamuchi et al., 1980,
13.	Polyphenol-oxidase	S	Lilyana and Yanko, 1968.
14.	Ascorbic acid oxidase	S	Chinoy et al., 1970; 1970 a.

Table 11.1. Contd.

Sr. No.	Enzyme	Stimulation (S) / Inhibition (I)	References
15.	Invertase	S/I	Saxena, 1969.
16.	Acid phosphatase (algai)	S	Godeh et al., 1981.
17.	Alkaline phosphate (animals)	S	Chinoy N.J., 1978; Chinoy N.J. et al., 1982.
18.	Xanthine oxidase	I (by Cu but reversed by AA)	Doisy et al., 1955; Bergel & Bray, 1959; Feigelson, 1952; Pasquier & Polonovski, 1957.
19.	Pyruvate decarboxylase	I	Sabri and Mohan Rao, 1967.
20.	Succinate dehydrogenase (animal)	S	Chinoy N.J., 1978; Chinoy N.J. et al., 1982.
21.	ATPase (animal)	S (by cysteine, glutathione or AA)	Ziff, 1944; Engelhardt, 1946.

Table 11.1. Contd.

Sr. No.	Enzyme	Stimulation (S) / Inhibition (I)	References
22.	Reduction of Cytochromes	S	Lewin, 1976.
23.	Esterase (Liver)	S	Same as under 4.
24.	Lipase (plants)	S	Chinoy et al., 1973; Saxena, 1969; Mehta, 1972; Shah S.K., 1972; Vyas and Patel, 1969; Yuldashev et al., 1971.
25.	AA-FR-special peroxidase (plants)	S	Chinoy et al., 1970; 1970 a, b; 1973 a; Jani, 1973; Mehta, 1973; Shah S.K., 1972.
	AA-FR-special peroxidase (animals)	S	Chinoy N.J., 1978; Chinoy N.J. et al., 1982.
26.	RNAse	S	Same as under 25 S; Mathew et al., 1978.
27.	Phosphodiesterase (animals)	I	Moffat et al., 1972; Lewin, 1976.

Table 11.1. Contd.

Sr. No.	Enzyme	Stimulation (S) / Inhibition (I)	References
28.	Aconitase (Plants)	S (by Fe^{2+} + AA)	Dickman & Speyer, 1954; Herr et al., 1956; Dickman & Cloutier, 1950; 1951.
29.	Nitrate reductase	No effect	Chauhan and Srivastava, 1980
30.	Phospho-hexokinase (plants)	S	Lardy, 1962.
		I (by thiol group inhibitors)	Axelrod et al., 1952; Melnikova and Neifakh, 1954.
31.	Glutamine Synthetase	S	Levine et al., 1981.
32.	Malate dehydrogenase NAD - diaphorase succinate (in fibroblasts and WBC of regenerating skin)	I I I	Aytandiloy et al., 1979.
33.	Carbonic anhydrase (Bovine)	I	Hirose & Kidani, 1980.

Table 11.1. Contd.

Sr. No.	Enzyme	Stimulation (S) / Inhibition (I)	References
34.	NADH- oxidase	S	Kern & Racker, 1958; Kersten et al. 1958
	Microsome catalyzed oxidation of NADH	S (by AA-FR)	Iyanagi & Yamazaki, 1969.
35.	Homogenentisate oxidase (microorganisms, Liver, Kidney)	Activation potentiated by Fe^{2+} + AA; GSH, other enediols)	Schepartz, 1953; Mason, 1957; Crandall, 1955; 1955 a.
		I (by salting out)	Suda and Takeda, 1950; 1950 a; Suda et al., 1951.
		I (by EDTA, O-phenanthroline, Diethyl-dithiocarbamate.	Schepartz, 1953.
		Reversal of inhibition by Fe^{2+}	Schepartz, 1953 a; Suzuki et al., 1952.
36.	β-dopamine hydroxylase (animals)	S (by AA-FR)	Blumberg et al., 1965.
37.	Tryptophan 5-hydroxylase	S	Cooper, 1961.

Table 11.1. Contd.

Sr. No.	Enzyme	Stimulation (S) / Inhibition (I)	References
38.	Three transaminases of glutamate - glyoxylate; glutamate-oxaloacetate and alanine glyoxate	I	Srivastava and Sirohi, 1969.
39.	$NADH_2$/NAD enzyme systems	S	Lewin, 1976.
40.	Hydroxylations during formation of steroids (animals)	S	Agrawal & Laloraya, 1977; Chinoy N.J., 1978; Chinoy N.J. et al., 1982.

12. REGULATION

The role of AA in some regulatory mechanisms are discussed in this chapter.

1. GENOTYPIC DIFFERENCES IN AA CONTENT

In the earlier stages of work on genetic aspects of AA, attempts were made to establish varietal differences in AA content of different cultivars of crop plants.

Determination of AA contents of a number of cultivars and strains of tomatoes revealed a variation of 13 to 44% among 98 cultivars grown at the same time (MacLinn et al., 1937). Currence (1939), on the other hand, showed that difference between tomato cultivars were indefinite and difficult to demonstrate statistically with samples obtained from a field experiment having three replications. There were significant differences in the AA content of the leaves of strains of snap beans, whereas, the differences in the AA content of fruits were not significant (Wade and Kanapaux, 1943). Pyke (1942), on the other hand, found no essential difference in the vitamin content of several cultivars of vegetables.

Significant differences in the AA concentration of various cultivars of spring crops of beet, pepper and pea were also recorded (Leonard et al., 1944). Similarly, Smith and Walker (1946) also found significant differences in AA content in cabbage cultivars in spite of fluctuations in the environment. Cultivars in the 'early pointed group' were in general higher in their AA content than those in the 'early round' and 'midseason round groups'. The 'midseason flat' cultivars and some of the 'late flat' ones had the lowest AA content. Seasonal differences in AA content were observed in a series of closely related strains. Smith and Walker (1946) also carried out a study in inbred lines of Wisconsin -All-Seasons, a cultivar which contained 51 to 56 mg per 100 g of plant material. In some of the selected lines AA content was not far removed from that of the present variety.

However, in one line, AA content was 80 to 91 mg/100 g. When individuals from this high yielding lines were crossed with individuals from the low yielding lines, AA content in the F_1 progenies was consistently intermediate between the two parents. On the basis of their work, Smith and Walker (1946) visualized the production of cabbage cultivars combining high nutritive value with high AA content, disease resistance and high productivity.

Walker and Foster (1946) made use of the high vitamin line of cabbage as parental material from which, by hybridization and selection, the factors for high AA content could be incorporated into standard commercial cultivars. They also studied the mode of inheritance of factors influencing high production of this vitamin in cabbage. The F_1 data from 23 crosses and the F_2 data from 8 of these crosses were taken into account. As it was not possible to secure complete data from the F_3 generation, which admittedly would have added much to the genetic interpretation of the inheritance of AA content, only the F_1 and F_2 data were studied. In the 23 F_1 hybrid progenies from crosses between an inbred line of cabbage high in AA content and the standard yellow resistant cultivar, which has a much lower range of AA content, the mean value of AA was intermediate between mean values of the corresponding parent progenies. In all out one case the mean of the F_1 was not far removed from the arithmetic and geometric means of the parents. In the 8 F_2 progenies studied, the frequency distribution of the AA values of individual heads approximated a normal curve. The range of AA content was wider for the F_2 than for parent progenies, when these were tested comparatively. The F_2 progenies varied in the difference between their standard deviations and of the respective parent progenies, indicating that the parent plants selected differed in the genes governing AA content. Walker and Foster (1946) concluded that AA content was an inherent quantitative character and its level attainable in a given line or individual was determined by the combination of genes present which governed this character. The expression of the character was influenced to some extent by environmental factors. The studies of crosses between the high AA-inbred and standard yellow resistant cultivars indicate that the average AA value of these and presumably other cultivars may be raised considerably with retention of the major horticultural characteristics of the cultivar concerned by a process of controlled crossing and selection.

Grover, Chinoy and Nanda (1958) have shown that fairly high concentrations of AA (284×10^{-5} M to 508×10^{-5} M) stimulated elongation of coleoptile sections during the first two hours (ref. Fig. 4.18). After that there was retardation in growth and actual contraction in section length. In lower concentrations however (28.4×10^{-5} to 113.10×10^{-5} M), continuous increase in section length occurred upto varying periods of growth. The data of extension growth of Avena coleoptile sections was treated by the classical method of enzyme kinetics proposed by Michaelis and Menten (Foster et al., 1952; McRae and Bonner, 1953). The reciprocals of Avena section length (1/v) attained during the first two hours were plotted against reciprocals of the five concentrations of AA solutions (1/S). A linear relationship was obtained between the (1/V) and (1/S) which indicates that AA reaction on growth of coleoptile sections follow the principles of enzyme kinetics like auxins and that it could be considered to have a reaction similar to auxin (ref. Fig. 4.18).

A number of workers (Chinoy et al., 1958; 1961; 1969; 1969 b; Garg et al., 1958; Zemlyanukhin, 1956; Chailakhyan, 1956) have also studied at length the effect of exogenous application of AA on plant growth and in most of the cases, a promotive effect was obtained.

2. CORRELATION BETWEEN AA TURNOVER AND METABOLISM, GROWTH, DEVELOPMENT AND HEREDITY.

Various aspects of the chemistry, metabolism and uses of AA in plants and animals have been discussed at length recently (Seib and Tolbert, 1982).

Correlations between metabolism, growth and development of wheat have been extensively studied with a view to elucidating their heredity (Chinoy, 1958; 1966; 1969; 1970). Generally there was a close correlation between the rate of development and rates of production and utilization of regulatory substances. The most important and universally present of these regularoty substances was found to be AA whose rates of production and utilization governed the rates of development and consequently rates of mineral and organic metabolism, enzymic activities and growth of different organs of the plant.

It was found that the rates of synthesis of AA and ASG were considerably enhanced by long photoperiod and vernalization treatment in both the early (E) and late (L) flowering genotypes of wheat (Chinoy, 1970). The rate of synthesis of AA was higher and that of ASG lower in the early genotype under normal day control (NDC) compared to those in Bledsoe (L) under the same environment (Chinoy et al., 1969 b). The mean ASG content of the reproductive organs was generally lower compared to that of the vegetative apex. Under long photoperiod and vernalization treatment (LDV), AAU was considerably enhanced during the period of reproductive differentiation in both the genotypes. The AAU was also significantly higher in N.P. 718 under NDC in early stages of its reproductive differentiation compared with that of Bledsoe (Chinoy et al., 1969 b).

The presence of AA - MM charge transfer complex (CTC) has now been well established in actively metabolizing cells of plants and animals (Chinoy, 1962; 1970; Chinoy et al., 1965; Chinoy N.J., 1978). The rate of CTC of AA with macromolecules was also accelerated under LDV in both the genotypes. Mean values of AA - MM complexing were significantly higher during the period of reproductive differentiation of both the genotypes compared with the corresponding values during vegetative differentiation.

Further substantiation of the above mentioned fact is derived from the work of Gurevich (1963) on AA - FR - peroxidase which catalyzes the formation of the free radical of AA and is almost universally present in actively metabolizing tissues of plants and animals (Chinoy, 1962; Chinoy et al., 1965; 1969 b, c, d; Chinoy N.J., 1970; 1973; 1978), (please see Chapters 1, 2).

Experiments on early and late genotypes of wheat have shown that the rate of AA - FR peroxidase and catalase activities were considerably enhanced during the reproductive phase of both the genotypes and especially under LDV condition (Chinoy et al., 1969 b). The participation of the free radicals in enzymic reactions has been shown by Commoner et al. (1954; 1957; 1958). Numerous other workers have also elucidated the importance of free radicals in biological processes (Chinoy, 1962, 1970; Chinoy et al., 1969 b, c, d; Krasnovsky and Brin, 1946; Blois et al., 1961; Zholkevich et al., 1965; Kols et al., 1966; Bidzilya and Shakhov, 1967; Burlakova, 1967; Komentiani and Cagan, 1967; Kovarskii et al., 1967; Teichler-Zellen and Levine, 1967; Vernon et al., 1967; Raymond and Purves, 1968). Work cited here clearly brings out the fact that the intensity of EPR signals was enhanced during the period of reproductive differentiation compared to those found during the vegetative phase. During the entire period of reproductive differentiation also, the various metabolic activities were at their peak (Chinoy, 1970).

These studies necessitated the formulation of a new concept of heredity (Chinoy, 1958; 1966; 1969, 1970). Three components of the "heredity quantum" were recognized: (i) the structural complex - "Nucleon" (n); (ii) the Energy Receptor complex - "Energon" (ε); and (iii) The developmental Complex - "Metabolism" (m). Considering the above mentioned experimental evidence Chinoy (1970) formulated the energy trapping potential of the energon in terms of electron energy of the free radical of AA and charge transfer complex thus:

$$\varepsilon \alpha R.C.$$

$$\therefore \ \varepsilon = \frac{1}{F} R.C.$$

where R is the rate of formation of the free radicals, C is the rate of formation of CTC, and 1/F (a constant) is the rate of development of a genotype or a hybrid segregate.

The rates of DNA and RNA synthesis were also considerably enhanced in both the genotypes under LDV treatment. DNA was synthesized at a faster rate during the earlier stage of reproductive differentiation, whereas, RNA was synthesized at a faster rate throughout the period of reproductive differentiation (Chinoy et al., 1969 b). Presumably faster rates of DNA and RNA synthesis under LDV as well as in the early genotype were due to the enhanced rate of electron flow from free radical and CTC.

'Nucleon', one of the components of the "Heredity Quantum" is visualized as an "activated complex" of macromolecules of which the primer DNA constitutes a part (Chinoy, 1970). It provides the molecular basis for the trapping and utilization of energy by the nucleotides. On the basis of the foregoing, it was postulated that AA constituted "the primary product of the nucleon" (Chinoy, 1958).

3. GENE ACTIVATION BY PHYTOCHROME MEDIATED AA

Schopfer (1966) found that a dark grown seedling of white-seeded mustard (Sinapis alba L) contained an appreciable amount of AA. Its content increased under the influence of light. This effect was considered to be due to phytochrome mediation and not due to photosynthesis. On the basis of Mohr's (1966) assumption that the function of phytochrome in the case of "positive" photoresponses involves a differential gene activation of "potentially active" genes, Schöpfer (1967) advanced the following working hypothesis: Phytochrome induced accumulation of AA will lead to a separation of DNA-histone complexes in the range of "potentially active" genes. This makes possible the DNA dependent synthesis of m-RNA at those sites which are lastly responsible for the initiation of 'positive' photoresponses. The evidence presented is considered to support the view that AA exerts a function in connection with the regulation of gene activity.

4. ACTION OF ASCORBIC ACID ON NUCLEIC ACIDS

There is no doubt that AA profoundly affects the composition and biosynthesis of nucleic acids. Sharma and Datta (1956) were able to induce mitosis in polyploid cells and somatic reduction in diploid cells of onion root tips by treatment with AA. This result was ascribed to the increased production of nucleic acids leading to the disturbance in their balance in the cell. Bogatyreva and Znamenskaya (1962) found that properties of DNA were altered after its reaction with AA. Berneis (1963) found that an aqueous solution of AA degraded the DNA molecule in the presence of molecular oxygen and this degradation was completely arrested by the addition of catalase or of peroxidase. Venetianier and Straub (1963) used DHA in enzymic reactivation of reduced ribonuclease. Price (1966) was able to observe an increased biosynthesis of RNA by placing isolated wheat nuclei in a medium containing AA. An appreciable enhancement in RNA content of seedlings of Sinapis alba occurred by exogenous application of AA (Schöpfer, 1967). The interaction of nucleic acids and AA have also been discussed in detail in Chapter I.

Exogenous application of AA has been shown to enhance the biosynthesis of RNA as well as stimulate various enzymic activities (Chapter 11) in the germinating seeds of Cicer arietinum L. cv. Chafa (Chinoy and Saxena, 1969; 1972) and other crop plants (Chinoy et al., 1969). On the basis of this evidence, these workers postulated that "ascorbic acid controls the coding mechanism of a plant by influencing differentially the production of m-RNA and thereby controlling the biosynthesis as well as the reaction velocities of different enzymes".

Shah and Pandey (1978) reported that the basal cells of the pro-embryo of Limnophyton obtusifolium was conspicuous by a gradual increase in nuclear DNA during development. Concomitantly, increase in the content of AA, RNA and -SH occurred. These authors also established a precise correlation in time between the levels of AA, RNA and proteins containing -SH groups on the one hand and the chromosomal cycle of meiosis in developing pollen grains of some monocots on the other (Shah and Kulkarni, 1979). During organogenetic study in tissue culture of Solanum melongena cv. Florida market, Fassuliotis et al. (1981) found that pith callus cells when cultured in MS agar + IAA + γγ-isopentenyl adenine failed to form organized shoot. However, transfer of this tissue to a similar culture medium supplemented with AA (0.1 M) alone or in 49 μ M, 2 i P caused a development into leafy shoots.

DNA, RNA and nucleotides have a profound effect on the formation and decay of free radical of AA and formation of AA - MM complex (Chinoy et al., 1974) (ref. Chapter 1). Addition of DNA and RNA to AA in the presence of H_2O_2 resulted in the enhanced formation of the free radical of AA and its earlier and faster decay. Enhancement in the signal height varied with the concentration of the nucleic acid and so did the rate of decay. During the reaction of AA with DNA or RNA in the presence of H_2O_2, it was observed that the AA disappeared from the medium. However, on hydrolysis of the reaction mixture with 5% HPO_3 for 10 minutes at 80° C, AA was released from the complex and could be determined quantitatively. It was also seen that the rate of decay of the signal with time was correlated with the amount of DNA or RNA and AA present in the complexed form. Complexing was also proportional to the concentration of both AA and of RNA. Further, complexing of macromolecules with AA could take place only after the free radical of AA was generated in the medium. Even bases and nucleotides,especially thymine and uracil brought about an acceleration in the formation of the free radical of AA (ref. Chapter 1 for details).

Boosting of the AA - FR - generation by DNA and RNA may be due to the localization of charges on the AA molecule as a result of complex formation. The localization of the charges as well as formation of a complex, both will tend to disrupt the electronic configuration of the molecule, causing the hydrogen bond to break, thus, bringing about the separation of the two DNA chains, exposing new sites for m-RNA synthesis, which in turn regulates the rate of biosynthesis of enzymes and subsequently other cell constituents leading to growth and differentiation, as also reported by Hendry et al. (1977), Stich et al. (1976) and Jamaluddin et al. (1981) (ref. Chapter 1).

The nucleon can, therefore, be visualized as follows:

$$n \propto M.N.a.$$

where n is the nucleon, M, "the active mass of molecules" (primer included), a, the rate of biosynthesis of the primary product of the nucleon, which in all probability is AA, and N, the rate of production of 'excited' nucleotides.

$$n = \varepsilon\ (M.\ N.\ a.),\ \text{where } \varepsilon \text{ is a constant}$$

$$\therefore \text{Thus} \quad n = \frac{RC\ (MNa)}{F}$$

Chinoy (1970) has reported that the rates of biosynthesis of metabolites like DNA, RNA, proteins and -SH groups, including enzymatic reactions like RNase, protease as well as amylase and invertase were found to be decidedly faster in the genotype having an inherently faster rate of development compared with that of Bledsoe. Further, under long day and vernalization treatments, these rates were accelerated in both the genotypes leading to the quickening of the rate of development. Such results have been repeatedly obtained for early and late genotypes of a number of crop plants (Chinoy et al., 1969 a, b; Patel, 1967; Acharya, 1968; Pandya, 1969; Abraham, 1969; Dave, 1969). There is now sufficient evidence to show a close correspondence between the rates of enzymatic reactions, biosynthesis of the major metabolites and the rate of development. All these evidences provide an experimental basis for the evaluation of metabolon

- the energy utilizing component of the heredity quantum in quantitative terms.

The metabolon (m) of the heredity quantum represents the potential of an organism for the utilization of energy for metabolic activity on the basis of a given nucleon, the visible manifestation of which is growth, differentiation and development. On the basis of what has been said, the rate of DNA replication, the rate of biosynthesis of m-RNA and rates of enzymatic reactions amongst others will represent the potentiality of the metabolon in a given environment.

Thus:

$$m \;\alpha\; d.r.z.$$

$$\therefore \; m = n\,(d.r.z.)$$

$$= \frac{RC\ (MNa)}{F} \;.\; drz$$

where d is the rate of DNA replication, r, the rate of biosynthesis of m-RNA, z, the overall rate of enzymic reactions and n, the nucleon, which is a constant.

There are multifactorial relationships between the three components of heredity quantum, rates of production of cell metabolites, rates of production of AA, free radicals and CTC, rates of enzymic reactions, photosensitivity as well as the photo-thermic quanta of different genotypes (Chinoy, 1970). These are diagramatically presented (Fig. 12.1). As the photosensitivity of an early flowering genotype is greater compared with that of the late one, the photothermic energy needed for carrying on the developmental process is less. θ_1, and θ_2, represent rates of utilization of electronic energy of free radical of AA and CTC of AA with macromolecules as well as rates of enzymatic reactions of early and late genotypes respectively. In a similar manner θ_1 and θ_2 represent rates of biosynthesis of metabolites such as AA, RNA, proteins and others in early and late genotypes respectively. The magnitude of nucleon, n (Black circle), the energon, (white semicircle) and metabolon, m (dotted

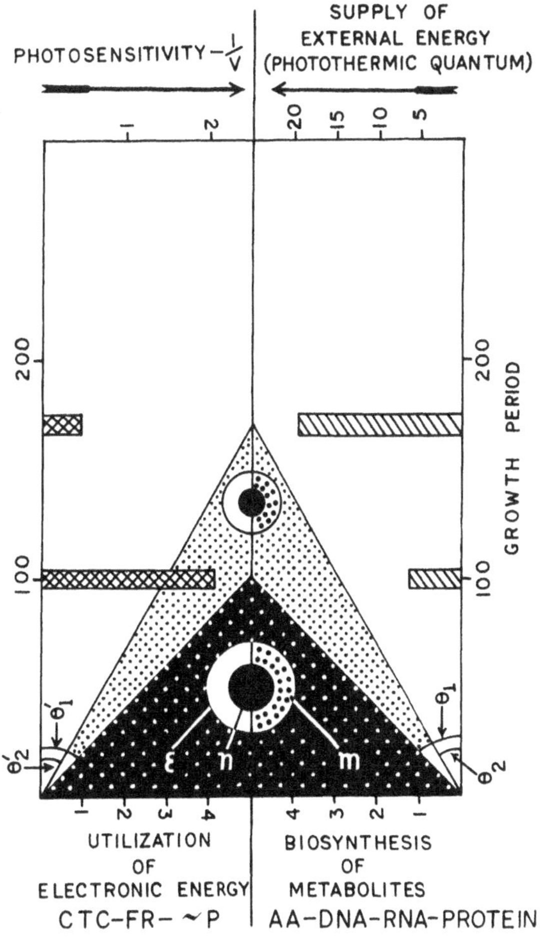

FIGURE 12.1. Multifactorial relationships between the three components of heredity quantum, rates of production of cell metabolites, rates of formation and utilization of ascorbic acid, FR and CTC, rates of enzymic reaction, photosensitivity as well as photothermic quanta of two genotypes (Chinoy, 1970).

semicircle) are represented by the areas of the heredity quantum they cover for the two genotypes. The area of smaller black triangle (with white dots) and that of the larger white triangle (with black dots) represent the magnitude of growth achieved by early and the late genotypes respectively (Fig. 12.1). There is ample evidence to show that early cultivars of a number of crop plants including wheat possess high photosensitivity and low requirement for photothermic energy, faster rates of formation of regulatory substances like AA, which on monovalent oxidation produces its free radical, MDHA and a complex between itself and a macromolecule (Chinoy, 1970). Rates of production of various metabolites are also faster in an early compared with a late genotype. Correlation between the rate of flow of electron energy (or free radical of AA and CTC), rates of biosynthesis of cell metabolites and the rate of development are not disturbed even by subjecting a genotype to differential photoperiodic and vernalization treatments. When a late cultivar of wheat or other thermophobe is subjected to LDV condition, it not only flowers early but also shows all the characters of growth, metabolism and energy regulation of an early flowering cultivar. In other words, it becomes its phenocopy. Even the segregating hybrid populations show the same type of correlation between growth and development. In other words, a late flowering genotype or a hybrid becomes a phenocopy of an early cultivar (or a hybrid) when its developmental rate is accelerated under LDV to synchronize with that of the early one.

Finally, the molecular and sub-molecular aspects of heredity were visualized in terms of reaction rates and the rate of flow of energy from the free radical of AA as well as from breaking up of AA - macromolecule complexes (Chinoy, 1970).

The excitation of the energon by photons is extranuclear, and is most probably concerned with the photosynthetic apparatus in which regulatory substances like AA are produced which profoundly influence the nucleon (Fig. 12.2).

A fact worthy of note is that exposure to radiation of moderate intensity augments AA turnover, the activity of AA - FR peroxidase and rate of complexing of AA with macromolecules (CTC). From the above facts it could be inferred that the photothermic quantum of a homozygous genotype or a hybrid

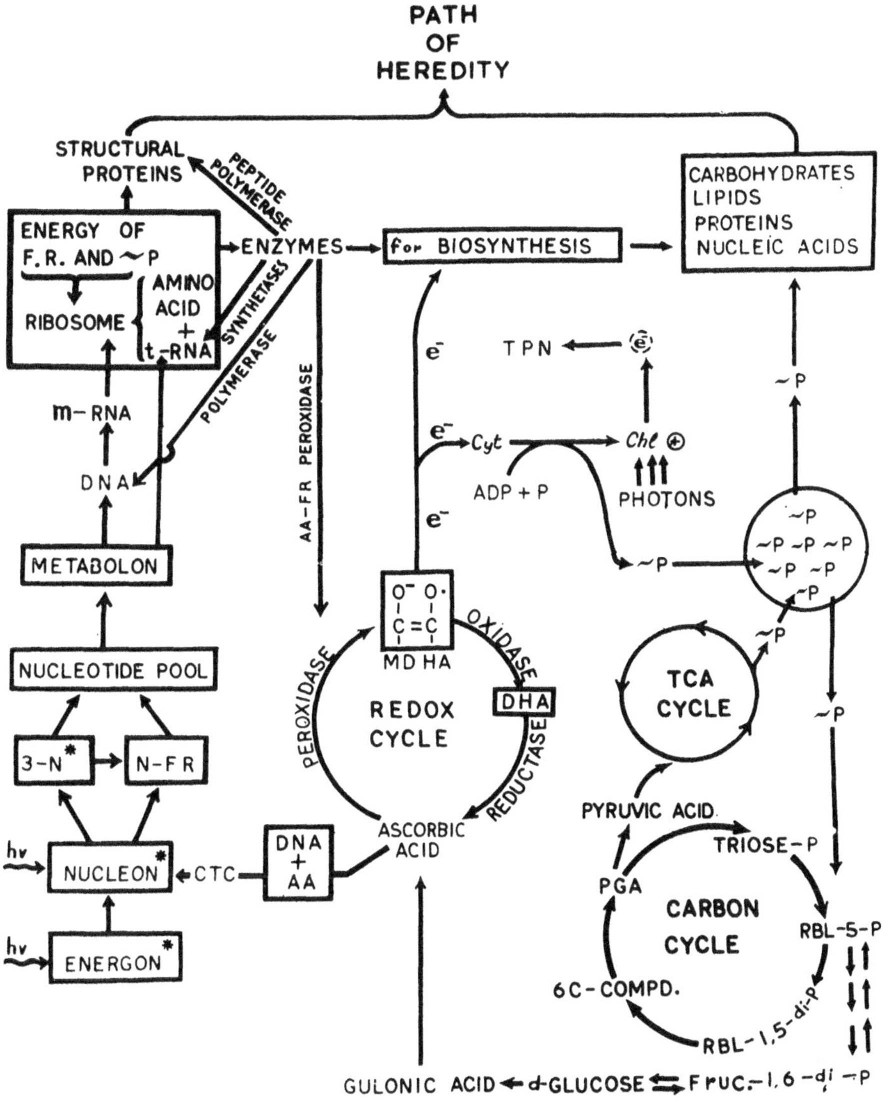

FIGURE 12.2. Molecular and submolecular aspects of heredity visualized in terms of reaction rates and the rate of energy flow from the free radical of AA as well as from the breaking up of the AA - MM complexes (Chinoy, 1970).

segregate determines the rate of production of AA - FR as well as that of the CTC formation (Fig. 12.2).

The postulated photon excitation of the energon is transmitted to the nucleon of the heredity quantum which can also be excited by photons of various spectral ranges. The macromolecules of the nucleon form charge transfer complexes with AA. On breaking up, these charge transfer complexes yield nucleotide triplets (3- N^0) as well as free radical of nucleotides (N - FR). As already stated earlier, localization of charges and AA – MM - complex formation tend to disrupt the electronic configuration of the macromolecules causing hydrogen bonds to break, thus bringing about the separation of the two strands of DNA, exposing new sites for the synthesis of m-RNA, which in turn would regulate the rates of biosynthesis of cell constituents and rates of enzymic reactions leading to a certain pattern of growth and differentiation.

The rate at which the free radical of AA is formed will depend upon the rate of reaction of the special AA - FR peroxidase. Thus the rate of electron flow will depend upon the above mentioned reaction. Further, the enzymes catalysing the conversion of MDHA $\xrightarrow{\text{oxidase}}$ DHA $\xrightarrow{\text{reductase}}$ AA will also partly determine the rate of electron flow. The carbon cycle will also contribute substantially to the production of AA by the genotypes thereby accelerating the formation of CTC with macromolecules to generate electron energy and contribute to the generation of free radicals of nucleotides thus participating in the replication of DNA.

5. SUMMARY

Genotypic differences in AA content of various crop plants have been reported by a number of workers. On the basis of a study of F_1 and F_2 populations of a number of cabbage crosses, it was concluded that AA content was inherited as a quantitative character. The dark grown seedlings of white seeded mustard (Sinapis alba) L. contained an appreciable amount of AA, but its concentration increased under the influence of light. This effect was considered to be due to phytochrome mediation and not due to photosynthesis.

Generally close correlations were found between the rates of production and utilization of AA on the one hand and rates of organic and mineral metabolism, rates of enzymic activities, rate of growth and development of the plant on the other. Synthesis of AA and ASG as well as AAU and AA - MM - complexing were considerably enhanced by long photoperiod and vernalization both in the early and late flowering genotypes of wheat. All the four above mentioned entities gave higher values during the period of reproductive differentiation compared with those obtained during the vegetative phase. The activity of a special isozyme of peroxidase which catalyzes the formation of free radical of AA, the AA - FR peroxidase was enhanced during the reproductive phase and especially under LDV condition. The intensity of EPR signals was also enhanced during the period of reproductive differentiation compared with those found during the vegetative phase.

On these bases the molecular and submolecular aspects of heredity were visualized in terms of reaction rates and the rate of flow of electron energy from the FR of AA as well as from the breaking up of AA - macromolecule complex (CTC).

REFERENCES

1. Abraham PG 1969. Ph.D. Thesis, Botany Department, Gujarat University, Ahmedabad - 380009, India.
2. Acharya UH 1968. Ph.D. Thesis, Botany Department, Gujarat University, Ahmedabad - 380009, India.
3. Berneis K 1963. Helv Chim Acta 46: 57.
4. Bidzilya NI, Shakhov AA 1967. In: Exposure of Plants to light impulses (Russ) Shakhov AA (ed) Timiryazev Inst Plant Physiol Acad Sci USSR.
5. Blois MS, Brown HW, Lemmon RM, Lindblom RO, Weissblutth M (eds) 1961. Free radicals in Biological systems, New York: Academic.
6. Bogatyreva SA, Znamenskaya MP 1962. Dokl Acad Nauk SSSR, Biochem Sec 140: 153.
7. Burlakova YeB 1967. Biofizika 12: 91.
8. Chailakhyan MKh 1956. Dokl Akad Nauk SSSR 111: 894.
9. Chinoy JJ 1958. Proc Sem "Modern Dev Plant Physiol" Uni Delhi, Delhi, India, Aug 26-30, 1957, p 130.
10. Chinoy JJ 1962. Indian J Plant Physiol 5: 172, FG Gregory Memorial Volume.
11. Chinoy JJ 1966. J Indian Bot Soc XLV: 150.

12. Chinoy JJ 1969. Proc XI釿ternl Bot Congr Seattle Washington Aug 24-Sept 2, Abst p 31.
13. Chinoy JJ 1970. Indian J Plant Physiol 13 (2): 115.
14. Chinoy JJ, Saxena OP 1969. Science and Culture 35: 606. Report on National Seminar on Plant Physiology and Crop Production.
15. Chinoy JJ, Saxena OP 1972. Proc 16th Int Seed Test Assoc (ISTA) Congr Washington DC June 7-12, 1973 37: 903.
16. Chinoy JJ, Grover R, Nanda KK 1958. Proc Sem "Modern Dev Plant Physiol" Delhi Univ, Aug 26-30, 1957, p 12.
17. Chinoy JJ, Grover R, Nanda KK 1961. Flora 150: 1.
18. Chinoy JJ, Shah CK, Patel HT, Suthar HK 1965. Proc.释nternl Sem "Tissue Culture" Baroda Jan 21-28, Ramakrishna CV (ed) p 377.
19. Chinoy JJ, Patel HD, Jani BM, John D, Vyas AV, Raj KPS 1969. Report PL 480 Research Project FG-In-208 (A7-CR-108), Botany Department, Gujarat University, Ahmedabad - 380009, India, p 1-456.
20. Chinoy JJ, Sastry K, Pandya RB, Gurumurti K, Abraham PG, Saxena OP, Shah PN 1969 a. J Biol Sci 12: 1.
21. Chinoy JJ, Saxena OP, Abraham PG, Pandya RB, Dave IC, Gurumurti K, Sastri K 1969 b. Report PL 480 Research Project FG-In-182 (A7-CR-87), Botany Department, Gujarat University, Ahmedabad - 380009, India, p 1-350.
22. Chinoy JJ, Saxena OP, Dave IC, Gurumurti K, Pandya RB, Abraham PG 1969 c. Curr Sci 38: 153.
23. Chinoy JJ, Saxena OP, Gurumurti K, Abraham PG, Pandya RB 1969 d. Sci & Culture 35: 690.
24. Chinoy JJ, Singh YD, Laloraya MM 1974. Vth Int Symp Mag Resonance Jan 14-18, Bombay, India, Abstr 17 A 64.
25. Chinoy NJ 1970. Stain Technol 45: 99.
26. Chinoy NJ 1973. Ind Natl Sci Acad 39 B: 119.
27. Chinoy NJ 1978. J Anim Morphol Physiol Silver Jubilee Volume: 68.
28. Commoner B, Townsend J, Pake GE 1954. Nature (Lond.) 174: 689.
29. Commoner B, Heise JJ, Lippincott BB, Norberg RE, Passonneau JV, Townsend J 1957. Science 126 (3263): 57.
30. Commoner B, Lippincott BB, Passonneau JV 1958. Proc Natl Acad Sci U.S. 44: 1099.
31. Currence TM 1939. Proc Amer Soc Hort Sci 37: 901.
32. Dave IC 1969. Ph.D. Thesis, Botany Department, Gujarat University, Ahmedabad - 380009, India.
33. Fassuliotis G, Nelson BV, Bhatt DP 1981. Plant Sci Lett 22: 119.
34. Foster RJ, McRae DH, Bonner J 1952. Proc Nat Acad Sci 38: 1014.
35. Garg OP, Chinoy JJ, Nanda KK 1958. Fyton 10: 17, III.
36. Grover R, Chinoy JJ, Nanda KK 1958. Proc Symp "Recent Advances in the Study of Plant Metabolism", Univ Allahabad, India, Jan 1958.
37. Gurevich AA 1963. Dokl Biol Sciences Sect. 145: 674
38. Hendry LB, Witham FH, Chapman OL 1977. Perspective in Biol & Med 21: 120.

39. Jamaluddin M, Rajalakshmi S, Sarma DSR 1981. Biochim Biophys Acta 653: 423.
40. Kols OR, Limarenko IM, Tarusov BN 1966. Dokl Akad Nauk SSSR 167: 956.
41. Komentiani Z, Cagan RH 1967. Biochim Biophys Acta 135: 1083.
42. Kovarskii AE, Konarskii VA, Kazantsev EF 1967. Dokl Biol Sci 175: 452.
43. Krasnovsky AA, Brin GP 1946. Compt Rend (Dokl) de l' Academic des Sci de l' URSS 53: 443.
44. Leonard P, Pepkowitz RE, Larson JG, Owens G 1944. Plant Physiol 19: 615.
45. McLinn WA, Fallers CR, Buck RE 1937. Proc Amer Soc Hort Sci 34: 543.
46. McRae DH, Bonner J 1953. Physiol Plantarum 6: 485.
47. Mohr H 1966. Z Pflanzenphysiol 54: 63.
48. Pandya RB 1969. Ph.D. Thesis. Botany Department, Gujarat University, Ahmedabad - 380009, India.
49. Patel HD 1967. Ibid.
50. Price CE 1966. Nature (Lond.) 212: 1481.
51. Pyke M 1942. J Soc Chem Ind 61: 149.
52. Raymond PL, Purves WK 1968. Plant Physiol 43: 454.
53. Schopfer P 1966. Planta (Berl.) 69: 158.
54. Schopfer P 1967. Planta (Berl.) 74: 210.
55. Seib A, Tolbert BM 1982 (eds) Ascorbic acid chemistry, metabolism and uses. Advances in chemistry series 200, Washington DC, American Chemical Society.
56. Shah CK, Kulkarni CD 1979. J Palynol 15: 91.
57. Shah CK, Pandey SN 1978. Phytomorph 28: 31.
58. Sharma AK, Datta A 1956. Fyton 6: 71.
59. Smith FG, Walker JC 1946. Am J Bot 33: 120.
60. Stich HF, Karim J, Koropatnick J, Lo L 1976. Nature: 260: 722.
61. Teichler-Zellen D, Levine RP 1967. Plant Physiol 42: 1643.
62. Venetianier P, Straub FB 1963. Acta Physiol Hung 24: 41.
63. Vernon LP, Ke B, Shaw ER 1967. Biochemistry 6: 2210.
64. Wade BL, Kanapaux MS 1943. J Agri Res 66: 313.
65. Walker JC, Foster RE 1946. Amer J Bot 33: 758.
66. Zemlyanukhin AA 1956. Fiziol Raste 3: 313.
67. Zholkevich VN, Chetverikov AG, Rogacheva AY 1965. Soviet Plant Physiol 12: 167.

SUBJECT INDEX

GPSR Compliance

The European Union's (EU) General Product Safety Regulation (GPSR) is a set of rules that requires consumer products to be safe and our obligations to ensure this.

If you have any concerns about our products, you can contact us on ProductSafety@springernature.com

In case Publisher is established outside the EU, the EU authorized representative is:

Springer Nature Customer Service Center GmbH
Europaplatz 3
69115 Heidelberg, Germany

Batch number: 10761599

Printed by Printforce, the Netherlands